Saved through Fire

Saved through Fire

The Fiery Ordeal in New Testament Eschatology

Daniel Frayer-Griggs

Foreword by
William R. Telford

☙PICKWICK *Publications* • Eugene, Oregon

SAVED THROUGH FIRE
The Fiery Ordeal in New Testament Eschatology

Copyright © 2016 Daniel Frayer-Griggs. All rights reserved. Except for brief quotations in critical publications or reviews, no part of this book may be reproduced in any manner without prior written permission from the publisher. Write: Permissions, Wipf and Stock Publishers, 199 W. 8th Ave., Suite 3, Eugene, OR 97401.

Pickwick Publications
An Imprint of Wipf and Stock Publishers
199 W. 8th Ave., Suite 3
Eugene, OR 97401

www.wipfandstock.com

PAPERBACK ISBN: 978-1-4982-0325-8
HARDCOVER ISBN: 978-1-4982-0327-2

Cataloguing-in-Publication data:

Frayer-Griggs, Daniel.

   Saved through fire : the fiery ordeal in new testament eschatology / Daniel Frayer-Griggs; foreword by William R. Telford.

   xx + 280 pp. ; 23 cm. Includes bibliographical references.

   ISBN 978-1-4982-0325-8 (paperback) | ISBN 978-1-4982-0327-2 (hardback)

   1. Fire—Biblical teaching. 2. Judgment Day—Biblical teaching. 3. Eschatology—Biblical teaching. I. Telford, William R. II. Title.

BS680.F53 F81 2016

Manufactured in the U.S.A.   03/31/16

For Jenn

Contents

Foreword by William R. Telford | ix
Acknowledgements | xiii
Abbreviations | xv

1. Introductory Matters | 1
2. The Functions of Fire in the Hebrew Bible | 25
3. The Fire of Judgment in Second Temple Texts | 70
4. John the Baptist and the Baptism in Fire | 131
5. The "Fire Words" of Jesus | 145
6. Saved through Fire | 199
7. Local Persecutions and the Cosmic Conflagration | 218
8. Concluding Remarks | 244

Bibliography | 249
Ancient Document Index | 267

Foreword

"But on the day when Lot went out from Sodom fire and sulfur rained from heaven and destroyed them all—so will it be on the day when the Son of man is revealed" (Luke 17:29–30).

"I came to cast fire upon the earth; and would that it were already kindled!" (Luke 12:49).

"For every one will be salted with fire" (Mark 9:49).

"I baptize you with water for repentance, but he who is coming after me is mightier than I, whose sandals I am not worthy to carry; he will baptize you with the Holy Spirit and with fire" (Matt 3:11=Luke 3:16).

"According to the grace of God given to me, like a skilled master builder I laid a foundation. . . . If the work which any man has built on the foundation survives, he will receive a reward. If any man's work is burned up, he will suffer loss, though he himself will be saved, but only as through fire" (1 Cor 3:10, 14, 15).

SAYINGS OF ESCHATOLOGICAL JUDGMENT such as these, and especially the so-called "fire" sayings of the New Testament, have filled scholars with a certain unease, and in consequence have tended to be relatively neglected in academic research. A fiery Jesus preaching hellfire and damnation is not a popular or comfortable one. The ultimate fate of the righteous, as well as of the wicked, is seemingly also called into question by these passages. Are these scriptural references to be understood literally or metaphorically? What function does this "fire" play at the endtime? Is its purpose merely punitive or does it have a purificatory role? It is no wonder, then, that each of these texts has been seen as a *crux interpretum* by New Testament scholars. Protestant scholars, in particular, have reacted against an interpretation of the Pauline passage (1 Cor 3:15) that sees adumbrated in it the Catholic doctrine of purgatory developed by early church fathers and theologians like Clement and Origen. It is these difficult questions, however, and these enigmatic New Testament texts, that Daniel Frayer-Griggs has chosen to engage in this strongly argued and lucidly written book.

Himself a skilled master builder, Frayer-Griggs has laid his exegetical work on a foundation established first of all by an MA thesis for Pittsburgh Theological Seminary in 2008, written under the supervision of Dale C. Allison Jr, and subsequently by a PhD thesis for the Department of Theology and Religion at Durham University in 2012, for which I, myself, was principal supervisor, with Professor Loren Stuckenbruck, Dr. Stephen Barton, and Professor Francis Watson acting (variously) as his second supervisor(s). Daniel came to us in Durham with a strong educational background, significant teaching experience and impressive references. In addition to having graduated first in his Master's class at PTS (Theology and Religious Studies, 2006–8), his qualifications included a BA in Religion and English (Hope College, 2000), an MEd in Secondary Education (Aquinas College, 2004) and an MA in English Literature (McNeese State University, Lake Charles, 2002–4). One of his referees described him as one with "the creative spark necessary to make a genuine contribution to his discipline," and the Synoptic eschatological "fire" sayings, first studied for his Master's thesis, so ignited his enthusiasm that, in his doctoral studies, he extended his work to encompass the relevant data in the Old Testament, the intertestamental literature, the remainder of the New Testament, the patristic literature and even Stoicism. To this study, he has brought his numerous skills and competencies, whether in ancient (Aramaic, Greek, Hebrew, Latin) or modern (French, German) languages, or in the other tools necessary for contemporary biblical research (source criticism, redaction criticism, textual criticism, literary, historical, and contextual analysis, history of interpretation, etc.). Built on these foundations, this book is the product of these doctoral efforts and it is a great pleasure to see it in print. With published articles already to his credit in refereed periodicals and edited publications, including encyclopedia entries, Daniel is a young man who is clearly making a valuable contribution to the Biblical Studies field. Among the honours and awards that he has been garnering, for example, was his selection for a Society of Biblical Literature Regional Scholar Award in 2013 for a paper he wrote on John 9:6, and which he presented to the Eastern Great Lakes Region of the SBL.

Saved by Fire: The Fiery Ordeal in New Testament Eschatology is the first full-length treatment of the New Testament eschatological "fire" sayings in the English language, and Pickwick Publications are to be congratulated for bringing it to a wider audience. As such, it will be of interest to scholars, graduate, postgraduate, and seminary students, and clergy and informed laypersons who are attentive to the latest developments in biblical exegesis. In his first chapter, Frayer-Griggs lays out his aims and agenda, as well as the scope of his investigation, setting the exegetical context, reviewing previous research, articulating the outstanding questions and issues (including literal or symbolic interpretation) and outlining his methodology. The sayings selected for analysis comprise Matt 3:11//Luke 3:16; Mark 9:49; Luke 12:49; Luke 17:26–32; Luke 23:31; *Gos. Thom.* 82; 1 Cor 3:10–15; 1 Pet 1:6–7; 4:12–13 and 2 Pet 3:10. Each of these "fire" sayings has a bearing on the subject of eschatological judgment in its manifestation as an ordeal or test affecting the human race *in toto*, both the righteous and wicked, at the endtime. They are distinguishable, hence, from those texts that describe the "fire" of hell or Gehenna, reserved by the New Testament writers, exclusively and eternally, for the wicked following the last or final judgment. Frayer-Griggs' purpose is to show that the "fiery ordeal" of which these particular texts speak has a dual function, the eschatological "fire" visited upon humanity being understood as being purificatory or even purgatorial as well as punitive. This is an exercise in motif-analysis, therefore, his aim being to identify and demonstrate a distinctive, widespread or recurrent thread of eschatological belief which, he claims, is inherent in the Christian tradition much earlier than has hitherto been thought.

To set the context for his argument, Frayer-Griggs devotes two chapters to the Jewish literary tradition before the New Testament, examining, in chapter 2, the functions of fire in the Hebrew Bible (the Pentateuch, the Prophets, and the Writings) and, in chapter 3, notions of judgment by fire in Second Temple apocalyptic texts (First Enoch, the Dead Sea Scrolls, the Septuagint, the Sibylline Oracles, Fourth Ezra). Herein, fire is shown not only to accompany the presence of God in theophanies or (in combination often with water) to have a punitive or destructive role with respect to the wicked, but also to perform a purifying, refining, testing, or purgative function with regard to the righteous. This last feature of eschatological judgment, whether manifest in a localized fiery ordeal or cosmic conflagration, takes on a greater significance in Second Temple Judaism, and so provides a fitting backdrop to Frayer-Griggs' analysis on the New Testament texts in the central chapters of the book. The readers will find in chapters 4, 5, and 6 of the book a lucid, balanced, and illuminating examination of these perplexing New Testament "fire" sayings, beginning appropriately with John the Baptist (chapter 4) and his (textually variant) prophecy of a "coming one" who will baptize "with the Holy Spirit and with fire." They will also be offered a well-argued case for the identity of the "coming one," for the original form of the saying, and for the meaning of "baptism with fire" in John's (and Jesus') first-century apocalyptic context.

Chapter 5 moves on to consider the "fire" sayings found on the lips of Jesus himself, John's disciple, in the Synoptic (and Thomas) tradition.

Building on the work of J. D. G. Dunn, T. J. Baarda, and others, Frayer-Griggs thoughtfully explores the text-critical issues, the history of interpretation, the sources, and the historical plausibility of each of these sayings, and cautiously opts in favour of, at least, the essential consonance of the sayings with the eschatological message of Jesus (and John). Taken together, he argues, they attest to a recurrent theme, namely, "that in the face of the coming of God's eschatological judgment all, including Jesus and his own disciples, must face a period of fiery testing or trial and that entry into the kingdom is in some sense contingent upon passing that ordeal" (198). In chapter 6, the enigmatic words of the apostle Paul in 1 Cor 3:15, which give the book its title, are considered, and the case for the purifying and soteriological function of fire in this hotly debated passage (1 Cor 3:10–15) is strongly made. Building on, but also himself testing and refining the work of A. N. Kirk, and noting here the allusion to Malachi 3, the temple, and the "refiner's fire," Frayer-Griggs adds additional weight to the interpretative case for the purificatory (if not purgatorial) function of (eschatological) fire in Paul's declaration. Chapter 7 completes the investigation by turning to the Petrine epistles where fiery ordeal and cosmic conflagration motifs are highlighted and adeptly discussed, the first (1 Pet 1:7; 4:12) employing the "refining fire" and temple imagery of Malachi as a metaphor for present suffering, the latter (2 Pet 3:10–15) drawing not only on the Jewish eschatology previously explored but also, Frayer-Griggs interestingly claims, on Stoic cosmology and the doctrine of *ekpyrosis*. Concluding remarks in chapter 8 summarize the argument, explore its implications, and make suggestions for further research.

This is a bold piece of work which not only attempts to come to grips with a series of individually difficult and enigmatic passages, but, in isolating and identifying the eschatological "judgment by fire" motif, with its dual punitive and purificatory dimensions, also seeks to establish a thread of continuity between them. "The detailed treatment of this string of texts could give the impression of walking through a minefield," claimed one of Frayer-Griggs' examiners, and that is undoubtedly true. Minefields produce their own fiery ordeals or cosmic conflagrations. Frayer-Griggs' work has been built on good foundations, but, to return to Paul, it is sure to be tested through the fire (of academic review). I am sure he will survive, and will receive his reward.

Thursday, March 3, 2016

Dr William R. Telford
Visiting Fellow
St John's College
Durham

Acknowledgments

THIS BOOK IS A revision of my doctoral thesis, submitted to Durham University in September 2012. As does almost anyone who embarks on such a long-term endeavor, I have relied on the assistance and encouragement of others throughout my research and writing. I would like first to thank my parents, Tom and Mary Griggs. Without their generosity and support, I would not have been able to pursue doctoral studies at Durham. They have assisted me and my family in countless ways, and I owe them a debt of gratitude I will never be able to repay. I can only strive to be as devoted and loving to my own children as they have been to me.

I would also like to thank Professor Dale C. Allison, Jr. at Princeton Theological Seminary. He read and commented on a draft of this thesis, for which I am extremely thankful. His continued interest in my research is much appreciated. I also wish to thank my friend and colleague, Tucker Ferda, for reading and commenting on select chapters of this book.

No one has contributed more to my writing of this thesis than my supervisor, Dr. William R. Telford, who graciously read and re-read initial drafts, revised drafts, and the penultimate draft of the thesis, offering insightful comments and suggestions at every stage. Even though I was not in residence at Durham during the writing of my doctoral thesis, I always felt that I had Bill's undivided attention. I recall one occasion when I completed a chapter and submitted it to Bill via email late one evening. Feeling quite pleased with myself, I thought I had won myself a week or so of rest before returning to writing and revising, but to my surprise I woke the next morning to find a marked-up draft of my recently submitted chapter in my inbox. Bill's thorough comments in the margins of drafts have brought to my attention several infelicities and have helped to strengthen my arguments considerably. He has been nothing but encouraging throughout this lengthy process, and I am glad to have had him as my supervisor. I am particularly grateful for his willingness to see the thesis through to its completion beyond his retirement.

I am grateful to Koninklijke Brill NV for permission to re-publish "'Everyone Will Be Baptized in Fire': Mark 9.49, Q 3.16, and the Baptism of the Coming One," *JSHJ* 7 (2009) 254–85, sections of which now appear in chapters 4 and 5 and to Cambridge University Press for permission to reprint material first published in "Neither Proof Text Nor Proverb: The Instrumental Sense of διά and the Soteriological Function of Fire in 1 Cor 3.15," *NTS* 59 (2013) 517–34, and which now appears in chapter 6.

Thanks are also due to Calvin Jaffarian and Robin Parry for their work in helping bring may manuscript to publication.

Finally, my wife, Jenn, has been my constant source of support, not only through my four years of doctoral studies, but over the past fourteen years of our marriage, during which time her love and friendship have made my life immeasurably richer. She has done more than her fair share for our family while working, pursuing her own graduate studies, and being an amazing mother to our sons, Jonah and Levi. It is to her that this book is dedicated.

Abbreviations

AB	Anchor Bible
ABD	*Anchor Bible Dictionary.* Edited by D. N. Freedman. 6 vols. New York, 1992
ACW	Ancient Christian Writers
AGJU	Arbeiten zur Geschichte des antiken Judentums und des Urchristentums
AnBib	Analecta biblica
ANF	*Ante-Nicene Fathers*
ANRW	*Aufstieg und Niedergang der römischen Welt Geschichte und Kultur Roms im Spiegel der neuern Forschung.* Edited by H. Temporini and W. Haase. Berlin, 1972–
ANTC	Abingdon New Testament Commentaries
APSP	*American Philosophical Society Proceedings*
ASNU	Acta seminarii neotestamentici upsaliensis
ATDan	Acta theological danica
BBB	Bonner biblische Beiträge
BDAG	W. Bauer, W. F. Arndt, and F. W. Gingrich (3rd ed.; rev by F. W. Danker), Greek-English Lexicon of the New Testament
BECNT	Baker Exegetical Commentary on the New Testament
BNTC	Black's New Testament Commentaries
Bib	*Biblica*
BN	*Biblische Notizen*

BTS	Biblical Tools and Studies
BZ	*Biblische Zeitschrift*
BZAW	Beihefte zur Zeitschrift für die alttestamentliche Wissenschaft
CBQ	*Catholic Biblical Quarterly*
CBQMS	Catholic Biblical Quarterly Monograph Series
CC	Continental Commentaries
CEJL	Commentaries on Early Jewish Literature
CGTC	Cambridge Greek Testament Commentary
ConBNT	Coniectanea biblica: New Testament Series
COQG	Christian Origins and the Question of God
CrQ	*The Crozer Quarterly*
CTL	Crown Theological Library
DSD	*Dead Sea Discoveries*
EDNT	*Exegetical Dictionary of the New Testament*
EKKNT	Evangelisch-katholischer Kommentar zum Neuen Testament
EH	Europäische Hochschulschriften
EvJ	*Evangelical Journal*
ExpTim	*Expository Times*
FC	Fathers of the Church
FRLANT	Forschungen zur Religion und Literatur des Alten und Neuen Testaments
GTJ	*Grace Theological Journal*
HCOT	Historical Commentary on the Old Testament
HNT	Handbuch zum Neuen Testament
HR	*History of Religions*
HSem	Horae semiticae. 9 vols. London, 1908–12
HTR	*Harvard Theological Review*
IBC	Interpretation: A Bible Commentary for Teaching and Preaching
ICC	International Critical Commentary
ITQ	*Irish Theological Quarterly*

JBL	*Journal of Biblical Literature*
JEA	*Journal of Egyptian Archaeology*
JETS	*Journal of the Evangelical Theological Society*
JPSTC	Jewish Publication Society Torah Commentary
JQR	*The Jewish Quarterly Review*
JSHJ	*Journal for the Study of the Historical Jesus*
JSJSup	Journal for the Study of Judaism: Supplement Series
JSNT	*Journal for the Study of the New Testament*
JSOT	*Journal for the Study of the Old Testament*
JSOTSS	Journal for the Study of the Old Testament: Supplement Series
JTS	*Journal of Theological Studies*
KEK	Kritisch-exegetischer Kommentar über das Neue Testament (Meyer-Kommentar)
LCL	Loeb Classical Library
LNTS	Library of New Testament Studies
LXX	Septuagint
MT	Masoretic Text
NCB	New Century Bible
Neot	*Neotestamentica*
NETS	*The New English Translation of the Septuagint*
NHMS	Nag Hammadi and Manichean Studies
NHS	Nag Hammadi Studies
NICNT	New International Commentary on the New Testament
NICOT	New International Commentary on the Old Testament
NIGTC	New International Greek Testament Commentary
NovT	*Novum Testamentum*
NSBT	New Studies in Biblical Theology
NTAbh	Neutestamentliche Abhandlungen
NTApoc	*New Testament Apocrypha*. 2 vols. Rev. ed. Edited by Wilhelm Schneemelcher. English trans. ed. Robert Mcl. Wilson. Cambridge: Clarke; Louisville: Westminster John Knox, 2003

NTD	Das Neue Testament Deutsch
NTL	New Testament Library
NTS	*New Testament Studies*
Numen	*Numen: International Review for the History of Religions*
OBO	Orbis biblicus et orientalis
OTL	Old Testament Library
OTP	*Old Testament Pseudepigrapha*. Edited by James H. Charlesworth. 2 vols. New York: Doubleday, 1983, 1985
PBM	Paternoster Biblical Monographs
PG	Patrologia graeca [=Patrologiae cursus completus: Series graeca]. Edited by J.-P. Migne. 162 vols. Paris, 1857–86
PL	Patrologia latina [=Patrologiae cursus completus: Series latina]. Edited by J.-P. Migne. 217 vols. Paris, 1844–64
PNTC	Pillar New Testament Commentaries
POS	Pretoria Oriental Series
RB	*Revue biblique*
RevQ	*Revue de Qumran*
SB	Sources bibliques
SBLDS	Society of Biblical Literature Dissertation Series
SBLMS	Society of Biblical Literature Monograph Series
SBT	Studies in Biblical Theology
SE	*Saeris Erudiri*
SHJ	Studying the Historical Jesus
SiBL	Studies in Biblical Literature
SNT	Supplements to Novum Testamentum
SNTSMS	Society for New Testament Studies Monograph Series
SPB	Studia postbiblica
ST	*Studia theologica*
STDJ 56	Studies on the Texts of the Desert of Judah
SUNT	Studien zur Umwelt des Neuen Testaments
SVF	*Stoicorum veterum fragmenta*. H. von Arnim. 4 vols. Leipzig, 1903–24

SVTP	Studia in Veteris Testamenti pseudepigraphica
TDNT	*Theological Dictionary of the New Testament*
TDOT	*Theological Dictionary of the Old Testament*
TENTS	Texts and Editions for New Testament Study
Theol	*Theology*
ThH	Théologie historique
THKNT	Theologischer Handkommentar zum Neuen Testament
TNTC	Tyndale New Testament Commentaries
TQ	*Theologische Quartalschrift*
TSAJ	Text und Studien zum antiken Judentum
TZ	*Theologische Zeitschrift*
VCSup	Vigiliae christianae: Supplement Series
VT	*Vetus Testamentum*
WBC	Word Biblical Commentary
WC	Westminster Commentaries
WMANT	Wissenschaftliche Monographien zum Alten und Neuen Testament
WTJ	*Westminster Theological Journal*
WUNT	Wissenschaftliche Untersuchungen zum Neuen Testament
ZNW	*Zeitschrift für die neutestamentliche Wissenschaft und die Kunde der älteren Kirche*

All ancient author abbreviations and biblical references conform to the guidelines set forth in:

>Alexander, Patrick, et al., editors. *The SBL Handbook of Style: For Ancient Near Eastern, Biblical, and Early Christian Studies*. Peabody, MA: Hendrickson, 1999.

CHAPTER 1

Introductory Matters

"For everyone will be salted with fire" (Mark 9:49). This saying has vexed more than a few interpreters of the New Testament over the last two millennia. One possible source of the bewilderment evoked by this pronouncement may be its odd juxtaposition of the two distinct and remarkably connotative images of salt and fire.[1] Perhaps more perplexing, however, is the fact that this verse presumes that *everyone*, regardless of his or her standing before God, will face the self-same fire. This assumption, whether it was made on the part of the historical Jesus, the Markan evangelist, or some early Christian prophet or author, raises several questions. Foremost among them is this: what effect is this fire believed to have upon those who must endure it? Does the saying presume that the fire will mean destruction for all? Does it indicate that it will refine all who pass through it? Or should we suppose that it will test each individual, perhaps purifying the repentant while punishing the wicked?

Further, of what fire does the pronouncement speak? While the fire of hell or Gehenna (e.g., Mark 9:43-48) predominates in the Synoptic Gospels, particularly in Matthew and Mark, this does not appear to be what Mark 9:49 has in mind, for nowhere else in the tradition does Jesus consign *everyone* to hell. Does Mark 9:49, then, refer to the fire of judgment that breaks into history, like the fire that rained down on Sodom and Gomorrah (Gen 19:24) or that Elijah called down upon the messengers of Ahaziah (2 Kgs 1:10-12)? Does the verse presume knowledge of the Zoroastrian or Stoic

---

1. For the many uses of salt in antiquity see Pliny, *Natural History* 31.34-45; Latham, *The Religious Symbolism of Salt*. On the multiple functions of fire in the Bible see Hamp, "אש"; Lang, "πῦρ, κτλ." Latham concludes his exhaustive survey of the religious symbolism of salt in the ancient world with the observation that Mark 9:49 "is too rich in symbolism to settle on any one meaning. This is especially evident because the two symbols, salt and fire, contain, each one, contrary meanings" (239).

ideas about the cosmic conflagration that consumes the entire universe?[2] Is it evidence for the Roman Catholic notion of purgatory? Or is this the fire that attends the Day of the Lord (Zeph 1:18; Mal 3:2; 4:1)? These questions concern not only the exegesis of Mark 9:49 but of other obscure New Testament and non-canonical texts as well, including Q 3:16; Luke 12:49–50; *Gos. Thom.* 82; 1 Cor 3:10–15; 1 Pet 1:7; 4:12; 2 Pet 3:10 and others, which previous interpreters have rarely studied together.

The aim of this book is to illumine verses such as these that anticipate passing through fire of some sort as an aspect of the eschatological destiny not only of the wicked, but of the whole of humanity, and to contribute to the understanding of the fire of eschatological judgment as an ordeal or test, as distinct from the fire of hell or Gehenna, in the New Testament and earliest Christianity. More specifically, this investigation seeks to demonstrate that belief in the dual (purificatory as well as punitive) function of fire is attested to in the Christian tradition at a date earlier than is frequently recognized.[3] Our first task will be to examine the context of Jewish apocalyptic within which this tradition originates, noting the shift within the prophetic and apocalyptic traditions from the notion that the function of eschatological fire is strictly punitive to the more nuanced view that the judgment by fire comprises not only punishment for the wicked but testing and even purification for the righteous or repentant. The second task of our investigation will be to discern elements of this motif, according to which both the righteous and the wicked will face an eschatological ordeal by fire, within the preaching of John the Baptist and the historical Jesus, particularly with regard to John's proclamation of "baptism in fire" and several of Jesus' more

---

2. The question of whether Zoroastrianism had any influence on biblical eschatology is discussed in detail in Mayer, *Die biblische Vorstellung vom Weltenbrand*. Mayer ultimately concludes that biblical concepts regarding the fire of eschatological judgment developed independently and that Zoroastrian influence cannot be demonstrated. The opposite view is presented in Cohn, *Cosmos, Chaos, and the World to Come*. On the Stoic concept of the cosmic conflagration as it relates to our subject see especially van der Horst, "The Elements Will Be Dissolved with Fire." While the present study does not purport to be an exercise in the history of religions (*Religionsgeschichte*), the Stoic doctrine of *ekpyrosis* will receive attention in our treatment of 2 Peter 3:7, 10–14. All other matters pertaining to issues of *Religionsgeschichte* will be relegated to the footnotes. For a history of religions approach to our subject, see Lang, "Das Feuer im Sprachgebrauch der Bibel"; Lang, "πῦρ, κτλ."

3. In Christian tradition, the notion of a purifying fire is most frequently associated with the idea of purgatory, the origin of which is often attributed to the second and third century Alexandrian theologians, Clement and Origen. For discussion of these so-called "fathers of Purgatory," see Anrich, "Clemens und Origenes"; Eno, "The Fathers and the Cleansing Fire"; Le Goff, *The Birth of Purgatory*, 52–95.

enigmatic sayings. Lastly, we shall trace the contours of the development of this motif in the Pauline and Petrine Epistles.

Exegetical Context

Several of the texts to be considered herein were of particular interest to patristic exegetes—such as Clement of Alexandria, Origen, Lactantius, and Ambrose—who read them as adverting to a fiery test or ordeal at the last judgment.[4] The teachings of these notable church fathers are of great importance, not only because their interpretations helped lay the groundwork for the later doctrine of purgatory, but also because they provide us with some of the earliest exegesis of these texts, much of which resembles the arguments of this book. I wish here to contextualize our discussion by noting three points of historical development in the interpretation of the passages under consideration. First, we may observe that a number of early exegetes believed some of these texts to suggest that fire of some sort would attend the final judgment and that this fire would serve a refining and purifying function for the repentant but a destructive role for inveterate sinners. Second, these early interpretations contributed to the development of the later

---

4. For further discussion of the fiery ordeal in patristic interpretation, see Bulhart, "Ignis sapiens"; Edsman, *Le Baptême de Feu*; van Unnik, "The »Wise Fire« in a Gnostic Eschatological Vision." These scholars draw attention to the motif of "wise fire" in patristic and gnostic literature. Many of the relevant texts' depictions of the last judgment envision a river of fire through which all must pass. The fire burns the wicked but does not harm the righteous and is thus deemed "wise." While the gnostic literature falls outside the scope of this study, it is worth noting here that the "wise fire" motif also occurs in *Pistis Sophia* 45. The "river of fire" motif appears widely in both Jewish and Christian literature. See, for instance, Dan 7:10; 1QH XI.29–32; *1 En.* 14.19; 67.13; *Sib. Or.* 2.196–205; 252–54; 3.54; 84–87; *4 Ezra* 13.10–11; *2 En.* 10.20; *Apoc. Pet.* 6; 12; *T. Isaac* 5.21–25; *Apoc. Paul* 31–36; *Hist. Jos. Carp.* 13.9; 22.1; *Pistis Sophia* 145; 26.5; *2 Apoc. John* 26; *3 Apoc. John* 5, 13; *Encomium on Saint John the Baptist* 13; *Teaching of Apa Psote* 3b. The lesser known works of *2 Apoc. John* and *3 Apoc. John* can be found in Court, *The Book of Revelation and the Johannine Apocalyptic Tradition*, while *Encomium on Saint John the Baptist* and *Teaching of Apa Psote* appear in Budge, *Coptic Texts*, vols 3 and 5.2 respectively. See also the rabbinic texts cited in Schwartz, *Tree of Souls*, 158–59. For discussion of the river of fire in some of these texts and its probable Zoroastrian roots, see Cumont, *Lux Perpetua*, 224–28; Zandee, *Death as an Enemy*, 307–10. The closest parallel to the river of fire in the Zoroastrian literature is Bundahishn 30.19–20: "Afterwards the fire and halo melt the Shatvairo in the hills and mountains, and it remains on this earth like a river. Then all men will pass into that melted metal and become pure; when one is righteous, then it seems to him just as though he walks continually in warm milk; but when wicked, then it seems to him in such manner as though, in the world, he walks continually in melted metal" (trans. E. W. West in F. Max Müller [ed.], *The Sacred Books of the East*, V.125–26).

doctrine of purgatory. And third, Protestant commentators and theologians reacted with varying degrees of contempt to the doctrine of purgatory, a consequence of which has been substantial reluctance, even among contemporary commentators, to see any element of purification or refinement in passages where fire features in the final judgment.

While the great Alexandrian theologians never developed a comprehensive doctrine of purgatory, Clement and Origen are often considered the "fathers of purgatory," for they believed that the fire of the final judgment would be cathartic for some, thus contributing to the evolution of the later doctrine.[5] Clement, for instance, describes the fire that poured down on Sodom as τοῦ φρονίμου πυρὸς ἐκείνου "that wise fire"[6] and suggests that the fire is "an image of salvation," which may indicate that it functions remedially (*Paed.* 3.8.44.2.).[7] Elsewhere he writes of a fire "capable of exposing and curing superstition" and goes on to describe pagan temples destroyed by "the discerning [σωφρονοῦν] fire" (*Protr.* 4.53.2).[8] While the destruction of Sodom and the burning of pagan temples may appear solely punitive, in a passage on sacrifices Clement clearly attributes two functions to fire, suggesting that there are in fact two distinct fires: one is "the all-devouring earthly fire" but another is that which "sanctifies." The latter is a fire "of wisdom [φρονίμον], which pervades the soul passing through the fire" (*Strom.* 7.6.34.4).[9] Most significant for our study, when commenting on Matt 3:11–12//Luke 3:16–17, where John the Baptist preaches concerning the Coming One's baptizing with Spirit and fire and burning the chaff from the threshing floor with fire, Clement describes fire "not as being evil or bad, but as strong and capable of cleansing away evil. For fire is conceived as a good force and powerful, destructive of what is baser, and conservative of what is better. Wherefore this fire is by the prophets called wise [φρονίμον]" (*Ecl. Proph.* 25.4).[10] The "wise fire" that Clement describes in these texts discerns

---

5. See Anrich, "Clemens und Origenes als Begründer der Lehre vom Fegfeuer."

6. Van Unnik ("Wise Fire," 279) suggests that the presence of the demonstrative ἐκείνου ("that") may indicate the expression "wise fire" was well known.

7. My translation. According to the interpretation of Ramelli ("*Stromateis* VII and *Apokatastasis*," 254–55), "This means that salvation passes through punishment if necessary, and that punishment in turn aims at salvation."

8. *ANF* 2:187.

9. *ANF* 2:532.

10. *ANF* 8:46. Since nowhere do the Hebrew prophets identify fire as "wise," it may be that Clement here numbers among the prophets Heraclitus, who according to Hippolytus spoke of "wise fire" (*Ref.* 9.10.7). See Reinhardt, "Heraklits Lehre vom Feuer," 25.

between the just and the wicked; it knows when to mete out punishment and when to provide purification.

Similarly, when commenting on John the Baptist's proclamation of the Coming One's baptism in fire, Origen writes:

> In the Jordan river, John awaited those who came for baptism. Some he rejected, saying, "generation of vipers," and so on. But those who confessed their faults and sins he received. In the same way, the Lord Jesus Christ will stand in the river of fire near the "flaming sword." If anyone desires to pass over to paradise after departing this life, and needs cleansing, Christ will baptize him in this river and send him across to the place he longs for. But whoever does not have the sign of earlier baptisms, him Christ will not baptize in the fiery bath. (*Hom. Luc.* 24)[11]

Origen describes immersion in the river of fire as a means of eschatological cleansing; it serves a purifying function that enables those who receive immersion to enter into paradise. Notably, this "fiery bath" is something to be desired: only those who have already been baptized in water and Spirit will be blessed with the cleansing baptism in fire. Elsewhere, when discussing Paul's statement that one might be saved "through fire" (1 Cor 3:15), Origen speculates that this ordeal is universal: "We must all, I think, come to that fire. Even if one is a Paul or a Peter, he must still come to that fire" (*Hom. Ps.* 3.36).[12] In Origen's view the righteous will pass through the fire unharmed while sinners would experience its blazing flame as a purgative remedy.[13]

Origen's views on the purifying function of fire are likewise manifest in his comments on Romans 1, when he writes "This recompense of error weighs more heavily on them in their being delivered up to shameful affectations than if they were purified in the fire of wisdom [τῷ φρονιμῷ πυρί]." Here again we encounter the "wise fire" motif, which we have already observed in Clement. As in Clement, the motif here indicates that the fire may be punitive or remedial as the circumstance necessitates. Origen, however, stresses the remedial aspect, explaining how the fire is "a help towards

---

11 Trans. Lienhard, *Homilies on Luke*, 103.

12. Trans. Daly in Von Balthasar (ed.), *Fire and Spirit*, 354.

13. Origen believed in *apokatastasis*, the restoration of all things, and taught that all would eventually be saved. In his view, after the fiery ordeal some may be sent to hell, but their torment there would not be eternal; they would eventually return to the eternal contemplation of God (see *Princ.* 2.10.4.5). For discussion of this topic, see Daley, *The Hope of the Early Church*, 57–59. Indeed, such was Origen's faith in the gracious and purifying aspect of divine fire that he believed the lake of fire in Revelation would have a purifying effect on even the devil. See Ramelli, "Origen's Interpretation of Violence in the Apocalypse," 59–62.

the purification of the evils of their error," subsequently citing and explicating Malachi 3: "For He cometh like a refining fire and like the fuller's herb, cleansing and purifying them that have need of such remedies" (*Or.* 29.15).[14]

When commenting on 1 Cor 3:12–15 in *Hom. Ezek.* 1:3, Origen returns to the "wise fire" motif:

> What is this fire, O Apostle, that tests our works? What is this fire [*ignis*] that is so wise [*sapiens*] that it guards my "gold" and shows forth my "silver" more brilliantly, that it leaves undamaged that "precious stone" that is in me, so that it consumes only the evil I have done, the wood, hay, and straw that I have built up? [see 1 Cor 3:12–15]. What is this fire? "I came to cast fire upon the earth; and how I wish that it were kindled! [12:49]. Jesus Christ says: "How I wish that it were already kindled!" For he is good, and he knows that if this fire were kindled, wickedness will be consumed.[15]

Strikingly, Origen links the fire of 1 Cor 3:10–15 with the fire of Luke 12:49, a connection that is significant for our purposes for it brings together two key texts that will be examined in this book. Further, following the saying about casting fire upon the earth in Luke 12:49, Jesus goes on to speak of baptism, a juxtaposition which echoes John the Baptist's proclamation concerning the Coming One's baptism in fire. And as we have seen, Origen associates that baptism in fire with the cleansing and chastising fire of eschatological judgment. Significantly, Origen elsewhere associates 1 Cor 3:10–15 with the cosmic conflagration, probably with 2 Peter 3 in mind (*Contr. Cels.* 5.15). Thus, at least in the mind of Origen, several of the texts we are to consider belonged to a single category of eschatological thought.

According to Lactantius, at the last judgment God tries both the wicked and the righteous by means of fire. Like Clement and Origen, Lactantius believes that the righteous will not feel the flame, but he apparently does not see the ordeal as purifying the wicked:

> The same divine fire, therefore, with one and the same force and power, will both burn the wicked and will form them again, and will replace as much as it shall consume of their bodies, and will supply itself with eternal nourishment. . . . But when He shall have judged the righteous, He will also try them with fire. Then they whose sins shall exceed either in weight or in number, shall be scorched by the fire and burnt: but they whom full justice and maturity of virtue has imbued will not perceive the fire; for

---

14. Trans. O'Meara, *Origen: Prayer, Exhortation to Martyrdom*, 124.
15. Trans. Scheck, *Homilies 1–14 on Ezekiel*, 30.

they have something of God in themselves which repels and rejects the violence of the flame. So great is the force of innocence, that the flame shrinks from it without doing harm; which has received from God this power, that it burns the wicked, and is under the command of the righteous. (*Div. Inst.* 7.21)[16]

While the phrase "wise fire" is not here employed, we are again presented with a fire that discerns between the wicked and the righteous.[17] The fire appears to function punitively with regard to the former, for as the fire consumes the bodies of the wicked it regenerates for them new bodies, only so that they may be consumed eternally. The notion is not one of purification but of eternal punishment. The fate of the righteous, however, is much different. They are immune to the fires of judgment. They pass the ordeal on account of their virtue.

Ambrose envisions a fiery ordeal at the last judgment as well, though it is unclear whether his views are closer to those of Clement and Origen or to those of Lactantius:

> Before the resurrected lies a fire, which all of them must cross. This is the baptism of fire foretold by John the Baptist, in the Holy Ghost and the fire, it is the burning sword of the cherub who guards the gate of heaven, before which everyone must pass: all shall be subjected to examination by fire; for all who want to return to heaven must be tried by fire. (*In Ps. CXVIII*, sermo 20)[18]

Notably, Ambrose brings together several aspects of the most significant texts we shall consider. He envisions a baptism in fire (see Matt 3:11//Luke 3:16); he interprets this as a trial or a test (see 1 Cor 3:10–15); and he believes that *everyone* will face this fiery trial (see Mark 9:49). Regarding this last element, Ambrose believes that "even Jesus, the apostles, and the saints first had to pass through the fire before entering Heaven" (see Luke 12:49–50).[19]

While this early and important strand of tradition interpreted many of these texts through the paradigm of the final judgment and assigned the

---

16. ANF 6:217. While Lactantius does not in this context explicitly cite the *Oracle of Hystaspes*, Hultgård ("Persian Apocalypticism," 75) notes, "[t]his is precisely the character of the eschatological fire that is described in the Iranian apocalyptic texts." See Hinnells, "The Zoroastrian Doctrine of Salvation," 134.

17. One subtle distinction worth noting is that Lactantius seems to emphasize the character of the individual over the intelligence of the fire, for it is on account of the "force of innocence," rather than the wisdom of the fire, that "the fire shrinks."

18. Trans. Le Goff, *Birth of Purgatory*, 59.

19. Le Goff, *Birth of Purgatory*, 59.

fire of the eschatological trial a testing and refining function, most modern commentators exhibit some aversion to the potentially purifying function of fire, presumably owing to the influence of these texts and passages like them in the historical development of the doctrine of purgatory and the Catholic-Protestant debates that subsequently ensued. For my present purposes I am less interested in the fully developed doctrine of purgatory and more interested in the reactionary stance of some of the most prominent early Protestant scholars and commentators, whose influence continues to be felt.

The center of gravity in these debates has always been the interpretation of 1 Cor 3:15. When commenting on this passage, Luther writes, "They turn this fire also into a purgatory, according to their custom of twisting Scripture and making it mean whatever they want."[20] The comments of John Calvin are in the same vein: "anyone who fouls the golden purity of God's Word with this filth of purgatory must undergo the loss of his work."[21] More vehement are the popular commentators of the 1600s. John Trapp, for instance, insists that the fire of v. 15 is "[n]ot of purgatory (a popish fiction), but of the Holy Ghost,"[22] while Matthew Poole offers the following: "For the fire of purgatory, it is a fiction, and mere imaginary thing, and of no further significancy than to make the pope's chimney smoke."[23] One finds much the same in Matthew Henry: "On this passage of scripture the papists found their doctrine of purgatory, which is certainly hay and stubble: a doctrine never originally fetched from scripture, but invented in barbarous ages, to feed the avarice and ambition of the clergy."[24] While this sort of anti-papal rhetoric is thankfully absent from most recent commentaries, reticence regarding the potential purificatory quality of fire persists. My contention is that the majority of Protestant exegetes who have revolted against the use of texts such as 1 Cor 3:15 as proof texts for purgatory have been overly skeptical regarding the possibility that fire may play a purifying function in these texts. While I do not wish to argue that these passages ought to serve as proof-texts for purgatory, I do wish to suggest that in some cases

20. "The Thirty-Seventh Article," in *Luther's Works*, 32:95. It is worth noting, however, that on the same page Luther writes, "The existence of a purgatory I have never denied. I still hold that it exists, as I have written and admitted many times, though I have found no way of proving it incontrovertibly from Scripture or reason."

21. Calvin, *Institutes*, 1:681. For his part, Calvin acknowledges the distinction between the early exegeses of the church fathers cited above, who associated the fire of 1 Cor 3:15 with the fiery trial at the last judgment, and the later development of the doctrine of purgatory. He is, however, also at odds with the interpretations of Clement and Origen.

22. John Trapp, *Commentary on the Old and New Testaments*, 5:524.

23. Poole, *Commentary on the Holy Bible*, 3:548.

24. Henry, *A Commentary upon the Holy Bible*, 6:126.

Research since 1950

The New Testament texts relating to fire seem to be a subject of relative unpopularity in recent biblical scholarship. This state of neglect is no doubt a reflection of the theological unease that many feel concerning eschatological judgment in general and the subject of hell in particular, with which fire is most frequently associated.[25] Moreover, as we have already suggested, it may likewise result from the distaste Protestant exegetes have traditionally felt for the Roman Catholic idea of purgatory and their consequent discomfort with those New Testament texts in fire appears to assume a cleansing function.[26] There was a period, however, primarily in the middle of the twentieth century, during which a handful of scholars took up the subject of fire in the Bible as an area worthy of research. These studies were primarily doctoral theses and dissertations, which in succeeding years have been supplemented by encyclopedia entries and journal articles focusing on specific sayings or texts. The following survey of the most significant of these contributions will focus largely on conceptual categories and broadstroke historical developments while engagement with individual texts will be postponed until the detailed exegesis reserved for later chapters.

Friedrich Lang

Perhaps the most systematic treatment of fire in the Bible and related literature to date is that undertaken by Friedrich Lang in his doctoral thesis, "Das Feuer im Sprachgebrauch der Bibel,"[27] an abridgment of which was later published in his entries on πῦρ, κτλ. in the *Theological Dictionary of the New Testament*.[28] Lang's discussion includes the functions of fire in Greek

---

25. For a discussion of the decline of belief in hell see Allison, "Problem of Gehenna."

26. I do not wish to suggest that all Protestant interpreters have arrived at their conclusions for theologically tendentious reasons. However, even where a prejudice against Roman Catholic interpretations is not consciously harbored, there has been such a fixed pattern of reading these texts in the Protestant tradition that the possibility of assigning a purifying function to the fire has not been seriously considered. The influence of this pattern of thinking appears even in the work of Joachim Gnilka, who was a member of the Pontifical Biblical Commission and whose book, *1 Kor 3,10–15*, is discussed below.

27. Lang, "Das Feuer im Sprachgebrauch der Bibel."

28. Lang, "πῦρ, κτλ." Given the influence of Lang's *TDNT* article, our engagement

and Hellenistic thought, in Persian thought, and in the canonical and non-canonical religious texts of Judaism and Christianity through the second century CE. One of the most useful contributions of Lang's work is his categorization of the various kinds of fire in the Bible and related Jewish and Christian texts. Lang observes that in the Hebrew Scriptures fire is often associated with God and his divine activities. It appears in relation to God namely in theophanies (Exod 3:2; 13:21f.; 14:24; 19; 24:17; Judg 6:21; Neh 9:12, 19; Ezek 1:28; Daniel 7; 10:6), as a means of divine judgment (Gen 19:24; Exod 9:24; Lev 10:2; 2 Kgs 1:10; Mal 3:19), as a sign of gracious visitation (Gen 15:17; Lev 9:23f.; Judg 6:21; 1 Kgs 18:38; 1 Chr 21:26; 2 Chr 1:7), and as a term for God (Deut 4:24; 9:3; Isa 33:14).[29] All of these categories are relevant to our investigation, but it is the second category, fire as a means of divine judgment, that is our chief concern. It is this category that appears to hold the most promise for contextualizing sayings like Mark 9:49. It must be borne in mind, however, that these categories are not isolated from one another. As Lang observes, "[t]he close relation between images of judgment and theophany expresses the fact that fire is understood, not as a blindly raging natural force, but as an instrument of punishment in the hand of the divine Judge."[30] Nevertheless, these categories provide a useful heuristic for differentiating between the disparate *functions* of fire in the biblical tradition. Beyond this classification of the different functions of fire in the Hebrew Bible, Lang suggests that the category of fire as a means of divine judgment can be divided into three distinct subcategories, reflecting the three roles that fire was expected to play in the eschatological assize:

> 1. It is a sign of the day of Yahweh, Jl. 2:30. 2. Yahweh will execute the judgment of eschatological destruction with fire on all His enemies, Mal. 3:19; Is. 66:15 f.; Ez. 38:22; 39:6. 3. The damned fall victim to eternal torment by fire ... see Is. 34:10; Jdt. 16:17; Sir. 21:9 f.[31]

Regarding the subcategory that he terms "eternal torment by fire," Lang avers that in the Old Testament this does not yet refer to hell.[32] Moreover, he finds no obvious reference to a cosmic conflagration and insists that nowhere is the judgment of fire connected with the concept of the ordeal.[33]

---

will be primarily with that publication.
29. Ibid., 935–37.
30. Ibid., 936.
31. Ibid., 936–37.
32. Ibid., 937.
33. Ibid., 936, n. 46.

Upon turning to the New Testament, Lang likewise gives most of his attention to the category of fire as a means of divine judgment. In this discussion Lang rightly distinguishes between "the eschatological fire of judgment" and "the eternal fire of hell" which torments sinners. We must pause, however, to elucidate this terminology, for it is somewhat confusing given that "the eternal fire of hell" can also be understood as a means of eschatological judgment when the word *judgment* is taken in the punitive sense. Marius Reiser's discussion of the language of judgment is instructive here: As Resier observes, "The concept of '*eschatological judgment*' or '*final judgment*' incorporates two different models that, if possible, should be kept separate: the eschatological judgment as a court of punishment and destruction, or as a forensic judgment."[34] What Lang terms "the eschatological fire of judgment" we might take as shorthand for the fire of forensic judgment that tests, chastens, and perhaps even purifies individuals on the Day of the Lord, finding some righteous and others wicked, whereas what he calls "the eternal fire of hell" might be taken to indicate the eternal fire of punitive judgment in hell; it is the execution of judgment carried out upon those whom God has found guilty. It should be noted that whereas the former may be understood as a test, punishment, or perhaps even as purification, the latter is understood only as punishment in the New Testament.

While the sayings of hell fire are a topic deserving of scholarly attention in their own right, the New Testament sayings concerned with the fire of eschatological judgment, which are even more neglected than those concerning the fires of hell, are our chief concern.[35] Lang correctly observes that the idea of an eschatological judgment of fire is "the best starting-point for an interpretation of the difficult sayings Mk. 9:49 and Lk. 12:49" as well as John the Baptist's pronouncement concerning the Coming One's baptism with Holy Spirit and fire and the apocryphal logion now found in *Gos. Thom.* 82.[36] Likewise, according to Lang, some texts outside of the Gospels allude to the eschatological judgment of fire, namely 1 Cor 3:13; 15; 2 Thess 1:8; Rom 12:20 in the Pauline corpus and Heb 12:29; Rev 20:9; and 2 Pet 3:7.[37]

Lang's studies will prove useful to our exploration in several ways. First, his delineation of the different categories of fire in the Hebrew Bible, while they may not be mutually exclusive in every case, provide a viable

---

34 Reiser, *Jesus and Judgment*, 22.

35. For recent scholarship on the topic of hell see especially Allison, "Problem of Gehenna"; Bernstein, *Formation of Hell*; Powys, '*Hell*'; Turner, *History of Hell*.

36. Lang, "πῦρ," 936.

37. Regarding 2 Pet 3:7, Lang writes, "This late passage is the only one in the NT in which the doctrine of a world conflagration, current in Babylonia, Persia and Greece, is distinctly combined with the apocalyptic concept of judgment" (945).

heuristic for considering the varying *functions* of fire as it appears in several disparate texts. Secondly, with very few exceptions he assigns the New Testament texts we wish to consider herein to the category of "eschatological fire of judgment," thus suggesting a helpful context for understanding the texts with which this book is concerned. Finally, Lang's silence regarding a purifying or cleansing function of fire in the Bible invites further research and consideration.

Joachim Gnilka

In his doctoral dissertation, *Ist 1 Kor 3,10–15 ein Schriftzeugnis für das Fegfeuer?*, Joachim Gnilka focuses his attention on the Pauline pericope 1 Cor 3:10–15, taking up the particular question of whether or not this text is best interpreted as a scriptural witness to the doctrine of purgatory.[38] The bulk of his work is a collection of patristic quotations concerning the interpretation of this passage, which need not be entered into here. More immediately relevant are his introductory and concluding chapters in which Gnilka provides a discussion of the Jewish apocalyptic literature pertaining to fire in general and offers his assessment of the usefulness of this material for the exegesis of 1 Cor 3:10–15 in particular.

Gnilka rightly begins with the observation that the starting point for exegesis of the New Testament is an understanding of the thought world of the Hebrew Bible and the Jewish literature of the period. As does Lang, Gnilka observes that the image of fire functions variously throughout these texts and, though less systematically than Lang, he applies several of the same categories. Fire appears in theophanies and even metaphorically to describe God himself. It is frequently also an image of God's wrath (Jer 21:12; Ps 89:47; Ezek 22:31), and the divine fury is vividly portrayed in descriptions of YHWH with breathing blazing flames from out of his nose (Jer 15:14; Ps 18[17]:9; Deut 32:22).[39] However, Gnilka stresses that it is often difficult to discern whether the fire is to be understood literally or figuratively, whether it is a real phenomenon or simply a metaphorical expression indicating divine anger.

While he detects no trace of the cosmic conflagration in the Hebrew Scriptures, Gnilka does find that a universal judgment by fire is described in apocryphal texts. He notes that fire alone (*T. Isaac* 7; *Syr. Apoc. Baruch* 48:39) or fire paired with flood (LAE 49; *T. Lev.* 3) imagery sometimes

---

38. Gnilka, *1 Kor 3,10–15* (All translations are my own).
39. Ibid., 13.

appears as a judicial element.⁴⁰ Thus, according to Gnilka, at a relatively early period a universal judgment by fire was envisioned. The question that concerns Gnilka is whether or not this expectation could have been known and used by Paul. Gnilka identifies what he believes to be a new motif in the subsequent literature, including texts such as *T. Isaac* 7:3 and *Sib. Or.* 7:13. In *T. Isaac* 7:3 the seer envisions the following: "The river of fire is so wise that it does not damage the pious, only the sinner alone, which it burns." The dual function of this fire is apparent and is similarly attested to in *Sib. Or.* 7:13: "And then all shall pass through the blazing river and the unquenchable flame. All the righteous will be saved, but the wicked will be destroyed for all ages." The purificatory qualities attributed to the fire in these texts suggest to Gnilka that the authors of both of these documents may have been influenced by Origen's doctrine of *apokatastasis*, the ultimate restoration of all things to God, and thus, according to Gnilka's presuppositions, they can represent only later Christian concepts. Gnilka further notices, however, that a very close parallel to the river of fire motif exists in the flood of fire depicted in several Jewish apocalyptic texts (*T. Levi* 3: Jos., *Ant.* 1, 2, 3; *4 Ezra* 7:31 f.; 13:10f; *Ps. Sol.* 15:6. *1 En.* 52; *L.A.E.* 49), though these texts envision the stream on a more cosmic scale.⁴¹ Gnilka observes that in the non-canonical Jewish texts that feature the river of fire, a reference to Dan 7:10 is probable. He stresses, however, that while the fire tests and punishes, it rarely purifies. In Gnilka's view Origen's association of this motif with purification at last judgment is due to the influence of later Jewish apocalypses, which also impacted Origen's exegesis of 1 Cor 3:10–15.⁴²

Regarding the role that fire plays in the New Testament texts concerning the Day of Judgment, Gnilka comments that "[a]t only one point, 2 Peter 3, is the cosmic conflagration explicitly mentioned in connection with the general judgment."⁴³ Moreover, in reference to the Synoptic Gospels, Gnilka makes reference only to the apocalyptic discourses in Mark 13 and its parallels as being remotely concerned with the Day of Judgment, concerning which he is content to conclude only that the discourse is not concerned with fire. Thus, he apparently sees fire as playing no significant role as a test or ordeal in the Gospels. Regarding his chief concern, 1 Cor 3:10–15, Gnilka judges that the world judgment by fire is in no way envisioned. Moreover, he insists that the interpretation that the fire has a cleansing quality, "is not possible without a very loose treatment of the

---

40. Ibid., 14.
41. Ibid., 16, n. 16.
42. Ibid., 16.
43. Ibid.

biblical texts."⁴⁴ He traces this entire line of interpretation back to Origen's influence, concluding, "Origen speaks for the first time of a cleansing fire. He is rightly regarded as the founder of the doctrine of purgatory."⁴⁵ Gnilka contends that Origen's reading was unduly influenced by the Old Testament apocryphal literature, which contains elements that in later exegesis were inappropriately imported into 1 Cor 3:10–15. Since he is more concerned with proving that the fire of 1 Cor 3:10–15 does not witness to the doctrine of purgatory, Gnilka says very little about what it does represent until his conclusion. Even there, however, he minimizes the role of the fire, suggesting that it is primarily theophanic. Reading 1 Cor 3:10–15 in light of 2 Thess 1:8, he concludes that the fire is an aspect of God's coming in judgment, but is not instrumental to that judgment.⁴⁶

Our full engagement with 1 Cor 3:10–15 must wait for a later chapter. At the moment, however, it is necessary to make a few observations concerning Gnilka's proposals as they relate to our study. Gnilka is surely correct that the doctrine of purgatory is not evidenced by any New Testament texts and that it owes its developed form to later Christian theologians such as Clement and Origen. However, he has prematurely limited his focus to the categorical concepts of the developed Christian doctrine of purgatory and the Stoic notion of the cosmic conflagration. For instance, he completely disregards the more obscure texts in the Synoptic Gospels such as Q 3:16; Mark 9:49; and Luke 12:49–50; 17:26–32; 23:31 which envision fire as playing some instrumental role in eschatological judgment. Gnilka frequently and forcefully asserts that a cleansing quality is not attributed to fire prior to the influence of Origen. This claim will require further investigation on our part as we proceed through our consideration of the early Jewish literature. Specifically, we will have to scrutinize thoroughly Gnilka's assertion that the Jewish apocalyptic literature that may have influenced Origen's exegesis does not represent a worldview contemporaneous with the authors of the New Testament and hence does not provide a suitable context for understanding Jesus, Paul, and the later evangelists, for as we shall see presently, other exegetes have read the material differently. Lastly, Gnilka's connection of the stream of fire with the eschatological judgment will play an important role in our consideration of the idea of baptism in fire, as it does in the exegesis of James D. G. Dunn, to whose work we now turn.

---

44. Ibid., 118.
45. Ibid., 115.
46. Ibid., 126.

James D. G. Dunn

In a number of publications, James D. G. Dunn has shown a lively interest in John the Baptist's pronouncement that the Coming One would baptize "with the Holy Spirit and fire."[47] Dunn has consistently argued that the Baptist envisioned baptism in Holy Spirit and fire as a metaphor of judgment and, at the same time, a metaphor of mercy. That is, he understood baptism in Holy Spirit and fire to signify "the fearful effects of God's wrath on Israel's sin and presumption, purgative, purifying those who repented, destructive, consuming those who did not."[48] Notably, Dunn envisions the baptism in Holy Spirit and fire (ἐν πνεύματι ἁγίῳ καὶ πυρί) as a single baptism. The Baptist did not expect two distinct baptisms, one gracious baptism in Holy Spirit for the repentant and another wrathful baptism in fire for the unrepentant. Rather, Dunn argues:

> John held out before his hearers . . . a baptism which was neither solely destructive nor solely gracious, but which contained both elements in itself. Its effect would then presumably depend on the condition of its recipients: the repentant would experience a purgative, refining, but ultimately merciful judgment; the impenitent, the stiff-necked and hard of heart, would be broken and destroyed.[49]

Dunn contends that since fire could symbolize purification as well as judgment, John the Baptist's metaphor is best understood in the context of the apocalyptic image of "judgment, purification in *a river of fire*." More, spirit (רוּחַ), like fire, could denote both blessing and judgment and was used in connection with the river of judgment (Isa 30:27f.). Thus, Dunn proposes that John combined all three symbols—river, fire, and spirit—so as to depict the imminent judgment as "God's fiery breath [רוּחַ] like a great stream through which all men must pass before the new age could appear."[50] Further, Dunn believes,

> [John's] metaphor is best understood as *a variation on a prominent theme within apocalyptic expectation—viz. "the messianic woes,"* that is, the conviction that the ending of the old age and

---

47. See Dunn, *Baptism in the Holy Spirit*, 8–22; Dunn, "Spirit-and-Fire Baptism"; Dunn, "Birth of a Metaphor"; Dunn, *Jesus Remembered*, 366–69. The following comments will engage primarily with the articles "Spirit-and-Fire Baptism" and "Birth of a Metaphor," which contain the fullest expression of Dunn's views.

48. Dunn, "Birth of a Metaphor," 135.

49 Dunn, "Spirit-and-Fire Baptism," 86.

50. Dunn, "Birth of a Metaphor," 136.

> the introduction of the new would be marked by a period of severe distress, or universal affliction and tribulation, of intense suffering and anguish, "the birth pangs of the Messiah."[51]

Thus, those penitents who willingly accepted John's baptism in the Jordan would be preserved in and through the coming tribulation, conceptualized as a stream of fiery breath issuing forth from the throne of God (Daniel 7), while the obdurate would be destroyed. This is the very image of a fiery ordeal or test.

Perhaps most significant for continued research into this motif is Dunn's further suggestion that while Jesus may not have fully embraced John's rhetoric of judgment, there exists an "important but neglected strand of teaching in the Jesus tradition," which he calls "fire-words," namely, Mark 9:49; Luke 9:54; 12:49–50; and *Gos. Thom.* 82.[52] These, he notes, bear a striking resemblance to the message of John the Baptist. Significantly, with the sole exception of Luke 9:54, these and the Baptist's proclamation are precisely the sayings that Lang included in his category of "eschatological judgment of fire" texts in his treatment of the Gospels. These sayings of Jesus, particularly Luke 12:49–50, which similarly juxtapose the two images of baptism and fire, seem to echo deliberately the proclamation of John the Baptist.

Dunn, however, perceives a reinterpretation of the Baptist's preaching in the words of Jesus, for while in Luke 12:49 it appears that Jesus may have believed that he had come to administer the fiery baptism, he now foresees himself as being required to endure this baptism himself:

> That is to say, he saw himself not so much as the baptizer in Spirit and fire but as the one who would himself be baptized in Spirit and fire, who would himself (alone?) endure the birth pangs of the new age—presumably in the hope that the new age might be brought in fully thereby, and that others might either be spared the messianic tribulation or at least be enabled to pass through the river of fiery *ruah* [רוּחַ] for themselves (*see* He 2:5–10).[53]

Dunn perceives a subtle shift. Jesus is no longer the dispenser of the baptism in fire, but its recipient, perhaps its sole recipient.

Dunn has contributed much of substance to the discussion of the question at hand. First, he has pointed to the dual function of the baptism prophesied by John the Baptist, arguing that it is simultaneously purgative and destructive depending upon the status of the recipient. He also provides

---

51. Ibid., 135.
52. Ibid., 137.
53. Ibid., 138.

a viable context for understanding this particular saying of John the Baptist, namely the tribulation or messianic woes. He has further observed that Jesus was influenced by this message and took it up himself in several of his more enigmatic pronouncements having to do with fire. Lastly, he raises the intriguing question of whether or not Jesus reinterpreted the message of the Baptist. All of these insights and questions are sure to play an integral role in our consideration of the "fire-words" of Jesus and other New Testament traditions related to the fiery ordeal.

Aaron Milavec

In his provocative and stimulating essay "The Saving Efficacy of The Burning Process in *Didache* 16:5," Aaron Milavec argues that "*Did* 16.5 offers an overlooked testimony to the dual function of eschatological fire more than a century prior to Tertullian and Clement of Alexandria," who are typically identified as the originators of the concept of purgatorial fire in Christian thought.[54] Moreover, he argues that examples of "the dual process of burning" can also be discerned "within the Jewish prophetic writings and the early Christian apocalyptic literature, thereby demonstrating that 'purgatorial fire' was not an invention of the *Didache* but the expression of a minor stream of thought which had a long development prior to the third century."[55]

Milavec draws our attention to the important fact that "[w]ithin the Jewish prophetic tradition, alongside of the metaphor of destructive fire as portraying God's judgment, one also finds intimations of those who pass through or near this fire unharmed" (see Isaiah 33). It is in Ezekiel and Malachi, according to Milavec, where the theme of purifying fire is most fully developed. In these texts the image of smelting is employed as a metaphor for God's refining and purifying fire of judgment (Ezek 22:20; Mal 3:2–3; see Isa 1:24; Jer 6:29; Zech 13:9; Prov 17:3; Sir 2:5). As Milavec describes it, "[t]he process here is that of the purification of raw metals that are mined from the earth by firing them in a furnace in order to allow the pure metal to be melted and thereby separated from the rocks to which it adheres."[56] The book of Malachi is particularly striking, for in it the dual function of fire is clearly displayed, providing "the clearest prophetic images of the dual functioning of fire within an apocalyptic scenario."[57] It is here that "The day of his coming" is described as a "refiner's fire" that will function to "purify

---

54. Milavec, "Saving Efficacy." See also Milavec, *The Didache*, 811–28.
55. Milavec, "Saving Efficacy," 131.
56. Ibid., 146.
57. Ibid., 147.

the sons of Levi and refine them like gold and silver" (Mal 3:2–3). In contrast, the next chapter of the very same text indicates that the fate of "the arrogant and all evildoers" will be quite different: "the day that comes shall burn them up" (Mal 4:1). It is within this context that Milavec situates his interpretation of *Did.* 16.5.

Didache 16:5 teaches that at the end, "all humankind will come up to the fiery test [τὴν πύρωσιν τῆς δοκιμασίας]; many will stumble and perish, but those who remain in their faith will be saved by the curse itself [σωθήσονται ὑπ' αὐτοῦ τοῦ καταθέματος]." The "curse" in this text has typically been interpreted as an allusion to Gal 3:13, according to which Christ "became a curse [κατάρα] for us."[58] However, in Milavec's view, its antecedent may be "the fiery trial" mentioned in the preceding clause, in which case the elect would be saved by the fire itself, that is by its purifying and refining effect.[59] While Milavec's interpretation of "the curse" is not without its critics,[60] he is surely correct to emphasize that "the term δοκιμασία (testing) would appear to suggest that both positive and negative results are anticipated."[61] The fire to which "all of humankind" will be subjected will presumably find some righteous and others wicked. Milavec further notes that *Herm. Vis.* 4.1.10, which employs the image of gold being refined by fire to symbolize the fire which destroys all that is wicked in the world, simultaneously tests and refines the elect, thus providing further evidence for the dual function of fire in early Christian thought. Similarly, the Christian *Sibylline Oracles* (esp. 2.286–89; 315–17; 8.411) depict a final judgment with fire. "All of this indicates," writes Milavec, "that the dual function of the eschatological fire was the preferred metaphor for God's final judgment long before Clement of Alexandria and Origen gave it theological precision within the Christian message."[62]

Milavec concludes that *Did.* 16.5 represents one instance within a stream of thought in Jewish and Christian apocalyptic literature according to which fire serves a dual function, simultaneously purifying the elect while punishing the wicked. This tradition originates, at least in the Hebrew

---

58. See Niederwemmer, *Didache*, 221. It is worth noting, however, that the parallel is closer in the English translation than it is in the Greek. Behind the English word "curse" of each text lie two distinct, though related, Greek words. Whereas *Did.* 16.5 uses κατάθεμα, Gal 3:13 has κατάρα. If an allusion were intended, we might expect the same word.

59. Ibid., 137–51; see Draper, "Jesus Tradition in the Didache," 282.

60. See Pardee, "Curse that Saves." Although Pardee does not endorse Milavec's reading, she concedes that his proposal is "not without its merit" (158).

61. Milavec, "Saving Efficacy," 154.

62. Ibid., 150.

Bible, within the prophetic tradition, and its impact is evident in certain early Christian apocalyptic texts. However, while he allows that 1 Cor 3:10–15 and 1 Pet 1:5–7; 4:12–13 are ambiguous regarding the precise function of eschatological fire, Milavec is reluctant to discern such a motif in the New Testament itself. Curiously he is silent concerning sayings of John the Baptist (Q 3:16) and Jesus (Mark 9:49; Luke 12:49–50) that may be taken to attribute at the very least a similarly ambiguous or perhaps even dual function to fire.

As does Dunn in his treatment of John the Baptist's message, Milavec discerns in the fiery ordeal of *Did.* 16.5 the notion that fire may serve both to purify and to punish in the period of eschatological tribulation, thus providing possible evidence for a relatively early conception of purificatory fire. More importantly, he makes clear that although the very possibility that fire could have purificatory qualities is frequently ignored in the treatment of New Testament texts, it is by no means foreign to the thought world out of which these texts arose, namely the scriptures of Israel, and particularly the prophets. Milavec's suggestion that eschatological fire serves a dual function in the literature ranging from the prophets and into the earliest Christian writings outside the New Testament is highly suggestive and raises the question of whether there is any evidence of this tradition in the New Testament itself.

Outstanding Issues and Questions

The contributions surveyed above provide a rich and welcome context for considering the several New Testament texts such as Q 3:16; Mark 9:49; Luke 12:49–50; *Gos. Thom.* 82; 1 Cor 3:10–15 and others that speak enigmatically of the fire to which apparently everyone (or at least many, including the elect) must be subjected. The context that has been found most suitable for the exegesis of these texts is the conceptual category that Lang describes as the eschatological fire of judgment, which may also be associated with the final tribulation or messianic woes as suggested by Dunn and Milavec, and which we may include under the idea of the fiery ordeal.

The scholars surveyed above differ in their opinion as to whether there is any evidence in the New Testament for the notion of a cleansing fire in association with the final judgment. Most notable in this regard are Gnilka's negative judgment concerning 1 Cor 3:10–15 and Dunn's affirmative verdict regarding Q 3:16. To come to our own conclusions regarding this controversial issue, we must first turn our attention to a survey of the relevant texts in Hebrew Bible and the literature of Second Temple Judaism before proceeding to our exegesis of the New Testament texts in question.

Given our interest in the apparent dual function of fire in the New Testament, perhaps most suggestive for our study are Milavec's arguments 1) that the Hebrew Scriptures attest to the dual function of fire as is most clearly demonstrated by the simultaneously refining and destroying fires in Malachi, and 2) that some of the earliest Christian texts outside the New Testament attest to a similar function. Milavec himself is, however, reluctant to assert that any texts within the New Testament canon have both purifying and destructive functions, locating the earliest appearance of this motif in Christian literature in *Did.* 16:5. Hence, the principal question we seek to address is this: can a dual function similarly be discerned in any New Testament texts?

Dunn has offered a positive response to this question in at least one instance, namely in the message of John the Baptist. Dunn's work thus provides an excellent starting point for our consideration of New Testament texts. This is especially so not only because he presents the strongest argument in favor of the dual function of fire in the New Testament to date, but because he locates the origin of this tradition in the preaching of John the Baptist, whose ministry antedates that of Jesus, Paul, and naturally the authorship of the New Testament texts we wish to take up. Dunn, moreover, has linked several of Jesus' "fire-words" to the preaching of the Baptist, suggesting that many of these sayings deliberately echo the language of John. What Dunn does not address explicitly, however, is the question of whether or not the "fire-words" of Jesus suggest a dual function in the same way that he believes John's proclamation of the baptism in the Holy Spirit and fire did. Arguing that Jesus did in fact adopt this teaching from the Baptist will thus be a key concern in the following argument.

Lastly, several of the exegetes considered above, particularly Gnilka, expressed considerable reservations about taking the language of eschatological judgment with fire literally and were more content to interpret it purely as a metaphor for God's wrath and fury. Similarly, Dunn and Milavec were willing to interpret the fire motif as a metaphor for the tribulation or the messianic woes without clarifying whether or not the literal element of fire was an expected feature of this period of eschatological travail. In light of the recent debates over the nature of apocalyptic language and whether it is to be understood literally or metaphorically, we cannot ignore the question of whether this particular apocalyptic motif was employed as a figurative expression or represents a literal expectation.[63]

---

63. For discussion see Wright, *The New Testament and the People of God*, 94–98, 202–9, 320–68; Allison "Jesus and the Victory of Apocalyptic"; Adams, *Stars Will Fall*.

Symbolism, Imagery, and Realism

Fire is an unusually potent image.[64] Well aware of this, the Greek theologians Clement and Origen, whose influence has already been noted with regard to the history of the interpretation of our texts, asked important questions regarding the nature of the fire described in passages such as Matt 3:11//Luke 3:16, Luke 12:49-50, and 1 Cor 3:10-15. In his *Stromata*, Clement differentiates between a fire that "sanctifies . . . sinful souls" and "the all-devouring vulgar fire" (7.6).[65] On Le Goff's reading of the evidence, "For Clement of Alexandria, the 'intelligent' fire that enters into the sinner's soul was not a material thing . . . , but neither was it a mere metaphor: it was a 'spiritual' fire."[66] Similarly, Origen notes the dual function of fire, observing that fire may both "enlighten" and "burn" (*Hom. Exod.* XII.4), and he interprets the fire that Jesus came to cast upon the earth as an enlightening—and thus presumably metaphorical—fire. Elsewhere, Origen ponders whether the fire of judgment in 1 Cor 3:10-15 ought to be taken σωματικῶς "literally" (*Con. Cels.* 4.13; see 6.71) but ultimately favors a more figurative interpretation.

The question of whether a particular biblical image or symbol ought to be taken literally or figuratively remains a lively one in contemporary biblical scholarship, particularly with regard to eschatological language. In the concluding chapters of his book *The Language and Imagery of the Bible*, G. B. Caird proposes, "myth and eschatology are used in the Old and New Testaments as metaphor systems for the theological interpretation of historical events."[67] In contrast to the *Konsequente Eschatologie* of Weiss and Schweitzer, which took end-of-the-world language quite literally,[68] Caird argues that while the biblical writers did believe that at some point in the future the world would come to a literal end, "[t]hey regularly used end-of-the-world language metaphorically to refer to that which they well knew was not the end of the world."[69] This thesis has been taken up with vigor by N. T. Wright, Caird's former student, who emphatically asserts that the early Christians used apocalyptic language to refer to the climactic events

---

64. On the religious symbolism of fire in general see Edsman, "Fire," in *The Encyclopedia of Religion*.

65. *ANF* 2:532.

66. Le Goff, *The Birth of Purgatory*, 55.

67. Caird, *Language and Imagery of the Bible*, 219.

68. Weiss, *Jesus' Proclamation of the Kingdom of God*; Schweitzer, *Quest of the Historical Jesus*.

69. Caird, *Language and Imagery of the Bible*, 256. Caird was no doubt influenced by the views of C. H. Dodd, his teacher, whose case for "realized eschatology" was set forth in Dodd, *Parables of Jesus*.

unfolding in their time but without reference to "the end of the space time universe."[70] Wright's thesis, in turn, has been forcefully countered by Dale C. Allison Jr., who contends that Wright has misrepresented the perspectives of Weiss and Schweitzer—namely that they spoke of "the end of the space time universe"—and urges that texts such as Mark 13 and 2 Pet 3:10–13, among others, ought to be taken literally.[71] Most recently Edward Adams, in his thorough study of "cosmic catastrophe" texts in antiquity, argues that "cosmic catastrophe language cannot be regarded as symbolism for sociopolitical change; writers who use this language have in view a 'real' catastrophe on a universal scale."[72]

This debate concerning the nature of eschatological language with regard to biblical imagery is of significance to our subject. If the New Testament writers use end-of-the-world language metaphorically, is it also true that the texts under consideration refer to eschatological fire figuratively, as a metaphor that points to something else? Or, on the contrary, if they used end-of-the-world language literally, should we conclude that all references to eschatological fire likewise ought to be taken literally? We must exercise caution here, for it is not only possible, but altogether likely, that various interpretations of a single image or symbol coexist within the New Testament. One author may employ the image of fire quite literally while another might apply it figuratively. John the Baptist may have envisioned a literal baptism in fire while Luke the Evangelist took the same image symbolically (Matt 3:11//Luke 3:16). Paul may have envisioned a literal fire on the Day of the Lord (1 Cor 3:10–15; see 2 Thess 1:8) whereas the author of 1 Pet 4:12 clearly interpreted his community's present trials as a figurative "burning" (πῦρωσις).

As I am inclined to read eschatological language in a literal sense, I am also generally inclined to interpret the fire of eschatological judgment literally, all the while acknowledging that the fire of judgment transcends its mundane, earthly counterpart. Fire in the Bible is frequently depicted as a weapon of the LORD when God intervenes in history, whether in the destruction of Sodom and Gomorrah (Gen 19:24), Nadab and Abihu (Num 3:4), or Korah's men (Num 16:35). Since fire is taken literally as a divine weapon when God's judgment manifests itself in the course of history, there is good reason to suspect that fire also ought to be taken literally when it features in eschatological judgment. There are, of course, some exceptions. Luke, for instance, has a tendency to historicize his eschatology and to describe events already taking place with eschatological language, as when he

---

70. Wright, *New Testament and the People of God*, 321.
71. Allison, "Jesus and the Victory of Apocalyptic," 126–42.
72. Adams, *Stars Will Fall*, 3.

interprets Jesus going to his crucifixion as green wood prepared for the fire (23:31).[73] We perceive a similar technique in 1 Pet 4:12, where the author describes his community's "fiery ordeal" as already taking place.[74] In general, I shall adopt the following approach: when perceived as an element in some future eschatological event, the image of fire should be taken literally as an instrument of God's judgment just as it appears in the texts from the Hebrew Bible mentioned above. On the other hand, when the image is applied to the author's present situation or when the comparative context is unequivocal, the image of fire may be interpreted figuratively. Each text, of course, must be examined in its own right, and so it is to the relevant texts in the Hebrew Bible and literature of Second Temple Judaism that we will soon turn.

Methodology

Since we are dealing with texts from a number of distinct genres, no single methodology will be appropriate for every passage considered herein. The critical tools we will draw on include source, form, and redaction criticism, historical analysis, and history of interpretation. Each text will be subjected to critical exegesis, which will attempt to understand the saying or pericope within its immediate textual framework as well as its wider literary context, particularly in light of the relevant passages from the Hebrew Bible and other Jewish texts and often with reference to the history of interpretation. In regard to our treatment of Jewish sources, only those texts that are commonly believed to antedate or coincide with the composition of the New Testament texts under consideration will fall under our purview. When we are confronted with competing versions of the text under investigation, the insights from source, form, and redaction criticism will aid us in reconstructing its earliest form insofar as such a task is possible. With regard to the historical Jesus, the traditional criteria (e.g., embarrassment, [double] dissimilarity, multiple attestation, coherence, et al.) will be applied to the sayings of Jesus in an attempt to discern their historical plausibility.[75] These criteria have recently come under attack, and some criticism is warranted.[76]

---

73. On Luke's historicized eschatology see especially Conzelmann, *Theology of St. Luke*, 135. Because the history of interpretation of Luke 12:49–50 reflects the problems inherent in eschatological language in general, our exegesis of that text will provide an opportunity to examine these matters in greater detail.

74. See chapter 6 below.

75. One of the clearest articulations of all of the major criteria is that of Meier, *Marginal Jew*, 1:167–95.

76. On the criterion of dissimilarity, see Theissen and Winter, *Quest*. On multiple attestation see Eve, "Meier, Miracle and Multiple Attestation." On all of the major

There are certainly instances where scholars, despite their using the same criteria, have come to wildly different conclusions.[77] Thus, the criteria cannot be taken as *guarantors of historical authenticity*; however, when used together, they are effective *indicators of probability regarding the historical plausibility* of a certain saying, complex of sayings, or event.[78] In light of these contemporary criticisms, a more holistic approach may also be appropriate for our investigation, for whereas most of the sayings under consideration are singly attested, together they constitute a widespread or recurrent motif in the Jesus tradition—and throughout earliest Christianity—which is anticipated in the preaching of John the Baptist and represented in several early Christian texts. Thus, even if the authenticity of no individual saying can be guaranteed, the recurrence of this motif points in the direction of a historical memory.[79]

---

traditional criteria, see Allison, "How to Marginalize the Traditional Criteria."

77. Consider the very different Jesuses of John Dominic Crossan and John P. Meier.

78. I am in essential agreement with the remarks of Webb, "Historical Enterprise," 60–75. More recently, see the comments of Martin, "Jesus in Jerusalem," 12, n. 15: "I am under no illusion that the traditional, modern criteria render the kind of confident certainty imagined by some scholars in previous generations. But the simple argument that a saying or event passed along by different independent sources has a better chance of being historical is a sound one. Also, the argument that a saying or event that goes against the 'tendency' of the relevant text or the pieties of early Christianity renders that saying or event more likely historical seems to me still to be a valid point. I am not, that is, advocating a rigorous use of a particular 'method' designed to produce 'facts'; I am simply mounting what I take to be persuasive arguments."

79. On the criterion of recurrent attestation see Allison, *Constructing Jesus*, 15–22.

CHAPTER 2

The Functions of Fire in the Hebrew Bible

Introduction

IT IS A JUSTIFIABLY accepted axiom of biblical scholarship that the Hebrew Bible is an indispensable background for understanding the New Testament. On nearly every page, the latter quotes, alludes to, and assumes familiarity with the former. It is necessary, therefore, to begin our investigation of the fiery ordeal in the New Testament with a survey of relevant background texts from the Hebrew Bible. As the primary concern of this study is by no means an exhaustive treatment of the functions of fire in the Hebrew Bible, the goal of this chapter must be modest. It will be necessary to limit our investigation to popular narratives, widespread motifs, and stock phrases concerning fire as well as specific texts to which New Testament authors may allude. In so doing, we are not concerned primarily with the texts' tradition histories, but with their influence on the conceptions present in the New Testament period. Although the canon of the Hebrew Bible remained undefined during the first century CE, the texts under consideration here are those found in the books that were widely considered authoritative among both first-century Jews and early Christians, as can be seen by their presence at Qumran and by the fact the authors of the New Testament appeal them as scripture.[1] We shall examine the most relevant texts in the Pentateuch, the Prophets, and the Writings of the Hebrew Bible.

---

1. Esther is the only text from the Hebrew Bible not accounted for in the Dead Sea Scrolls.

We must first begin by limiting our discussion to that which Laughlin terms "holy fire," namely "any use of fire in the Old Testament which portrays the presence and/or activity of God."[2] Laughlin's definition is useful in distinguishing "holy fire" from the many mundane uses of fire. However, Laughlin's three-fold distinction of the manifestations of "holy fire" in theophanies, in the Levitical texts pertaining to the fire upon the altar, and as weapon of God's punishment and discipline is too limited in scope and too focused on the cultic altar fire for our purposes. Lang's divisions are slightly more useful. He identifies four functions of fire in relation to God: fire in theophany, fire as a means of divine judgment, fire as a sign of gracious visitation, and fire as a term for God. Even Lang's categories, particularly his category of "Fire as a Means of Divine Judgment," require some refining. For while he appropriately notes that this category must be subdivided to include both "Yahweh's judicial intervention in the course of history" and "the fire of eschatological judgment," he completely overlooks the fact that in certain eschatological texts in the Hebrew Bible—particularly in those texts that employ the imagery of smelting to evoke the refining function of fire (see Isa 1:25–26; Mal 3:2–3)—the divine fire is applied to the elect or the righteous and has a positive effect. Note Lang's silence regarding these texts in his discussion of eschatological fire:

> Fire has especially three roles in the eschatological drama. 1. It is a sign of the day of Yahweh, Jl. 2:30. 2. Yahweh will execute the judgment of eschatological destruction with fire on all His enemies, Mal. 3:19; Is. 66:15 f.; Ez. 38:22; 39:6. 3. The damned fall victim to eternal torment by fire.[3]

He is content to focus on the punitive functions of the fire of eschatological judgment to the exclusion of any possible refining functions of fire on the Day of the Lord.

The refining function of fire, which is systematically overlooked in discussions of eschatological fire, is precisely the category we believe will most illumine several of the enigmatic New Testament texts we shall take up in later chapters. We must, therefore, give them attention here, not least because they have been so frequently ignored in previous discussions. In this chapter, therefore, we shall take a thematic approach to our texts, exploring not only the categories discussed by Laughlin and Lang, but especially those suggested by Milavec, namely fire as a means testing and refining. Thus, a more comprehensive approach will include the consideration of not only

---

2. Laughlin, "Holy Fire," xiii, underlining removed.
3. Lang "πῦρ," 936–37.

the theophanic and punitive functions of fire, but also the purgative function of "holy fire."

The Theophanic Function of Fire

Fire and related phenomena, such as smoke and lightning, are recurring motifs in many of the Hebrew Bible's theophanies.[4] The earliest fire theophany occurs in Gen 15, when God cuts his covenant with Abram. According to the Lord's command, Abram brings a heifer, a ram, a goat, a turtledove, and a pigeon, and—with the exception of the two birds—cuts them in two, "laying each half over against the other" (v. 10). Then, "[w]hen the sun had gone down and it was dark, a smoking fire pot [תַנּוּר] and a flaming torch [וְלַפִּיד] passed between these pieces" (v. 17).[5] It appears that these elements represent the Lord's involvement in the covenant cutting ceremony. "And yet," von Rad cautions, "one can detect a certain reticence, for the narrator avoids simply identifying Yahweh with those strange phenomena."[6] It is no doubt true that the fiery phenomena are not unequivocally identified with the Lord; however, given that in ancient Near Eastern custom it was the party promising to keep a covenant who passed between the pieces of the sacrificed animals (see Jer 34:18–19), it appears that the "smoking fire pot" and the "flaming torch" are at the very least symbolic representatives of the Lord, if not indicative of the presence of the Lord himself. Whatever the relationship between the Lord and the fiery phenomena, it is clearly a close one.

Some, however, have asked why both symbols appear in this text, for if providing fire and smoke for the theophany were their only purpose, either one alone would have sufficed. According to Henri Cazelles, the torch (לַפִּיד) symbolizes the positive, protective fire of Yahweh and the furnace (תַנּוּר) foreshadows the destructive, dangerous side.[7] This is corroborated by John Ha's observation that both the destructive consuming power and the purifying aspect of fire are represented by the flaming torch and the furnace elsewhere in the Hebrew Bible: "The former [לַפִּיד] is found in Is. 62:1 and

---

4. Scriba (*Die Geschichte des Motivkomplexes Theophanie*, 28–31) discerns three distinct functions of fire in theophanies in the Hebrew Bible: 1) fire manifests the כָּבוֹד of God; 2) God or his mediator breathes fire out of his mouth as a means of destruction; and 3) fire is used as a metaphorical expression for the wrath of Yahweh. However, we shall treat Scriba's second category below under the heading of "the punitive functions of fire."

5. Laughlin ("Holy Fire," 5) reports, "both phenomena [the fire pot and the flaming torch] are recorded in Akkadian texts."

6. Von Rad, *Genesis*, 188.

7. Cazelles, "Conexions et Structure," 343. See Laughlin, "Holy Fire," 6.

Zech. 12:6 where it expresses the salvation of Jerusalem (and Judah) as well as the destruction of her enemies. Mal. 3:19f employs the latter [תַנּוּר] to depict the burning on the Lord's day that results in the destruction of the evildoers and the liberation of those who fear YHWH."[8] Thus, already in the earliest theophany in the Hebrew Bible fire plays a central role, perhaps already suggesting its simultaneously destructive and purificatory power.

Concerning the תַנּוּר "smoking fire pot," von Rad has cautiously registered the observation that "[s]ince *tannūr* (actually an oven) had the shape of a hollow clay cylinder tapering toward the top [. . .], it has been thought that this phenomenon could be understood as a mysterious preview of God's fiery mountain, and thus as a reference to the conclusion of a covenant with Moses on Sinai."[9] Others have speculated that the "smoking fire pot" and the "flaming torch" respectively anticipate the pillar of cloud and the pillar of fire that led the Israelites out of Egypt (Exod 13:21; 14:24; Num 14:14). While these links are by no means certain, they are indeed suggestive.

Before turning our attention to the Sinai theophany, however, let us first consider the well-known account of the call of Moses at Mount Horeb:

> There the angel of the Lord [מַלְאַךְ יְהֹוָה] appeared to him in a flame of fire out of a bush; he looked, and the bush was blazing [הַסְּנֶה בֹּעֵר בָּאֵשׁ], yet it was not consumed. Moses said, "I must turn aside and look at this great sight, and see why the bush is not burned up [מַדּוּעַ לֹא־יִבְעַר הַסְּנֶה]." When the Lord saw that he had turned aside to see, God called to him out of the bush, "Moses, Moses!" And he said, "Here I am." Then he said, "Come no closer! Remove the sandals from your feet, for the place on which you are standing is holy ground [אַדְמַת־קֹדֶשׁ]." He said further, "I am the God of your father, the God of Abraham, the God of Isaac, and the God of Jacob." And Moses hid his face, for he was afraid to look at God." (Exod 3:1–6)

Moses encounters the angel of the Lord (מַלְאַךְ יְהֹוָה) in the burning bush in v. 2;[10] the voice calling to Moses out of the bush is attributed to God (אֱלֹהִים) in v. 4b; and the pericope ends with Moses turning away from the burning bush, "for he was afraid to look at God [אֱלֹהִים]" in v. 6. All of these are clear indications that we are dealing with a theophany.

---

8. Ha, *Genesis 15*, 56.

9. Von Rad, *Genesis*, 188.

10. For the argument that "the angel of the Lord" is "God himself in human form," see von Rad, *Genesis*, 193–94. For a discussion of the angel of the Lord in relation to God in post-biblical Jewish literature, see Fossum, *Name of God*.

Three aspects of the fire in this theophany are striking. First, as is often noted, the fire does not consume the bush. David Noel Freedman observes that translators often conceal the apparent contradiction of the Hebrew, in which the same verb is used to indicate that הַסְּנֶה בֹּעֵר בָּאֵשׁ ("the bush was burning with fire") and that Moses looked on to see מַדּוּעַ לֹא־יִבְעַר הַסְּנֶה ("why the bush is not burning"). The latter phrase is usually translated as "why the bush was not consumed." While the translation is on the whole appropriate, the paradox inherent in the Hebrew text does serve to highlight the curious nature of the fire.[11] It is obviously an understatement that the fire here is no mundane fire, but has supernatural qualities. Second, it is significant that God instructs Moses to remove his shoes, for he is standing on holy ground (אַדְמַת־קֹדֶשׁ). This raises the question: is the ground inherently holy, or has it become holy from the fire, or more specifically from the presence of God in the fire? To be sure, the location of the burning bush on Mount Horeb, the mountain of God (v. 1), would indicate that the ground would have been considered holy already. However, the divine presence made visible by the fiery bush seems to impart a greater degree of holiness to the immediate location.[12] It would be going too far to insist that the holiness of the place derives from the fire itself and not from the divine presence within the fire; however, as we shall see, the fire and the presence are one. Third, the theophany inspires fear in Moses.[13] No doubt it is not the fire alone that arouses this fear. Indeed, the verse states, "he was afraid to look at God" (v. 6). However, implicit is Moses' understanding that to look at the fire is to look at God.[14] Thus, the burning bush theophany provides us with an instance of holy fire that, to some degree, imbues holiness and fear, for it is intimately associated, if not identified, with the Lord.

11. Freedman, "Burning Bush."

12. Noth (*Exodus*, 39) suggests that within the text as it stands the burning bush "is assumed to be the permanent feature of the place in question." If this is the case, the holiness attributed to the place is not a recently acquired quality, but would also be a "permanent feature."

13. Laughlin ("Holy Fire," 17) notes an intriguing parallel in the myth of Horus at Edfu, which describes Horus as "a flame . . . , inspiring fear(?), which rageth on a hillock of brushwood." See Blackman and Fairman, "Myth of Horus at Edfu," 16. The parallel is indeed striking: both gods are manifested as burning bushes and thus elicit the fear of bystanders; however, whether one narrative was dependent upon the other is impossible to tell and, at any rate, beyond the scope of the present investigation.

14. Regarding Moses' hiding his face, see Alter, *Five Books of Moses*. Alter writes, "[t]he gesture reflects the reiterated belief of biblical figures that man cannot look on God's face and live. What should be noted is how God's manifestation has shifted from Genesis. God spoke to Abraham face to face in implicitly human form. Here He speaks from fire, and even that Moses is afraid to look on" (319).

Exodus 19:16–18 narrates what is perhaps the most famous theophany in the Hebrew Bible, the Lord's appearance to Moses on Mount Sinai:[15]

> On the morning of the third day there was thunder and lightning [קֹלֹת וּבְרָקִים], as well as a thick cloud on the mountain [וְעָנָן כָּבֵד עַל־הָהָר], and a blast of a trumpet so loud that all the people who were in the camp trembled. Moses brought the people out of the camp to meet God. They took their stand at the foot of the mountain. Now Mount Sinai was wrapped in smoke [עָשַׁן], because the Lord had descended upon it in fire [בָּאֵשׁ]; the smoke went up like the smoke of a kiln, while the whole mountain shook violently [וַיֶּחֱרַד כָּל־הָהָר מְאֹד].

The motif of fire is employed again on several different occasions in relation to the Sinai theophany. Exodus 24:17 reads, "Now the appearance of the glory of the Lord was like a devouring fire [כְּאֵשׁ אֹכֶלֶת] on the top of the mountain in the sight of the people of Israel." In Deut 4:12 Moses tells the Israelites, "The Lord spoke to you out of the fire [מִתּוֹךְ הָאֵשׁ],"[16] and in 5:22–23 he reminds them once more:

> These words the Lord spoke with a loud voice [קוֹל גָּדוֹל] to your whole assembly at the mountain, out of the fire [מִתּוֹךְ הָאֵשׁ], the cloud and the thick darkness [הֶעָנָן וְהָעֲרָפֶל], and he added no more. He wrote them on two stone tablets, and gave them to me. When you heard the voice out of the darkness [הַקֹּל מִתּוֹךְ

---

15. It is sometimes suggested that the burning bush is related to the Sinai theophany, for the Hebrew word used for bush, סְנֶה (seneh), shows some resemblance to the name סִינַי (Sinai); see *Pirke Eliezer*, ch. 41. Haupt ("Burning Bush," 38–39) asserts "Sinai means covered with senna shrubs" (364), emphasis removed. While Noth insists the name Sinai cannot derive etymologically from the Hebrew סְנֶה, he allows for the possibility that "when the story was later incorporated into the framework of the Moses tradition the word *seneh* was felt to contain an allusion to the name Sinai. This could also be the reason for the phenomenon of the fire, which is reminiscent of the features which accompany the theophany of Sinai."

16. Wilson (*Out of the Midst of the Fire*, 60) notes that the phrase "out of the midst of the fire" occurs ten times in Deuteronomy and that eight of these texts (4:12, 15, 33; 5:4, 22, 26; 9:10; 10:4) refer to the Lord speaking to the Israelites. Wilson concludes, "[i]f YHWH is represented as speaking 'out of the midst of' a fire, then that would seem to suggest, *a priori*, that he is considered to be present within that fire." Deuteronomy 5:4 is particularly striking and indicative of the divine presence within the fire: "The Lord spoke with you face to face at the mountain, out of the fire." As Laughlin ("Holy Fire," 51) observes, the Deuteronomist "made use of theophanic fire as it was already known in traditions and added nothing new to them. The place that holy fire had come to occupy in the traditions is indicated by the fact that the author(s) of Deuteronomy, concerned as he was with image prohibitions, especially in chapters four and five, could not bring himself to remove these parts of the tradition which speak of Yahweh appearing in fire." This further indicates how essential the fire motif was to the Sinai tradition.

הַחֹשֶׁךְ], while the mountain was burning with fire [בֹּעֵר בָּאֵשׁ], you approached me.

Jörg Jeremias observes that each of these accounts, although attributed by source critics to differing strands of the tradition, indelibly links the theophany to the presence of fire on Mount Sinai:

> According to Exod 24:17 P the glory of Yahweh appears in the form of fire on the mountain top, according to Exod 19:18 J Yahweh comes down to Sinai in fire, according to Deut 4:12; 5:23 Yahweh speaks from out of the midst of the fire; in all of these texts of the Sinai tradition, fire is an accompanying feature or the means of the appearance of Yahweh at his coming.[17]

The fire in the Sinai theophany is no peripheral element. Indeed, it is perhaps second in importance only to the voice (הַקֹּל) of the Lord.

Some have urged that it was the Israelites' observations of volcanic activity that stand behind the fiery phenomena described in the Sinai theophany.[18] Noth, for instance, writes, "[t]he description of the circumstances speaks of three manifestations, the smoke on the whole mountain, the fire which is connected with the descent of Yahweh and the great quaking of the whole mountain. . . . The manifestations mentioned point quite clearly to a volcanic eruption."[19] Others, however, have disputed this claim, not least due to the fact that no volcano has ever been found that could correspond to any possible location for Mount Sinai. More probable, therefore, is the position of Frank Moore Cross, who argues that the tradition of the Sinai theophany "surely rests not on a description of volcanic activity, but upon hyperbolic language used in the storm theophany."[20] Cross adduces one significant text that antedates the Sinai tradition, namely 2 Sam 22:8–16 = Ps 18:7–15, which exhibits remarkable parallels to the Sinai theophany:[21]

> Then the earth reeled and rocked;
> the foundations of the heavens trembled
> and quaked, because he was angry.
> Smoke [עָשָׁן] went up from his nostrils,
> and devouring fire [וְאֵשׁ . . . תֹּאכֵל] from his mouth;

17. Jeremias, *Theophanie*, 108.
18. Ibid., 103–5.
19. Noth, *Exodus*, 159.
20. Cross, *Canaanite Myth and Hebrew Epic*, 167.
21. Cross cites only 2 Sam 22:9 [8] = Ps 18:8 [7]; however, the full text quoted here more fully illustrates the argument. The text cited above is from 2 Sam 22:8–16. The divergences in Ps 18:7–15 are minor.

glowing coals flamed forth [גֶּחָלִים בָּעֲרוּ] from him.
Then he bowed the heavens, and came down;
    thick darkness was under his feet.
He rode on a cherub, and flew;
    he was seen upon the wings of the wind.
He made darkness around him a canopy,
    thick clouds, a gathering of water.
Out of the brightness before him
    coals of fire flamed forth [בָּעֲרוּ גַחֲלֵי־אֵשׁ].
The Lord thundered from heaven;
    the Most High uttered his voice.
He sent out arrows, and scattered them
    —lightning [בָּרָק], and routed them.
Then the channels of the sea were seen,
    the foundations of the world were laid bare
at the rebuke of the Lord
    at the blast of the breath of his nostrils.

It is striking that in this poetic text we have several of the same elements—shaking mountains, smoke, a descent from heaven, dark clouds, and the voice of the Lord—that recur in the Sinai theophany. This is surely strong evidence that the Sinai theophany is best understood as a storm theophany rather than as deriving from the observation of volcanic activity. What is more, if a volcano were being described, one would not expect the fire to *descend* from heaven as Exod 19:18 explicitly states (יָרַד עָלָיו יְהוָה בָּאֵשׁ); rather, *ascending* fire would be more appropriate.

    Understanding the Sinai tradition as standing firmly within the tradition of storm theophanies has implications for our later discussion of the punitive function of fire, for the storm theophany depicts the Lord as a divine warrior and fire as one of his weapons of wrath. This connection between the theophanic and punitive functions of fire serves to remind us that there is no hard and fast separation between the many functions of fire in the Hebrew Bible, a point that deserves reiteration throughout this chapter. Having discussed some of the most significant instances of theophanic fire in the Pentateuch, let us now consider some of the key prophetic texts, particularly those found in 1 Kings, Isaiah, and Ezekiel, as well as the apocalyptic writing Daniel, that employ fiery phenomena in their theophanies.

    It is for good reason that the prophet Elijah has been called "the man of fire *par excellence*."[22] In a memorable confrontation with the prophets

---

22. Robinson, "Elijah, John and Jesus," 30.

of Baal, Elijah calls down fire from heaven to consume his offering and to demonstrate the Lord's superiority over Baal (1 Kgs 18:30–39), he later calls down fire on the men of King Ahaziah of Samaria when the king inquires of Baal-Zebub (2 Kgs 1:9–12), and at the end of his life he ascends to heaven in a fiery chariot (2 Kgs 2:11). It is no surprise, then, that in the later memory of the authors of such books as Malachi and Ben Sirach the mention of Elijah often provokes thoughts of the fire of divine judgment. We shall consider several of these Elijah traditions over the course of this chapter and the next. Let us first, however, examine the texts that gave rise to these later traditions. While the ascension of Elijah into heaven in a fiery chariot (2 Kgs 2:11) cements the association of fire with the figure of Elijah, and 2 Kgs 1:9–12 demonstrates the destructive element of fire, it is the theophanic fire in 1 Kgs 1:9–12 that demands our attention here.

In 1 Kgs 18 Elijah finds himself alone in a confrontation with the four-hundred-and-fifty prophets of Baal, whom he challenges to a contest to see whose deity will send fire to consume the sacrifices being offered upon the altar on Mount Carmel. Elijah proposes, "you call on the name of your god and I will call on the name of the Lord; the god who answers by fire is indeed God" (v. 24).[23] After the prophets of Baal are unsuccessful in their attempt to summon fire from Baal, Elijah commands that his own sacrifice be doused with water until the water flows down and fills the trench surrounding the altar. Then in a demonstration of power "the fire of the Lord fell [וַתִּפֹּל אֵשׁ־יְהוָה] and consumed the burnt offering, the wood, the stones, and the dust, and even licked up the water that was in the trench" (v. 38). The phrase אֵשׁ־יְהוָה "the fire of the Lord" coupled with the verb נָפַל "to fall" clearly identifies the source as the heavenly abode of the Lord, thus indicating the divine nature of the fire in question.

The entire premise of the confrontation is to demonstrate to the people that the Lord, not Baal, is the true God, as is clearly indicated by Elijah's admonition in v. 21b: "If the Lord is God, follow him; but if Baal, then follow him." Indeed, there is a direct correlation between the misplaced faith of the prophets of Baal and their inability to call down fire to consume their sacrifice, and Elijah's appropriate faith in the Lord and his success in summoning fire from heaven. The opening of Elijah's prayer requesting that the Lord send his fire as a sign of divine acceptance demonstrates this very fact: "O Lord, God of Abraham, Isaac, and Israel, let it be known this day that you are God in Israel, that I am your servant, and that I have done all these

---

23. De Vries (*1 Kings*, 228) comments, "both Baal and Yahweh claimed the power of fire (for Yahweh, see Gen 15, 19; Exod 3, 19; Judg 6, 13; Amos 1:4, 7, 10, 12, 14; 2:2, 5; Zech 2:5; Mal 3:2). The power actually to send fire will decide which of the two is really God."

things at your bidding" (v. 36b). The fire will justify the faith of Elijah and demonstrate that the Lord is the true God as is affirmed in the following line of Elijah's prayer: "Answer me, O Lord, answer me, so that this people may know that you, O Lord, are God, and that you have turned their hearts back" (v. 37).

It is essential to note that the fire that consumes the sacrifice is understood as indicating the Lord's acceptance of Elijah's sacrifice and is therefore a sign of his pleasure, and not his wrath. As Cogan indicates, "[t]he fiery apparition often symbolizes YHWH's presence (e.g., Exod 3:2; 19:18; 24:17), and, with reference to the altar inaugurations (Lev 9:24; 2 Chr 7:3), the fire attests divine approval and acceptance of the worshiper's act (1 Chr 21:26)."[24] This observation will prove important when we consider other contrasting texts in which fire falls from heaven and clearly plays a destructive role. However, there is no clear indication that the fire Elijah calls down from heaven is indicative of the divine presence. In this regard, we may question whether the present fire is truly theophanic. It is at the very least quite different from the theophanic fire we have observed thus far. To be sure, the Elijah tradition may intentionally be distancing itself from the notion that the divine presence dwells within the fire found in other previously-discussed theophanies, for in 1 Kgs 19:12, when Elijah experiences stormy phenomena on Mount Horeb, the text explicitly states that לֹא בָאֵשׁ יְהוָה "the Lord was not in the fire." Nonetheless, the fire in the present passage functions to demonstrate the power of the one true God over against Baal.[25] And, as previously noted, the premise of Elijah's contest with the prophets of Baal is their dispute over the identity of the true God in Israel. The response that the fire from heaven elicits from the people makes this clear: "When all the people saw it, they fell on their faces and said, 'The Lord indeed is God; the Lord indeed is God'" (v. 39). Thus the fire that consumes the sacrifice is bound up with the identity of the God of Israel even if according to this tradition the Lord is not present in the fire.

Turning from the prophet Elijah to the prophet Isaiah, we see that fire continues to exercise its influence as a significant motif in theophanic visions. The call of Isaiah, narrated in Isa 6, employs fiery elements, though in a much more subtle manner than we have seen thus far. The relevant section of the chapter is as follows:

> In the year that King Uzziah died, I saw the Lord sitting on a throne, high and lofty; and the hem of his robe filled the temple.

---

24. Cogan, *1 Kings*, 443.

25. See Ibid., 444. Cogan describes the fire as "an impressive show of YHWH's might."

> Seraphs [שְׂרָפִים] were in attendance above him.... The pivots on the thresholds shook at the voices of those who called, and the house filled with smoke [עָשָׁן]. And I said: "Woe is me! I am lost, for I am a man of unclean lips, and I live among a people of unclean lips; yet my eyes have seen the King, the Lord of hosts!" Then one of the seraphs flew to me, holding a live coal [רִצְפָּה] that had been taken from the altar with a pair of tongs. The seraph touched my mouth with it and said: "Now that this has touched your lips, your guilt has departed and your sin is blotted out." (Isa 6:1–7)

The presence of the seraphim may be related to the fire motif, for the word seraphim derives from the root שָׂרַף, meaning "burn."[26] More explicit, however, are the עָשָׁן "smoke" that fills the temple and the רִצְפָּה "live coal" taken from the altar, both of which are frequently recognized as representing the fiery element so common in biblical theophanies.[27] The smoke, as we have already seen was a significant motif in the Sinai theophany. And presumably the smoke implies the presence of fire. Interestingly, here it is also coupled with shaking foundations as it was in Exodus 19.[28]

In a striking scene, Isaiah laments that he is "a man of unclean lips," provoking one of the seraphim to approach Isaiah with a live coal held in a pair of tongs and press it to Isaiah's lips. The words proclaimed by the seraph are of profound significance, for they present us with evidence for the purifying function of fire within the context of a theophany: "Now that this has touched your lips, your guilt has departed and your sin is blotted out" (v. 7b). The burning coal has the power to cleanse that which is unclean and to blot out one's guilt. As Wildberger notes, "one might refer to similar rites which involved ... the purifying of metal by means of fire (Num. 31:22f.)"[29]

Even more so than with Isaiah's call, fire plays a significant role in the call narrative in Ezekiel 1, which presents us with a profoundly complex and symbolic description of Ezekiel's mystical vision of the throne chariot:

> As I looked, a stormy wind [רוּחַ סְעָרָה] came out of the north: a great cloud [עָנָן גָּדוֹל] with brightness around it and fire flashing forth continually [וְאֵשׁ מִתְלַקַּחַת], and in the middle of the fire [מִתּוֹךְ הָאֵשׁ], something like gleaming amber. In the middle of it was something like four living creatures.... In the middle of

---

26. See Laughlin, "Holy Fire," 54.
27. Lang, "Das Feuer," 56; Laughlin, "Holy Fire," 52–57.
28. Blenkinsopp, *Isaiah 1-39*: "The tectonic phenomena induced by the seraphic acclamation belong to the standard description of theophanies" (225).
29. Wildberger, *Isaiah 1-12*, 269.

> the living creatures there was something that looked like burning coals of fire [כְּמַרְאֵה־אֵשׁ בֹּעֲרוֹת], like torches [כְּגַחֲלֵי הַלַּפִּדִים] moving to and fro among the living creatures; the fire was bright, and lightning [בָרָק] issued from the fire. The living creatures darted to and fro, like a flash of lightning. (vv. 4–5a, 13–14)

The רוּחַ סְעָרָה "stormy wind," עָנָן גָּדוֹל "great cloud," and בָרָק "lightning" all indicate the influence of other storm theophanies from Israel's tradition.[30] Indeed, Steven Tuell links this vision with the Sinai theophany, for "only in Ezekiel's inaugural vision (1:4) and in Deuteronomy (4:12, 15, 33, 36; 5:4, 22, 24, 26; 9:10; 10:4) does the Lord speak from the center of the fire [מִתּוֹךְ הָאֵשׁ]: here to Ezekiel in Babylon; there to Moses on Sinai."[31] There is, however, another noteworthy parallel that highlights the dangerous element in this theophany. The phrase וְאֵשׁ מִתְלַקַּחַת, which the NRSV translates "and fire flashing forth continually," is a difficult and rare one in the Hebrew Bible, occurring only elsewhere in Exod 9:24 in reference to the plague of hail.[32] Thus, while one intertextual link positively associates the fire in Ezekiel's vision with the Sinai theophany, the allusion to the plague of hail in Egypt is a vivid reminder that the fiery theophany was no tame experience, but at once fascinating and terrifying.

These literary connections indicate both that the influence of the Sinai theophany, particularly its identification of fire with the divine presence, lived on into the period of Israel's exile, and also that the fiery phenomena continued to evoke the fearful awe that it had in the burning bush theophany, upon which Moses was afraid to look. Perhaps most striking in this passage, however, is that the fire attests to the enigmatic character of Ezekiel's vision. Indeed, the entire theophany is rife with circumlocutions. The seer adverts to "*something like* gleaming amber" (v. 4), "*something like* four living creatures" (v. 5), "*something that looked like* burning coals" (v. 13), and so on, culminating in the expression, "This was the appearance of the likeness of the glory of the Lord" (v. 28). It has long been observed that these qualifications attest to the enigmatic nature of the theophanic vision and Ezekiel's inability to articulate his mystical experience with human vocabulary. It is very possible that the fire itself plays a similar function here, for fire also has an enigmatic quality to it. Fire is a mysterious symbol that points towards the inarticulate.

Quite similar to Ezekiel's vision is the apocalyptic vision of Dan 7:9–10, in which the seer, Daniel, describes his vision of the Ancient of Days:

---

30. See the discussion in Zimmerli, *Ezekiel*, 119.
31. Tuell, *Ezekiel*, 14, boldface removed.
32. Greenberg, *Ezekiel 1–37*, 43.

his throne was fiery flames [שְׁבִיבִין דִּי־נוּר],
   and its wheels were burning fire [נוּר דָּלִק].
A stream of fire [נְהַר דִּי־נוּר] issued
   and flowed out from his presence.

Treating this text immediately after our discussion of Ezekiel's vision allows us to see quite clearly Daniel's dependence on Ezekiel for the content of his vision.[33] To be sure, Daniel does not merely copy Ezekiel's text without making it his own. However, several of the same motifs recur here. The fiery throne recalls the description of the throne in Ezek 1:27, the burning wheels almost certainly derive from the wheels described in Ezek 1:15-21, and Daniel's enigmatic "one like a son of man" probably has its origin in Ezekiel's "something that seemed like a human form" in 1:26.[34] Thus, here, as in Ezekiel, the fire is an expression of the glory of the Lord, as it is often employed in theophanies.

While largely influenced by Ezekiel's description of the divine throne, Daniel does introduce a new motif into the theophanic vision, a motif that would come to the fore in later apocalyptic literature and Merkavah mysticism, namely the stream of fire that flows from the throne of God. John J. Collins observes that the נְהַר דִּי־נוּר "recalls the river of molten metal that plays an important part in Persian eschatology." He then goes on to assert that "[i]n Daniel 7, however, the river does not have an eschatological function (e.g., as ordeal or as instrument of destruction), so any relationship to the Persian motif is doubtful."[35] Collins's comment is curious, however, for while the stream of fire in Daniel's vision appears to function primarily as a component of the theophany, it is very plausible that when v. 11 alludes to the beast who "was put to death, and its body destroyed and given over to be burned with fire" the author may be suggesting that the beast's body was cast into the stream of fire, thus functioning as an "instrument of destruction." Indeed, in the immediate context of v. 11 there is no better candidate for the fire that destroys the beast than the stream of fire itself.[36] Moreover, early interpretations of Daniel 7 support this reading. Already 4 Ezra 13, a text clearly dependent upon Daniel 7, depicts a figure described as "something like the figure of a man," who comes up from "the heart of the sea," and from this figure's mouth issues a stream of fire that consumes his adversaries.[37] If

33. See 1 Enoch 14.
34. Halperin, Faces of the Chariot, 76-77.
35. Collins, Daniel, 302-3.
36. Collins does not discuss this possibility and simply states, "The location of the fire [into which the beast is thrown] is not specified in Daniel 7" (304).
37. For further discussion of 4 Ezra 13, see chapter 3. That the author has in mind

this interpretation of the stream of fire in Daniel's vision is correct, it serves as a reminder once again that the theophanic function of fire ought not be distinguished completely from the punitive function of fire. Moreover, if the fire functions as an ordeal, the question of Persian influence cannot be ruled out.

Finally, several of the psalms in which fire serves a theophanic function deserve brief mention.[38] Psalm 97:3 may contain a notion similar to that which we find in Daniel where the fire is simultaneously theophanic and destructive: "Fire [אֵשׁ] goes before him, and consumes [וּתְלַהֵט] his adversaries on every side." It is both a sign of his presence and a weapon against his enemies. Psalm 104 offers a storm theophany in which fire plays a significant role. Verse 2 describes the Lord as "wrapped in light [אוֹר] as with a garment" while v. 4 states "flame and fire [אֵשׁ לֹהֵט] are your messengers." The Lord, moreover, is associated with רַעַם "thunder" (v. 7), and is described as the one "who looks upon the earth and it quakes [וַתִּרְעָד], who touches the mountains and they smoke [וְיֶעֱשָׁנוּ]" (v. 32). And in Ps 144:5–6 the psalmist's prayer contains elements of the storm theophany:

> Bow your heavens, O Lord, and come down;
>   touch the mountains so that they smoke [וְיֶעֱשָׁנוּ].
> Make the lightning [בָּרָק] flash and scatter them;
>   send out your arrows and rout them.

In all of these psalms fiery phenomena indicate the presence of the Lord, but they are not limited to their theophanic function. They are simultaneously weapons of the Divine Warrior, serving once again as a reminder that the functions of fire in the Hebrew Bible are not easily categorized, for the functions often overlap.

We have seen that fire is part and parcel of the theophanic genre in the Hebrew Bible. From the earliest theophany fire plays a central role, and in many texts it is as if the presence of the Lord and the fire are one. The fiery element in these theophanies served many functions: it invoked the awe of Moses; for Elijah it demonstrated that the Lord, and not Baal, is the true God; it played a purificatory role in the case of Isaiah's call; it served as a weapon of the Divine Warrior; and it took the form of a stream of fire in Daniel's vision. In all of these functions, however, the close relationship

---

the stream of fire when referring to the fire into which the beast is cast gains plausibility when one considers the interpretation of this text by later Jewish rabbis, who envision angels being born out of the river of fire daily to sing praises before the throne of the Lord and then returning to the river of fire at the day's end to be consumed by the same fire from which they originated (see *Hag.* 14a; *Gen. Rab.* 78:1).

38. Psalm 18:7–15 has already been discussed above.

with the Lord in the theophanic visions discussed above makes clear that however the fire functions it is an expression of the divine will. This point is essential to grasp as the starting point for our discussion of the punitive and purifying functions of fire below.

The Punitive Function of Fire

As noted by Lang, the fire of divine judgment at times intervenes in the course of history while elsewhere it plays a central role in the eschatological scenario. In what follows, we shall adopt Lang's observation that the fire of judgment occurs to destroy the enemies of the Lord both within history and at the conclusion of history with one caveat. As we shall see in our discussion of the purifying function of fire, the eschatological fire on the Day of Lord is not always destructive. Indeed, in some cases it may test or even purify and refine the righteous. In the following discussion of the punitive function of fire in the Hebrew Bible, we shall, when relevant, consider such matters as the source and scope of the fire, as well as the moral standing of the individuals subjected to fire, in order to determine whether the fire functions to test, punish, or destroy those towards whom it is directed. This approach will help us attain a greater understanding of the punitive function of fire in the Hebrew Bible. We shall first consider those texts dealing with the punitive fire that intervenes in the course of history and then take up those texts in which fire plays a role in the unfolding of the eschatological drama.

The Punitive Function of Fire in the Course of History

While we are primarily concerned with the *eschatological* functions of fire in the New Testament, there can be no question that the Hebrew Bible's narratives concerning the fire that intervenes in the course of history play a significant role in later eschatological concepts concerning the power and function of fire. Indeed, as we have already noted, the heuristic categories imposed on the functions of fire in the Hebrew Bible are by no means absolute. Theophanic fire and punitive fire cannot be considered in total isolation from one another. It is the same with the fire that intervenes in history and that which accompanies the end of history in eschatological thought. The one informs the other, and this is nowhere more obvious than in the case of Sodom and Gomorrah, where we shall begin our discussion of the punitive function of fire.

In Luke 17:28–32, Jesus likens the days of the coming Son of Man to the days of Lot and alludes specifically to the fire and sulphur that rained down on Sodom and Gomorrah. Elsewhere Jesus warns, "on the day of judgment it will be more tolerable for the land of Sodom than for [those who rejected Jesus]" (Matt 10:34; 11:24; see Luke 10:12). It is therefore quite apparent that the fiery event associated with Sodom and Gomorrah exercised considerable influence on the minds of the Gospel writers and probably upon that of Jesus himself. Moreover, the Genesis account of the destruction of Sodom and Gomorrah is perhaps the most well-known account of judgment by fire in the Hebrew Bible. Indeed, as Weston Fields notes, the description of the destruction of Sodom and Gomorrah "is so extraordinary, so striking, so exceptional and complete, that subsequent biblical accounts of destruction by fire are probably to remind the audience of this ancient momentous destruction."[39] Therefore, in addition to its canonical location in the book of Genesis, its potential influence on other passages dealing with judgment by fire in both the Hebrew Bible and the New Testament makes the account of the destruction of Sodom and Gomorrah an excellent place to begin our survey of the punitive fire that intervenes in history.

Genesis 18:20—19:28 narrates God's deliberations with Abraham concerning the destruction of Sodom and Gomorrah, the visitation of the angels to Sodom to rescue Lot, the near assault on these messengers by the citizens of Sodom, and finally the destruction of Sodom and Gomorrah by sulphur and fire that rains "from the Lord out of heaven" (19:24). The origin of the sulphur and fire that rains down on Sodom and Gomorrah is clearly divine. This is implicit at the very beginning of the narrative when the Lord says, "Shall I hide from Abraham what I am about to do . . . ?" (Gen 18:17) and becomes explicit in the following dialogue between the Lord and Abraham, in which Abraham asks, "Will you indeed sweep away the righteous with the wicked?" (Gen 18:23). When Abraham questions whether the Lord will not spare the city if fifty, forty-five, thirty, twenty, or ten righteous people are found there, the Lord's responses are equally clear. He says that if even ten righteous are found in Sodom, he will spare the city. Thus when the converse happens, it is apparent that it is a reflection of the divine will. The visitation of the two divine messengers who urge Lot to flee the impending destruction are likewise indicative of the divine origin of the destroying fire. Lastly, the description of the sending of fire makes the divine origin of the fire indisputable: "Then the Lord rained on Sodom and Gomorrah sulphur and fire from the Lord out of heaven" (Gen 19:24). The redundancy מֵאֵת יְהוָה מִן־הַשָּׁמָיִם ("from the Lord" and "out of heaven") makes it quite clear

---

39. Fields, *Sodom and Gomorrah*, 137.

that "in the mind of the biblical writer their devastation was due to no blind force of nature. It was due rather to the purposeful judgment of God whose weapon of destruction was very real fire."[40]

There is one inconsistency, however, that must be noted. In 19:13, the angels say to Lot, "*we* are about to destroy this place, because the outcry against its people has become great before the Lord, and the Lord has sent *us* to destroy it." Whereas the fire is later described as coming "from the Lord, from out of heaven" (19:24), here the narrative indicates that the angels who have entered the city are in some way responsible for the destruction of the city. Westermann indicates that this discrepancy is due to the combination of two originally distinct narratives.[41] While this observation does not contradict the divine origin of the destruction of fire, and a detailed discussion of the tradition history of our text would take us too far afield for our present purposes, it does demonstrate that according to some traditions intermediary figures such as angels may be the bearers of the fiery judgment.

On the one hand, the scope of the fire from heaven is clearly defined. It falls on Sodom and Gomorrah and the surrounding Plain (19:25). That the fiery judgment does not extend beyond this region is implied by the very possibility of Lot's escape with his two daughters. He is told to flee to the hills. While he fears that if he obeys the angels he will be overtaken by disaster, what disaster he fears is unclear. It appears to be something other than the impending destructive fire from heaven, for otherwise the messengers' admonition that he flee to the hills for safety makes little sense. Capitulating to Lot's appeals, the angels allow him to flee to the nearby city of Zoar where he will be safe. Had the destruction not been confined to the two cities and the plain, Lot would not have been able to flee to Zoar or the surrounding hills on foot without being destroyed. The fate of Lot's wife does not contradict this, for although she dies outside of the city limits, her fate is not attributed to the fire. Rather, as Westermann notes, we have here a widespread folk motif according to which one who is forbidden to look back on a scene of judgment does so with the result that he or she is transformed into a pillar of salt or a pillar of stone.[42] Her punishment is not a result of the fire from heaven. She is not consumed by fire but is transformed into a pillar of salt.

On the other hand, when one encounters Gen 19:31–32, a different understanding of the scope of the fire presents itself. In these verses, Lot's eldest daughter, commiserating with her younger sister, laments, "Our father is old,

---

40. Laughlin, "Holy Fire," 149.

41. Westermann, *Genesis 12-36*, 303.

42. Ibid., 304. See the many examples cited in Gaster, *Myth, Legend, and Custom*, 160–61.

and there is not a man on earth to come in to us after the manner of all the world. Come, let us make our father drink wine, and we will lie with him, so that we may preserve offspring through our father." Many ancient commentators, both Jewish and Christian, attributed the motives of Lot's daughters to their ignorance. Seeing the destruction of the two cities and the death of their mother, they mistakenly believed that the whole of humanity had perished as in the time of Noah (Philo, *QG*; Josephus, *Ant.* 1:205; Irenaeus, *Adv. Haer.* 4.31.2; *Gen. Rab.* 51:8; Ephraem, *Comm. on Gen.* 19:31; Jerome, *QG* 19:30).[43]

In quite a different mode of interpretation, however, modern commentators speculate that these verses are the vestiges of an earlier tradition according to which the fire that fell from heaven destroyed everyone except Lot's family in the same way the flood washed away all of humanity with the exception of Noah's family. Moreover, the narrative parallels certain themes of the aftermath of the great flood, when Noah became drunk and his honor was violated, perhaps sexually, by his offspring (Gen 9:18–27). Thus the narrative may presume an earlier tradition according to which a worldwide destruction by fire, equal in scope to the great flood, left only Lot and his daughters to repopulate the earth. Prior to the advent of modern criticism, however, the reading that attributes ignorance to Lot's daughters was probably the prominent reading and is thus more in keeping with the understanding the authors of the New Testament would have held. In light of this we will judge the scope of the fire that fell from heaven to be limited to the cities of Sodom and Gomorrah and the surrounding plain.

Two groups of people are identified in this narrative, and each group is assigned a different fate. The first group, the wicked of Sodom, faces judgment by fire from heaven. The second group, Lot and his family, is spared the punishment. The destruction of the wicked is quite straightforward. God says, "How great is the outcry against Sodom and Gomorrah and how very grave their sin" (18:20), indicating the reputation for wickedness that the inhabitants of Sodom have acquired. Abraham's question, "Will you indeed sweep away the righteous with the wicked?" (18:23) followed by God's reply that he will not if he finds only ten righteous in the city suggests that if there are any righteous in the city, their number is few. The attempted rape of the visiting angels by "the men of the city, the men of Sodom, both young and old, all the people [כָּל־הָעָם] to the last man" makes clear that the entire city, at least the entire male populace, are wicked and deserving of the destruction planned by God (19:4).

Apparently the only righteous people in the city are Lot and his family. They may be deemed righteous only by association with Abraham, or they

---

43. Cited in Kugel, *The Bible as it Was*, 194.

may be considered righteous due to the hospitality they have shown towards the angels. On the one hand, Lot's very decision to leave the virtuous Abraham to take up residence in Sodom, whose citizens were apparently known for their depravity, and his subsequent incest with his daughters lead one to question Lot's virtue. On the other hand, the protection he affords the angels clearly sets him apart from the people of Sodom who harbor ill intentions towards the visitors. Indeed, the question of Lot's moral standing was debated in second temple Judaism.[44] In many texts Lot is considered wicked and is saved only because of Abraham's righteousness (*Jub.* 16:6–8; Philo, *QG* 4:54). In other texts it is said that Lot is spared because he had learned hospitality from Abraham (Philo, *QG* 4:10; 107–14; *T. Ab.* (A) 1:1–2; *T. Jac.* 7:22). It seems most probable that Lot's righteousness—and that of his immediate family—is primarily derivative of Abraham's righteousness, and it is due to this righteousness that they are given the opportunity to flee the impending judgment. This is most evident in the statement, "God remembered Abraham, and sent Lot out of the midst of the overthrow" (19:29b). Thus, apart from his association with Abraham, Lot is a morally ambiguous character.

While Lot and his family are spared this fiery judgment, their survival can only be attributed to the fact that they escape in time to avoid the fire falling from the sky. The fire itself is indiscriminate and functions only to destroy that upon which it falls. Thus, there is apparently no possibility of survival within the city. Indeed, 19:25 states, "he overthrew those cities, and all the plain, and all the inhabitants of the cities, and what grew on the ground." The destruction is so complete that Lot's family's only hope is evacuation. Thus the function of the fire and brimstone is purely destructive, regardless of one's moral standing. There is no hint of testing or purification here. Hence the angels' warning, "Get up, take your wife and your two daughters who are here, or else you will be consumed in the punishment of this city" (Gen 19:15b).

Another striking narrative from the Pentateuch involving the fire of judgment is the strange and troubling story of Aaron's two sons, Nadab and Abihu. These two Aaronide priests, whose legitimacy as priests is never called into question, and whose presence in the temple cannot constitute their transgression, are consumed by fire from the Lord. Their fate is particularly unnerving, for it is difficult to discern whether or not they actually violated any explicitly stated commandment. What they are charged with is offering אֵשׁ זָרָה "strange fire" or "unholy fire" at the altar of the Lord, an opaque charge which many scholars have expended significant exegetical energy attempting to elucidate.

---

44. See especially the references in Ibid., 181–95.

According to Lev 10:2 the fire that consumes Nadab and Abihu וַתֵּצֵא אֵשׁ מִלִּפְנֵי יְהוָה "came out from the presence of the Lord" and it is later described as "the burning that the Lord has sent" (v. 6). Many commentators have noticed the similar expression is used in Lev. 9:24a: "Fire came out from the Lord [וַתֵּצֵא אֵשׁ מִלִּפְנֵי יְהוָה] and consumed the burnt offering and the fat on the altar." The location of מִלִּפְנֵי יְהוָה "the presence of the Lord" is not clearly specified, however. Is this a reference to the Lord in the heavens, or does it refer to the presence of the Lord residing within the holy of holies? This question has direct bearing on our exegesis and thus demands our attention. According to the *Sifra* on Leviticus, the fire that consumed Nadab and Abihu fell from heaven like the fire that fell from Sodom and Gomorrah.[45] This would indicate that the phrase "the presence of the Lord" is here used in reference to the Lord's presence in his heavenly abode. However, as Jacob Milgrom notes in his expansive commentary on Leviticus, in other instances in the Hebrew Bible when fire falls from heaven to consume a sacrifice, it is made explicit with verbs like נָפַל "to fall" (1 Kgs 18:38) and יָרַד "to descend" (2 Chr 7:3) or the phrase מִן־הַשָּׁמַיִם "from heaven" (1 Chr 21:26). But in both Lev 9:24 and 10:2 the verb וַתֵּצֵא "to come out, emerge" is employed. "Thus," Milgrom argues, "there can be no doubt that the fire emerged from the adytum [or restricted area, i.e., the holy of holies], in conformance with the Priestly theology that the Lord's *kabod* [כָּבוֹד], encased in cloud, would descend upon the Tabernacle and rest between the outspread wings of the cherubim flanking the Ark."[46] That this happens within the sanctuary where Nadab and Abihu are in close proximity to the כָּבוֹד lends credence to Milgrom's claim.

Milgrom is probably correct that the fire that consumes the sacrifice in 9:24 and the fire that smites Nadab and Abihu in 10:2 both proceed from the holy of holies, yet he oddly asserts that these fires are not the same fire. He bases this on the slender evidence that if the fire in 10:2 were the same as that in 9:24, in the former the definite article would be used, and thus it would be called הָאֵשׁ "the fire," not simply אֵשׁ "fire." If both proceed from the holy of holies, where the *kabod* dwells and manifests itself in the form of fire, however, it seems most plausible that the fire is the same fire. Indeed, the precise parallelism between 9:24 and 10:2, which both employ the formula וַתֵּצֵא אֵשׁ מִלִּפְנֵי יְהוָה וַתֹּאכַל . . . ("fire came out from the Lord and consumed . . ."), would suggest that both verses are referring to the same fire. Indeed, if the definite article were added before the word "fire" in 10:2, as Milgrom insists it must be in order for it to be identified with the fire of

---

45. See Levine, *Leviticus*.
46. Milgrom, *Leviticus 1–16*.

9:24, this parallelism would be disrupted. It could thus be argued that the author refrained from adding the definite article in order to maintain the parallelism and that it is precisely this parallelism that indicates that the fires are one and the same. The significance of this parallelism will become clear when we consider the function of the fire that consumed Nadab and Abihu.

Laughlin proposes a different reading. According to him, the fire that consumes the sacrifice in 9:24 and Nadab and Abihu in 10:2 did not derive from the holy of holies but from the altar upon which the perpetual fire burns (see Lev 6:12–13).[47] The problem with Laughlin's proposal is that both 9:24 and 10:2 state explicitly that the fire "came out from the Lord." While the fire on the altar may have had a divine origin, it is never equated with the Lord himself. Rather, the כָּבוֹד of the Lord, which is elsewhere identified with fire (Exod 24:17), was thought to dwell in the holy of holies (Exod 25:22; Pss 80:1; 99:1), and therefore Milgrom's argument that the fire that "came out from the Lord" emerged from the holy of holies prevails over Laughlin's proposal.

In the narrative about Aaron's sons, the scope of the fire of judgment is far more limited than it is in the story of Sodom and Gomorrah. Whereas the latter consumed everyone within the cities and on the surrounding Plain, the former is directed only towards the two individuals who offered אֵשׁ זָרָה "unholy fire." Thus the scope of the fire can be delineated easily: it was directed only towards the two sons of Aaron, Nadab and Abihu. On this topic little more can be said.

As with Lot, the moral standing of Nadab and Abihu is contested in the later Jewish literature. Both Philo and some of the Rabbis contend that they were righteous priests whose only error was accidental while some among the Rabbis accuse them of a variety of sins, including drunkenness, arrogance, failure to father children, and wearing improper attire.[48] This variety of accusations is indicative of the fact that no one is quite sure what Nadab and Abihu's offence was. All that the text states is that "they offered unholy fire [אֵשׁ זָרָה] before the Lord, such as he had not commanded them" (10:1b). What precisely is meant by "unholy fire" (זָרָה אֵשׁ) is unclear. Leviticus 1:7 states, "The sons of the priest Aaron shall put fire on the altar and arrange wood on the fire." This fire, it is later commanded, is to be kept burning night and day; it is to be a perpetual fire that never dies out (6:12–13). According to Milgrom, the altar fire was believed to have had divine origins.[49] Hence the verb נָתַן "put" in 1:7 and not בָּעַר "kindle." The fire placed on the

---

47. Laughlin, "Holy Fire," 562.
48. See Kirschner, "Rabbinic and Philonic Exegeses."
49. Milgrom, *Leviticus*, 157.

altar was presumably already burning. It was therefore only this fire that was to be used for offerings. This has led many to believe that the זָרָה "unholy" or "strange" fire, which Nadab and Abihu put in their censors and burned incense upon, was taken from another source and not from the altar.[50]

Mary Douglas proposes that the punishment of Nadab and Abihu by fire was a result of Aaron's sin in facilitating worship of the golden calf (Exodus 32), for in that text Yahweh's wrath is said to וַיִּחַר־אַפִּי "burn hot" (v. 10) as is Moses' (v. 19), and fire is the medium by which the calf is destroyed (v. 20).[51] If she is correct, then Nadab and Abihu were punished for the sins of their father and not for their own actions. However, this reading ignores the "unholy fire" offered by Nadab and Abihu and is therefore unsatisfactory. The only element within the narrative itself that points to any rationale for the consumption of Aaron's sons is the reference to the "unholy fire." The causal relationship between the "unholy fire" they offer and the holy fire that consumes them has more explanatory power, than the notion that it is a result of their father's sin. In the wake of this strange event, the Lord warns Aaron, "You are to distinguish between the holy and the common, and between the unclean and the clean" (10:10). This strengthens the claim that the failure to distinguish between the holy and unholy fire was the ritual sin of Nadab and Abihu. Thus, while perhaps not morally corrupt, Aaron and Nadab engage in ritual impropriety in their service in the temple, and this is the cause of Yahweh's judgment on them.

In the Second Temple period various functions were attributed to the fire that consumed Nadab and Abihu. This variety is probably due to the sparseness of detail offered by the text. All that is stated is that "fire came out from the presence of the Lord and consumed them, and they died before the Lord" (10:2). Against the typical meaning of the verb אָכַל "consumed" stand two striking details in the narrative, which garner the attention of later interpreters. First is the fact that Aaron's sons are not literally "consumed." While they are certainly struck dead by the fire, their corpses remain as is made abundantly clear by 10:4, in which Moses orders Mishael and Elzaphan, the sons of Aaron's uncle Uzziel, "Come forward and carry your kinsmen away from the front of the sanctuary to a place outside the camp." While the verb אָכַל typically means "consume, devour," Levine observes that "when it is used of fire, it may simply mean "to burn, blaze."[52] And so it is used here to indicate that the fire blazed forth and killed Nadab and Abihu, without literally consuming them.

50. Morgenstern, *Fire upon the Altar*, 6.
51. Douglas, *Leviticus as Literature*, 200–205.
52. Levine, *Leviticus*, 60.

The second is the detail that Mishael and Elzaphan "carried them [Nadab and Abihu] by their tunics [בְּכֻתֳּנֹתָם] out of the camp" (10:5). The tunics belong not to Mishael and Elzaphan, but to Nadab and Abihu. That Nadab and Abihu were burned by the fire but that their tunics remained sound struck Philo and many of the Rabbis as odd and requiring explanation, and rightly so. Indeed, according to Gerstenberger, "especially the idea that the still-present corpses were taken away from the camp in their own tunics (v. 5) is utterly incompatible with the miscreants' death by incineration."[53] Both of these details indicate that the function of the fire, though it clearly resulted in the deaths of Nadab and Abihu, was not as destructive as that which rained down on Sodom and Gomorrah and destroyed all that lay within the cities. It was, nevertheless, clearly a form of punishment for Nadab and Abihu's ritual improprieties, and it demonstrates that even the Lord's chosen priests, and not only the wicked outsiders, are susceptible to judgment by holy fire.

The story of Nadab and Abihu shares much in common with and may show some literary dependence upon the account of Korah's rebellion. In this narrative, Korah, a Levite, his cohorts Dathan and Abiram (and On),[54] and two hundred and fifty men from the assembly protest against Moses and Aaron's exclusive role as the Lord's representatives to the community. The charge they lay against Moses and Aaron is that of exalting themselves above others. Korah and his men protest the exclusive role of Aaron and his sons as priests in the temple, indicating that they also wish to participate in this cultic role. In response to Korah's insubordination Moses proposes a sort of ordeal, instructing Korah and his followers to take their censers filled with fire and incense to the tent of meeting. Korah and the two hundred and fifty men oblige Moses' request while Dathan and Abiram remain in their homes. In consequence of their insurrection, Dathan and Abiram are swallowed up by the earth and go down to Sheol. More significant for our purposes, however, is the fate of the two hundred and fifty men who accompany them, for "fire came out from the Lord [מֵאֵת יְהוָה] and consumed the two hundred fifty men offering the incense" (Num 16:35). The fate of Korah is never specifically stated; however, it is probable that since he was among the men who offered incense, he was likewise consumed by fire from the Lord just as the two hundred and fifty were.[55]

---

53. Gerstenberger, *Leviticus*, 120.

54. After a brief mention in 16:1, On is no longer mentioned.

55. Indeed, most scholars believe that two separate stories, one about Dathan and Abiram and the other about Korah and the two hundred fifty men, are here fused. See Milgrom, *Numbers*, 138–39, 314, n. 77; Levine, *Numbers*, 410–17.

As with the narrative about Nadab and Abihu, the fire that consumes the two hundred and fifty men is said to come מֵאֵת יְהוָה "from the Lord." In keeping with our interpretation of the phrase "from the Lord" in Lev 9:24 and 10:2, and given the men's presence in the tent of meeting, we would do well to conclude once again that the fire that consumes the men does not fall from heaven, but like that which consumed Nadab and Abihu, proceeded from within the tabernacle, wherein resided the glory of the Lord in the form of fire.[56]

The scope of the fire in Num 16:35 is broader than the fire in Lev 10:2 but narrower than that which appears in Gen 19:24. The narrative of Num 16, however, bears striking similarities to both of these texts. Judging by Num 16:20, it appears that the Lord's initial intention is to consume the entire congregation, save for Moses and Aaron, just as he destroyed all of the citizens of Sodom and Gomorrah with the exception of Lot's family. However, in the manner of Abraham, Moses pleads with the Lord on behalf of the congregation, "O God, the God of the spirits of all flesh, shall one person sin and you become angry with the whole congregation?" (v. 22). Consequently, the fire consumes only Korah and the two hundred and fifty just as the fire in Lev 10:2 consumes only Nadab and Abihu, who similarly took their censers and offered incense before the Lord.

Less mystery surrounds the question of what prompted the destruction of Korah and his men than that which brought about the deaths of Nadab and Abihu. Here the sin of those consumed by fire is clearly their rebellion against Moses and, by extension, God and their arrogant belief that they are worthy of offering incense before the Lord despite the fact that they were not of the proper priestly lineage. It is this arrogance that elicits Moses' reproach: "You Levites have gone too far" (16:7b). While the role of offering sacrifice was limited only to Aaron and his sons, Korah is explicitly identified as a Levite, and Moses' protestation indicates that at least Dathan and Abiram are as well. Thus whereas the presumed sin of Nadab and Abihu was their offering of unauthorized fire, the sin of Korah's men was the fact that they were unauthorized priests. Indeed, after the fire consumes the two hundred and fifty men, their bronze censers are hammered out into a covering for the altar, which is to be "a reminder to the Israelites that no outsider, who is not of the descendants of Aaron, shall approach to offer incense before the Lord, so as not to become like Korah and his company—just as the Lord had said to him through Moses" (16:40).

The fact that their sin was cultic illegitimacy is made all the more clear when their actions and fate are contrasted with those of Aaron, for Moses

---

56. See Milgrom, *Numbers*, 138.

commands Aaron to join Korah and his men in offering incense: "And Moses said to Korah, 'As for you and all your company, be present tomorrow before the Lord, you and they and Aaron; and let each one of you take his censer, and put incense on it, and each one of you present his censer'" (Num 16:16–17). Presumably, Aaron performs the same actions as Korah and his men do. However, Aaron is not consumed by the fire in v. 35 as they were. Thus the sin of Korah and his men is clearly that of arrogating to themselves the right of performing the priestly rituals reserved only for Aaron and his sons.

According to Num 16:35, fire consumed (וַתֹּאכַל) Korah and the two hundred and fifty men who offered incense in the tent of meeting. Unlike Leviticus, in the account of the two hundred and fifty men who followed Korah, there is no mention of any corpses that needed to be removed from the tent of meeting. Thus it appears that the verb אָכַל (consume, devour) is used in its literal sense. Indeed, it seems that the disappearance of the two hundred and fifty men is just as complete as that of Dathan and Abiram, who were swallowed up into the chasm that opened in the earth. In Num 16:36–39 all that is required is that Aaron's son Eleazar collect the censers that remained. This is a striking detail. What is even more interesting is what the Lord says to Moses about the censers: "Tell Eleazar son of Aaron the priest to take the censers out of the blaze; then scatter the fire far and wide. For *the censers of these sinners have become holy* at the cost of their lives. Make them into hammered plates as a covering for the altar, for they presented them before the Lord and *they became holy*." Despite the fact that the men who offered incense in their censers before the Lord were not authorized to do so, in the wake of their destruction by the consuming fire of the Lord, their censers are deemed sacred. While Korah and his men are consumed by the fire, their censers are made holy. While the means by which the censers were sanctified is not explicitly stated, Milgrom presents the possibility that they became holy "because they were touched by divine fire."[57] This suggestion is supported by the existence of a similar notion elsewhere in the Priestly legislation, according to which fire plays a purificatory role: "everything that can withstand fire shall be passed through fire, and it shall be clean" (Num 31:23a).

While the fire may play a purificatory role in relation to the censers, it is apparent that it would have indiscriminately consumed any human being in its path. The indiscriminate nature of the fire is indicated by the Lord's command to Moses and Aaron, "Separate yourselves from this congregation, so that I may consume them in a moment" (16:21). Like Lot and his

---

57. Ibid., 139.

family, those who wish to escape the fire from the Lord must be outside of the scope of the flame.

Similarly, in 2 Kgs 1:9–12 Elijah condemns King Ahaziah for having sent men to inquire of Baal-zebub. When the king sends a captain and fifty men to summon Elijah to him, Elijah responds by calling down fire upon them. Elijah's plea in v. 10 "let fire come down from heaven" is answered by the narrator's descriptive statement in v. 11 that "the fire of God came down from heaven." When the narrative repeats with a second captain and a second deployment of fifty men, the same stock phrases—Elijah's "let fire come down from heaven" (v. 12a) and the narrator's "Then the fire of God came down from heaven" (v. 12b)—are once again employed. These phrases function to identify unambiguously the source of the fire. It comes *from heaven*. More specifically, however, if there had been any question as to the precise heavenly origin of the fire, this uncertainty is dispelled by the fact that the fire is described as "the fire of God."[58] Its origin, like the fire that Elijah summoned from heaven to consume the sacrifice on Mount Carmel, is clearly divine.

The scope of the fire is rather limited. Each fire consumes a captain and his regiment of fifty men. These instances of fire falling from heaven directed towards a precise target parallel those texts we have already examined in the Pentateuch in which the fire proceeds from the holy of holies to consume a focused target, namely in the cases of Nadab and Abihu and Korah and his men. Here, too, the function is clearly destructive, for it consumes King Ahaziah's men; however, the fire also has a secondary function, for it validates Elijah as a "man of God." When calling down fire, Elijah says, "If I am a man of God, let fire come down from heaven" (vv. 10, 12). The fire of God (אֵשׁ אֱלוֹהִים) that falls from heaven confirms Elijah's standing as a man of God (אִישׁ אֱלוֹהִים).[59]

We must also give consideration to the moral standing of King Ahaziah's two captains and their regiments of fifty men. At first consideration, one may be compelled to the conclusion that these captains and their men are merely unwitting messengers and that their guilt before Elijah is only an extension of this sin of King Ahaziah, for in both cases the captains only come to Elijah and bid him to come down from his hill top at the king's command. It is the mere fact that they are relaying an order from the king that elicits Elijah's wrath. However, it cannot be insignificant that the third captain who comes to Elijah does not presume to tell Elijah what to do, but

---

58. Cogan and Tadmor (*II Kings*, 26–27) note, however, that אֱלוֹהִים may be used to express the superlative and thus translate אֵשׁ אֱלוֹהִים as "an awesome fire" See also Thomas, "Consideration of Some Unusual Ways."

59. See Bronner, *Stories of Elijah and Elisha*, 8.

instead prostrates himself before the prophet and pleads, "O man of God [אִישׁ הָאֱלֹהִים], please let my life, and the life of these fifty servants of yours [עֲבָדֶיךָ], be precious in your sight" (v. 13), and that he and his men are thus spared. Indeed, the words of the third captain make clear the sin of the first two captains, for the third refers to himself and his men as עֲבָדֶיךָ "servants of yours," thereby subordinating himself to Elijah and thus aligning himself with Elijah's God, the Lord. By contrast, the first two captains viewed themselves not as servants of Elijah, and thus of the Lord, but as subordinates of King Ahaziah, who had inquired of Baal-zebub, and therefore in the eyes of Elijah they are deemed as sinful as King Ahaziah and deserving of the fire of judgment.

Although both this passage and 1 Kgs 18 employ the same verb to describe the function of the fire—אָכַל "to consume"—the function of the fire on Mount Carmel that consumes Elijah's sacrifice to the Lord differs significantly from the fire that consumes the captains of King Ahaziah and their men. In the first case, the fire that consumes the sacrifice is to be understood as a sign of the Lord's acceptance of Elijah's sacrifice and is therefore a sign of the Lord's pleasure, and not wrath. As Cogan indicates, "[t]he fiery apparition often symbolizes YHWH's presence (e.g., Exod 3:2; 19:18; 24:17), and, with reference to the altar inaugurations (Lev 9:24; 2 Chr 7:3), the fire attests divine approval and acceptance of the worshiper's act (1 Chr 21:26)."[60] On the other hand, the fire that consumes the captains sent by King Ahaziah and the men who accompany them is clearly a sign of God's wrath. They are consumed not because they are a pleasing sacrifice to the Lord, but because they have offended the Lord through their allegiance to the king despite the fact that he has inquired of Baal-zebub. The fire from heaven serves as a means of divine judgment in the punitive sense.

The primary function in both cases, however, is to demonstrate the power of the Lord God over against Baal or Baal-zebub.[61] As previously noted, the premise of Elijah's contest with the prophets of Baal is their dispute over the identity of the true God in Israel. The response that the fire from heaven elicits from the people makes this clear: "When all the people saw it, they fell on their faces and said, 'The Lord indeed is God; the Lord indeed is God'" (v. 39). Similarly, the fire that falls from heaven upon the captains and the men sent by King Ahaziah seem to be in confirmation of Elijah's conditional "If I am a man of God . . . " (vv. 10, 12). The fire of God is sent in response to demonstrate that Elijah is a true man of God, and that, in turn, the Lord is the true God of Israel as Elijah has testified.

---

60. Cogan, *I Kings*, 443.

61. Ibid., 444. Cogan describes the scene as "an impressive show of YHWH's might."

Having considered a sampling of several of the more significant texts in which fire functions as a means of judgment intervening in the course of history, we can see that its primary function is destructive, as can be seen in the narratives about Sodom and Gomorrah, Nadab and Abihu, and Korah's Rebellion examined above. It appears that in every one of these cases the fire does not discriminate between the righteous and the wicked, and thus the righteous are forewarned so that they may escape the destructive fire. However, we did note that there is evidence in the Priestly writings that fire plays a purificatory role in relation to certain inanimate objects, namely the censers used by Korah and his men. It should also be noted that in every instance the fire is a literal fire, and whether it falls from heaven or proceeds from the holy of holies, its source is unambiguously God. Having considered these foundational texts in which fire intervenes in the course of history, let us consider the key texts in which the punitive function of fire features in the eschatology of the Hebrew Bible.

The Punitive Function of Eschatological Fire

In addition to the fire that intervenes within the course of history, several prophetic texts speak of an eschatological fire that attends the Day of the Lord. These texts are perhaps more significant for our study, for in the New Testament, fire is nearly always eschatological. In addition to texts that specifically refer to "the Day of the Lord," we shall also consider here several other prophetic texts that originally referred to impending destruction within the confines of history. For despite the fact that these prophecies originally pointed towards historical referents, it is generally recognized that many such texts took on an eschatological hue in later interpretation.[62] The eschatological fire takes several forms. Sometimes it is referred to as a "devouring fire." This motif is particularly popular among the prophets, who utilize it to depict the consuming wrath of the Lord; frequently in tandem with the motif of the devouring fire, the prophets use the metaphors of chaff, stubble, and thorns in reference to the wicked who are consumed by the destructive fire. Other times the fire appears in conjunction with the motif of water or rivers; sometimes the fire is even in the form of a river, similar to that we have seen in our discussion of the theophanic fire of Daniel 7. In other cases still eschatological judgment is described in language reminiscent of a forest fire that consumes the trees of the forest to their

---

62. For an excellent discussion of this tendency, see chapter 32, entitled "The Eschatological Reinterpretation of Prophecy" and the literature cited therein, in Blenkinsopp, *History of Prophecy*, 226–39.

very roots. These motifs, which sometimes overlap with one another, are too widespread for us to discuss exhaustively here, and so we must limit our discussion to a few key texts. We shall begin with the motif of the devouring fire that consumes chaff, stubble, and/or thorns.

Isaiah makes frequent use of the devouring fire that consumes chaff, stubble, and thorns as a metaphor for God's judgment. In Isa 5:24 the prophet states that those who "have rejected the instruction of the Lord of hosts" and "despised the word of the Holy One of Israel" will be like stubble consumed by a tongue of fire. Speaking to Jerusalem in 29:5–6, Isaiah proclaims the following:

> But the multitude of your foes shall be like small dust [כְּאָבָק דַּק],
> and the multitude of tyrants like flying chaff [וּכְמֹץ].
> And in an instant, suddenly, you will be visited by the Lord of hosts
> with thunder and earthquake and great noise,
> with whirlwind and tempest, and the flame of a devouring fire [וְלַהַב אֵשׁ אוֹכֵלָה].

Here elements of the storm theophany reside alongside the devouring fire motif. It is quite evident, however, that the visitation of the Lord is one of judgment. The oracle is enigmatic, but it appears that the object of the devouring fire is Jerusalem itself. This conclusion is anticipated by v. 2, in which Jerusalem is addressed as (אֲרִיאֵל) "Ariel," which means "altar hearth," and v. 2, which reads "Yet I will distress Ariel [לַאֲרִיאֵל]. And there shall be moaning and lamentation, and Jerusalem shall be to me like an Ariel [כַּאֲרִיאֵל]." The final clause, which likens Jerusalem to an Ariel (an altar hearth) has suggested to some interpreters that the city itself will become the altar upon which a burnt offering is sacrificed. It is unclear, however, whether it is the inhabitants of Jerusalem themselves or if it is the enemies of the Lord besieging Jerusalem, who are likened to מֹץ chaff, that will be the consumed by the אֵשׁ אוֹכֵלָה devouring fire.[63]

Isaiah 30:27–28 likewise calls upon the devouring fire motif to depict the wrath of the Lord, though this time the image is paired with water:

> See, the name of the Lord comes from far away,
> burning with his anger, and in thick rising smoke;
> his lips are full of indignation,

---

63. See Wildberger, *Isaiah 28–39*, 74. Wildberger entertains the possibility that "Yahweh will make Zion into the locale where the offering of his enemies will be made." He goes on to caution, however, that "[w]ith our limited knowledge at this point in time, it is better to be extra cautious about trying to say too much."

> and his tongue is like a devouring fire [וּלְשׁוֹנוֹ כְּאֵשׁ אֹכָלֶת];
> his breath is like an overflowing stream [וְרוּחוֹ כְּנַחַל שׁוֹטֵף]
> that reaches up to the neck—
> to sift the nations with the sieve of destruction [יְחֶצֶה לַהֲנָפָה גוֹיִם בְּנָפַת שָׁוְא]
> and to place on the jaws of the peoples a bridle that leads them astray.

Here, too, we clearly have a depiction of the coming judgment in which fire plays a destructive role. What is striking in this passage, however, is the pairing of the phrases וּלְשׁוֹנוֹ כְּאֵשׁ אֹכָלֶת "his tongue is like a devouring fire" and וְרוּחוֹ כְּנַחַל שׁוֹטֵף "his breath is like an overflowing stream." If these two images are to be read not as distinct elements, but as a hendiadys, as may be suggested by the fact that they are both proceed from the Lord's mouth, we may be dealing with the stream of fire motif we have already encountered in Daniel 7. This reading is made all the more probable a few lines later in v. 33, which describes the place of punishment prepared for the Assyrian: "For his burning place has long been prepared; truly it is made ready for the king, its pyre made deep and wide, with fire and wood in abundance; *the breath of the* Lord, *like a stream of sulphur* [כְּנַחַל גָּפְרִית], *kindles it.*"

It is also interesting to note that, according to the NRSV translation, the stream of devouring fire functions יְחֶצֶה לַהֲנָפָה גוֹיִם בְּנָפַת שָׁוְא "to sift the nations with a sieve of destruction" (v. 28). While the title "sieve of destruction" makes the destructive element more prominent, the notion of sifting with a sieve may suggest that the judgment taking place here is at least partially forensic and not exclusively punitive, and we are left to question whether the sieve sorts out anything or anyone that is not turned over to destruction. The sieve probably implies the separation of chaff from whatever produce is being sifted, and presumably the wheat, or whatever grain, from which the chaff was removed would not be destroyed. It should be noted, however, that others have translated the phrase quite differently. Blenkinsopp has "he will place on the nations a yoke that spells their ruin,"[64] and Wildberger suggests, "he whirls the nations around with a disaster-bringing whirling."[65] Due to the difficulties inherent in the translation of this passage, too much confidence should not be placed in the likelihood that the sifting entails a positive outcome for any involved parties. Indeed, the theme of destruction and judgment predominates.

---

64. Blenkinsopp, *Isaiah 1–39*, 422.
65. Wildberger, *Isaiah 1–12*, 185.

Perhaps more significant is Isa 33:10–12; 14–16, in which the prophet speaks an utterance of judgment by fire against the wicked in Israel, accusing them of conceiving chaff and bringing forth stubble:

> "Now I will arise," says the Lord,
> "now I will lift myself up;
> now I will be exalted.
> You conceive chaff [חֲשַׁשׁ], you bring forth stubble [קַשׁ];
> your breath is a fire [רוּחֲכֶם אֵשׁ] that will consume you.
> And the peoples will be as if burned to lime [מִשְׂרְפוֹת שִׂיד],
> like thorns [קוֹצִים] cut down, that are burned in the fire."
>
> . . .
>
> The sinners in Zion are afraid;
> trembling has seized the godless:
> "Who among us can live with the devouring fire [אֵשׁ אוֹכֵלָה]?
> Who among us can live with everlasting flames [מוֹקְדֵי עוֹלָם]?"
> Those who walk righteously and speak uprightly,
> who despise the gain of oppression,
> who wave away a bribe instead of accepting it,
> who stop their ears from hearing of bloodshed
> and shut their eyes from looking on evil,
> they will live on the heights;
> their refuge will be the fortresses of rocks;
> their food will be supplied, their water assured.

Since חֲשַׁשׁ "chaff" and קַשׁ "stubble" are the useless byproducts of threshing, they probably allude to the evil actions of the wicked, presumably the actions that the righteous refrain from in vv. 15–16. Interestingly, those who are accused of conceiving chaff and bringing forth stubble are in some sense also equated with these byproducts, for they share the same fate. Like chaff and stubble, they too shall be consumed by fire. Similarly, the peoples (the foreigners in Jerusalem) will be consumed as קוֹצִים "thorns;" however, their fate is even more dismal, for when the prophet proclaims that "they will be burned as if to lime [מִשְׂרְפוֹת שִׂיד]," this envisions complete destruction. Their bones will be burnt until all that is left is a powdery white substance resembling lime (see Amos 2:1b).[66]

There is some debate as to whether the MT's רוּחֲכֶם אֵשׁ (NRSV: your breath is a fire) is the best rendering of v. 11b. Against this reading, some assume a textual error and favor the proposed reading רוּחַ כְּמוֹ אֵשׁ (a wind/

---

66. Wildberger, *Isaiah, 1–12*, 286.

breath like fire) or רוּחִי כְּמוֹ אֵשׁ (my breath is like fire).[67] Whereas the MT suggests that the source of the fiery breath that consumes the wicked inhabitants of Jerusalem is the wicked themselves, in the alternative readings the source is the Lord who is coming in judgment. The evidence from the Targum and the Vulgate support the proposed alternative reading רוּחַ כְּמוֹ אֵשׁ (a wind/breath like fire), and given the fact that throughout Isaiah, the Lord is the source of the devouring fire, there is good reason to favor that interpretation here as well.

In vv. 14–16, "the sinners of Zion" ask, "Who among us can live with the devouring fire?" Despite the despairing tone of their inquiry, a positive answer is given in response to their question: the righteous can live in the midst of the fire. Yet no specific function, such as purification or testing, is attributed to the fire in regard to its effect upon the righteous. It seems that they simply endure the devouring fire without experiencing its destructive powers. They are like Daniel's friends—Shadrach, Meshach, and Abednego—whom King Nebuchadnezzar had thrown into the fiery furnace when they would not worship his gods, but who were preserved from any harm (Daniel 3).[68] In connection with this observation, Isa 43:2 is also worthy of mention:

> When you pass through the waters [בַּמַּיִם], I will be with you;
> And through the rivers [וּבַנְּהָרוֹת], they shall not overwhelm you;
> When you walk through fire [בְּמוֹ־אֵשׁ] you shall not be burned,
> and the flame shall not consume you.

As in Isa 33:14–16, in 43:2 those who belong to the Lord are preserved from the devouring fire. It is also noteworthy that here again we may encounter a hendiadys, for the מַיִם "waters" and the אֵשׁ "fire" may be one—together they may compose a river of fire such as we have already encountered elsewhere. Whether or not this is the case, independently they represent the extremes of physical danger, and in the verse the Lord promises to protect his chosen from such perils.

---

67. See *BHS*, 724, n. 11a. Blenkinsopp (*Isaiah 1–39*) prefers the former emendation; Gunkel and Kissane (cited in Wildberger, *Isaiah 28–39*, 279) prefer the latter emendation; and Wildberger prefers the MT as it stands, arguing "it is questionable whether one should alter the text merely to arrive at the thoughts one expects to find. Yahweh certainly rises up and enters the scene. But there is no contradiction, as far as OT thought is concerned, in the idea that the godless dig their own graves by their actions" (279).

68. Daniel 3 offers an interesting instance of fire in the Hebrew Bible. Since, however, it does not fall under the category of "holy fire," for it does not have a divine source, we shall not include that text in our discussion.

In Jeremiah and Ezekiel we encounter a motif similar to the fire that consumes chaff and stubble. Here, however, the metaphor is not agricultural, but draws on the devastating phenomenon of the forest fire that consumes all in its path. In Jer 21:14, speaking of his judgment on the house of the king of Judah, the Lord proclaims, "I will punish you according to the fruit of your doings, says the Lord; I will kindle a fire in its forest [וְהִצַּתִּי אֵשׁ בְּיַעְרָהּ], and it shall devour [וְאָכְלָה] all that is around it." Ezekiel 15:1–8 likens the inhabitants of Jerusalem to עֵץ־הַגֶּפֶן "the wood of a vine" that God has given up to be burned by a forest fire. The Lord speaks to Ezekiel, saying that a piece of vine wood is completely useless, but it will be even more so once it is consumed by fire.[69] So it is with Jerusalem. Comparing Jerusalem to a useless burnt vine suggests that total destruction will result from the ensuing judgment.

The motif recurs at greater length in Ezek 20:45–47:

> The word of the Lord came to me: Mortal, set your face toward the south, preach against the south, and prophesy against the forest land in the Negeb, Hear the word of the Lord: thus says the Lord GOD, I will kindle a fire in you [מַצִּית־בְּךָ אֵשׁ], and it shall devour every green tree in you and every dry tree [וְאָכְלָה בְךָ כָל־עֵץ־לַח וְכָל־עֵץ יָבֵשׁ]; the blazing flame shall not be quenched, and all faces [כָּל־פָּנִים] from south to north shall be scorched by it. All flesh shall see that I the Lord have kindled it; it shall not be quenched.

From his position in Babylon, Ezekiel is to speak against the south, that is Israel, specifically Jerusalem, and he is to proclaim its destruction by fire.[70] Two details are striking in this prophecy. First, it is the Lord who says מַצִּית־בְּךָ אֵשׁ "I will kindle a fire in you" (v. 47). While those who literally kindle the fire of war are Nebuchadnezzar's men, the ultimate source of the fire is the Lord, and he will use it against his own people in Judah. This is a stark reminder that the prophets did not reserve proclamations of judgment for the other nations alone, but could, and often did, direct their criticism towards their own people. Secondly, that the fire "shall devour every green tree [כָּל־עֵץ־לַח] in you and every dry tree [וְכָל־עֵץ יָבֵשׁ]" seems to envision complete destruction. Indeed, "all faces [כָּל־פָּנִים] from south to north shall

---

69. As Zimmerli (*Ezekiel*, 320) notes, "One use for vine wood is doubtless to be conceded: its use as fuel."

70. Eichrodt, *Ezekiel*, 287. Negeb is not to be taken as a literal reference to the sparsely settled, politically insignificant, dry and treeless geographical region but its figurative sense, simply meaning "south." From Ezekiel's location in Babylon, the most obvious referent is Jerusalem, the center of Israel's power and influence. The parallel oracle that follows in 21:1–7, which refers specifically to Jerusalem (v.2), confirms this.

be scorched by it" (v. 47b). While dry trees, like chaff and stubble, are fit for burning in the fire, one does not expect the consumption of the green, living tree. Yet no forest fire is selective, consuming only the dead and worthless dried out trees. Unlike the fire that consumes the chaff after it has been sifted out from the wheat, the forest fire does not discriminate. On the basis of Ezekiel's parallel allegory of the sword that follows in 21:1–7, particularly v. 3b, in which the Lord proclaims, "I am coming against you, and will draw my sword out of its sheath, and will cut off from you both righteous and wicked [צַדִּיק וְרָשָׁע]," it appears that the green tree and the dry tree are to be taken as figurative references to the righteous and the wicked respectively.[71] Strikingly, the Lord's judgment by fire will not discriminate between the righteous and the wicked; all will share the same fate of destruction.

A number of texts refer explicitly to "The Day of the Lord" or allude to it by some shorter formula. Isaiah 34:8–10 speaks of "a day of vengeance" and draws upon the motif of the river of fire as a metaphor of divine judgment:

> For the Lord has a day of vengeance [יוֹם נָקָם]
>   a year of vindication by Zion's cause.
> And the streams of Edom shall be turned into pitch [. . . וְנֶהֶפְכוּ לְזֶפֶת],
>   and her soil into sulphur [לְגָפְרִית];
> her land shall become burning pitch [לְזֶפֶת].
> Night and day it shall not be quenched;
>   its smoke shall go up forever.
> From generation to generation it shall lie waste;
>   no one shall pass through it forever and ever.

That this text looks forward to a יוֹם נָקָם "day of vengeance" suggests an eschatological reading; the day envisioned is to be identified with "the Day of the Lord." The language is strikingly reminiscent of the destruction of Sodom and Gomorrah (see Jer 49:17–18).[72] The brimstone in v. 9 (גָפְרִית) recalls that of Gen 19:24 (גָפְרִית) and it may also be significant that the cities of Sodom and Gomorrah were associated with the region of Edom during the time of Isaiah.[73] Moreover, as Gen 19:25 uses the verb "overthrew" (וַיַּהֲפֹךְ) to describe the destruction of Sodom and Gomorrah, Isa 34:9 uses the same verb (וְנֶהֶפְכוּ) in reference to turning the streams of Edom to pitch.[74] We can thus conclude that the destruction of Sodom and Gomorrah functions as a

---

71. See Block, *Book of Ezekiel*, 668.
72. See Beuken, *Isaiah II/2*: 286.
73. Wildberger, *Isiah 28–39*, 333.
74. Ibid.

prototype for the eschatological destruction of Edom. As with the destruction of Sodom and Gomorrah, the destruction of Edom by fire is absolute: nothing is preserved; no one is saved; all is left in ruin.

Malachi 4:1 (LXX 3:19) is also among the eschatological texts that speak of the Day of the Lord. Malachi's use of "the day" is synonymous with "the Day of the Lord," and he likens that day to an "oven" or "furnace" (see Hos 7:4):

> See, the day [הַיּוֹם] is coming, burning like an oven, when all [כָּל] the arrogant and all [כָּל] evildoers will be stubble [קַשׁ]; the day [הַיּוֹם] that comes shall burn them up, says the LORD of hosts, so that it will leave them neither root nor branch [שֹׁרֶשׁ וְעָנָף].

Interestingly, Malachi conflates two previously considered motifs. He begins by likening the evildoers to קַשׁ "stubble," a motif we have already noted in the prophets, one that is commonly paired with "chaff" to indicate the worthlessness of those being judged. But he concludes with the forest fire motif: the fire will leave neither root nor branch (שֹׁרֶשׁ וְעָנָף). The objects of the divine judgment are specifically identified as "all the arrogant and all evildoers." Malachi's use of כָּל "all" indicates the thoroughness of the judgment: none of the arrogant or evil doers will be spared; all will be consumed like stubble thrown into the oven. Moreover, their judgment will result in utter destruction. Turning to the image of the forest fire, Malachi proclaims that they will be left "neither root nor branch." The result for the wicked is utter destruction. Pieter Verhoef cites Job 18:16–18 to indicate the significance of a people being compared to a tree without root or branch: "Their roots [שָׁרָשָׁיו] dry up beneath, and their branches [קְצִירוֹ] wither above. Their memory perishes from the earth, and they have no name in the street."[75] Not only are they destroyed, but it is as if they had never existed.

It is very interesting to note that the subject here is (הַיּוֹם) "the day" itself: "the day that comes shall burn them up" (3:19). Incidentally, as Verhoef observes, "[t]he expression that the coming day will cause a fire to consume the evildoers is found only here in the OT."[76] Presumably "the day" functions as a circumlocution for the divine as is indicated by 4:3, which states "on the day when I act." Nonetheless, it presents an intriguing usage, one that is paralleled by 1 Cor 3:13, where ἡ ἡμέρα "the Day" functions as the subject of disclosing, accompanied by a revealing and testing fire, and is therefore worth noting at this juncture for the light it may shed on that text later in our investigation.

---

75. Verhoef, *Books of Haggai and Malachi*, 326.
76. Ibid.

Finally, our discussion of passages dealing with the eschatological judgment by fire in the Hebrew Bible would not be complete without a consideration of Isa 66:15–16:

> For the Lord will come in fire,
>> and his chariots like the whirlwind,
> to pay back his anger in fury,
>> and his rebuke in flames of fire.
> For by fire will the Lord execute judgment,
>> and by his sword, on all flesh [אֶת־כָּל־בָּשָׂר];
> and those slain by the Lord shall be many.

There are some theophanic elements here: the Lord comes with chariots as the Divine Warrior, and he is accompanied by stormy winds. It is no simple theophany, though, for the scene is one of judgment. It is significant that judgment is exercised "on all flesh." Strikingly, however, that the verse states that "those slain by the Lord shall be many" and not "all" has as its corollary that at least some are expected to survive this judgment by fire and sword. Westermann notes a parallel usage in Jer 25:21, where judgment is upon "all flesh" and concludes that "[s]ince there these words introduce the oracles against the foreign nations, there can be no doubt that here it also refers to God's judgment upon them."[77] God comes in judgment to rescue Israel from the nations who oppress her.

Just a few verses later in Isa 66:22–24 the phrase "all flesh" recurs, but here it refers to the righteous who worship the Lord:

> For as the new heavens and the new earth,
>> which I shall make,
> shall remain before me, says the Lord,
>> so shall your descendants and your name remain.
> From new moon to new moon,
>> and from sabbath to sabbath,
> all flesh [כָּל־בָּשָׂר] shall come to worship before me,
>> says the Lord.
> And they shall go out and look at the dead bodies of the people who have rebelled against me; for their worm shall not die, their fire shall not be quenched, and they shall be an abhorrence to all flesh [לְכָל־בָּשָׂר].

After the eschatological judgment and the creation of a new heaven and a new earth, the righteous (the כָּל־בָּשָׂר of v. 23) go out to see the wicked

---

77. Westermann, *Isaiah 40–66*, 421.

whose "fire shall not be quenched" (the כָל־בָּשָׂר of v. 16). Thus the fire of eschatological judgment that is exercised upon the nations becomes a perpetual fire that punishes them eternally. It is no wonder, then, that this text so influenced later notions of eternal punishment in hell (see Mark 9:40-48). Most significant for our study, though, is the fact that in this case the fire of judgment appears to be selective. It has destroyed the wicked and rebellious while sparing those who worship the Lord. Like the flood that only spared Noah and his family, the fire has purified the land of evil so that a new creation can come into existence untainted.

The punitive fire of eschatological judgment takes several forms. At times the motif of a fire that devours useless waste such as chaff, stubble, and thorns depicts the destruction of evildoers. Elsewhere judgment is depicted as a raging forest fire that consumes everything in its path. In other instances the fire that attends the day of the Lord is depicted as a river of fire or a fire paired with a sword that destroys the nations. In a handful of cases we noted that the fire of judgment can sometimes be selective or that the righteous can endure the fire unscathed. From these texts we turn to examine those passages in which fire implicitly or explicitly plays a purifying role.

The Purifying Function of Fire

We have noted several instances in which the fiery judgment is selective, destroying only the wicked while the righteous are preserved. However, we have not yet encountered any texts in which fire plays a positive function. Indeed, one motif that has received relatively little attention thus far in the scholarly literature is that of the refining fire, a motif scattered throughout the prophets Isaiah, Jeremiah, Ezekiel, Zechariah, and Malachi, which refers to the process of smelting and refining whereby precious metals such as silver were melted so that the dross and alloy could be removed, leaving behind a purified product. Before considering these texts, a brief word on ancient metallurgy is in order.[78] The extraction of silver from lead ores involved two stages. First, the ore was smelted, and second the silver was refined from crude lead. Smelting itself involved two processes. In the Mediterranean, the smelter most often began with a lead ore called galena, which probably contained, in addition to silver, other elements, such as copper, iron, and tin. The smelter would first roast the galena with blasts of hot air, causing the sulphur dioxide in the ore to turn into gas and escape, thus resulting in a mixture of lead sulphate and lead oxide (the former is also

---

78. In what follows I rely mainly on the excellent excursus on "The Cuppellation of Silver" in Holladay, *Jeremiah 1*, 230-32.

known as litharge). The roasting process was complete once most of the galena had become litharge. The second process, reduction, involved heating the litharge with charcoal at a higher temperature and in the absence of oxygen until the lead sulphate and lead oxide interacted to produce sulphur dioxide, leaving behind crude lead containing trace amounts of silver. After the ore was smelted, silver could be refined from the crude lead by a process called cupellation. The crude lead was placed in a furnace on a hearth of a porous substance such as bone ash. The lead was heated to between 900 and 1000 degrees centigrade and then blasted with air, causing the crude lead to oxidize and become lead oxide (litharge), some of which was subsequently absorbed by the bone ash. The remainder of the litharge flowed off through a small notch in the hearth. The litharge could also act as a flux, carrying off the oxides of other metals. The end result was a small amount of purified silver, typically 0.05 percent to 0.1 percent of the original galena.

The first to compare the Lord to a smelter and his judgment as the process of smelting and refining is the prophet Isaiah in 1:25–26:

> I will turn my hand against you;
> I will smelt away your dross as with lye [כַּבֹּר]
> and remove all your alloy
> And I will restore your judges as at the first [כְּבָרִאשֹׁנָה]
> and your counselors as at the beginning [כְּבַתְּחִלָּה].
> Afterward [אַחֲרֵי־כֵן] you shall be called the city of righteousness,
> the faithful city.

First, it must be noted that this is a scene of judgment as indicated by the opening line: "I will turn my hand against you." Despite the positive outcome, the Lord is acting against Jerusalem, specifically the ruling classes. Many commentators prefer an alternative reading for v. 25, proposing בְּכֹר "in a furnace" (see Isa 48:10, which has בְכוּר) instead of כַּבֹּר "as with lye."[79] This is a very plausible emendation that makes good sense in the context of smelting and refining. However, it may be unnecessary, for as Wildberger notes, "salts of lye play a role as a flux when silver is being extracted."[80] Either way, the metaphor of smelting clearly indicates the purification of Jerusalem, for after the process is completed, after the dross and alloy have been purged away, the city is restored and is called righteous. The prophet alludes to both stages of the process: the dross is smelted away and then the alloys are removed in the refining process. As a result of this process, a return to an idyllic time is envisioned, an idealized past without political corruption,

---

79. Kaiser, *Isaiah 1–12*, 39; Blenkinsopp, *Isaiah 1–39*, 187.
80. Wildberger, *Isaiah 1–12*, 60.

THE FUNCTIONS OF FIRE IN THE HEBREW BIBLE 63

as is indicated by the lines כְּבָרִאשֹׁנָה "at the first" and כְּבַתְּחִלָּה "at the beginning." Blenkinsopp conjectures that since the address is to Jerusalem, "the idealized past is presumably that of the city under Davidic rule."[81] Clearly the motif of smelting and refining is used metaphorically to point to the removal of the wicked inhabitants of the city with the hope that אַחֲרֵי־כֵן "afterward" the righteous will remain.

Drawing on the image of the refiner to depict the process of purification by means of fire, Jer 6:27–30 turns the metaphor on its head to depict a dire situation:

> I have made you a tester and a refiner [בָּחוֹן] among my people
>   so that you may know and test [וּבָחַנְתָּ] their ways.
> They are all stubbornly rebellious,
>   going about with slanders;
> they are bronze and iron,
>   all of them act corruptly.
> The bellows blow fiercely,
>   the lead is consumed by the fire;
> in vain the refining goes on [לַשָּׁוְא צָרַף צָרוֹף],
>   for the wicked are not removed.
> They are called "rejected silver" [כֶּסֶף נִמְאָס],
>   for the Lord has rejected them.

One of the most striking aspects of this passage is that the בָּחוֹן "tester" and "refiner" is not the Lord, but Jeremiah. As Holladay points out, while the noun בָּחוֹן occurs only in this verse in the entire Hebrew Bible, the verb בָּחַן "assay, test" appears in five other places in Jeremiah, and in each of those cases the Lord is the subject.[82] Jeremiah is thus given a special role acting in the Lord's stead to test and refine the rebellious people.

The wicked are depicted as bronze, iron, and lead that cling to silver in the second stage of purifying silver, the refining process.[83] Strikingly, however, for Jeremiah the smelting process is done in vain, "for the wicked are not removed." They are ultimately rejected because they have not been purified by the process. Interestingly, Holladay notes that if the refining process were conducted under too extreme heat, it would frequently prove unsuc-

---

81. Blenkinsopp, *Isaiah 1–39*, 187.
82. Holladay, *Jeremiah 1*, 229.
83. Allen (*Jeremiah*, 91) observes that while the extraction of silver involved two stages (smelting and refining), both of which are used metaphorically in Isaiah 1:21–26), "Jeremiah developed the second stage, while Ezekiel was later to develop the first (Ezek 22:18–22)."

cessful and the alloy could not be removed from the pure silver. Perhaps this passage may be suggesting that the Lord's anger burned too hotly in this case for any positive outcome to be expected. Nonetheless, the fact that the process results in only כֶּסֶף נִמְאָס "rejected silver" is clearly due to the impenitence of the wicked, and presumably the Lord's anger is proportionate to the wickedness of those being tested and refined.

As Jeremiah made use of the metaphor of refining silver unsuccessfully executed, Ezek 22:18–22 employs the metaphor of smelting to a similar end.

> Mortal, the house of Israel has become dross [לְסוּג] to me; all of them, silver, bronze, tin, iron, and lead. In the smelter they have become dross [סִגִים]. Therefore thus says the Lord GOD: Because you have all become dross [לְסִגִים], I will gather you into the midst of Jerusalem. As one gathers silver, bronze, iron, lead, and tin into a smelter, to blow the fire upon them in order to melt them [לְהַנְתִּיךְ]; so I will gather you in my anger and in my wrath, and I will put you in and melt you [וְהִתַּכְתִּי אֶתְכֶם]. I will gather you and blow upon you with the fire of my wrath, and you shall be melted [תֻתְּכוּ] within it. As silver is melted in a smelter, so you shall be melted [וְלֻנְתַּכְתֶּם] in it; and you shall know that I the Lord have poured out my wrath upon you.

In the process of smelting, סִיג "dross" is the refuse, but as Zimmerli notes, "the nature of Israel as dross is already affirmed at the beginning of the statement."[84] Given this state of affairs the whole process of smelting appears useless from the start. There is no expectation of purification, no hope of a successful outcome; the Lord's only intention is נָתַךְ "to melt" Israel. Eichrodt puts it quite forcefully: "Yahweh gathers up all the dross left by his previous attempts at smelting, forms them into the great smelting furnace of Jerusalem, and stirs up a huge fire underneath, but it is nothing more than a bonfire lit by the blaze of his rage, which will put an end to the continuous contempt for his holiness."[85] The motif has lost its purificatory sense and is simply an image of divine wrath.

The threat in 22:18–22 comes to fruition in Ezek 24:1–13 in Ezekiel's parable of the pot, in which the city of Jerusalem is likened to a rusty pot that cannot be cleansed.

> Stand it [the pot] empty upon the coals,
>   so that it may become hot, its copper glow,
>   its filth melt in it, its rust be consumed.

---

84. Zimmerli, *Ezekiel*, 464.
85. Eichrodt, *Ezekiel*, 314.

> In vain [תְּאֻנִים] I have wearied myself;
> > its thick rust does not depart.
> > To the fire with its rust!
> Yet, when I cleansed you [טִהַרְתִּיךְ] in your filthy lewdness,
> > you did not become clean [וְלֹא טָהַרְתְּ] from your filth;
> you shall not again be cleansed until I have satisfied my fury upon you
> [לֹא תִטְהֲרִי־עוֹד עַד־הֲנִיחִי אֶת־חֲמָתִי בָּךְ]. (vv. 11–13)

While not using the technical language of smelting and refining, the parable shares obvious points of continuity with Ezekiel's previous use of that metaphor. The persistence of the unclean elements is here as well, just as in the previous passage. Yet, dark as it is, and despite the fact that the Lord laments wearying himself תְּאֻנִים "in vain" or "with toil," this parable is not quite as hopeless as Ezekiel's previous use of the smelting metaphor. To be sure, this is no joyous pronouncement of the successful purification of the city, but neither is it a pronouncement of abject hopelessness as we have seen in Jeremiah and already in Ezekiel. For while it involves enduring his fury, the Lord's intention is טָהֵר "to cleanse" and despite Israel's recalcitrance (וְלֹא טָהַרְתְּ) the closing line (לֹא תִטְהֲרִי־עוֹד עַד־הֲנִיחִי אֶת־חֲמָתִי בָּךְ) at least holds the possibility of cleansing as the final outcome.

More optimistic than Jeremiah and Ezekiel is Zech 13:8–9, where the prophet depicts the purification of a remnant:

> In the whole land, says the Lord,
> > two-thirds shall be cut off [יִכָּרְתוּ] and perish,
> > and one-third shall be left alive.
> And I will put this third into the fire [וְהֵבֵאתִי אֶת־הַשְּׁלִשִׁית בָּאֵשׁ],
> > refine them [וּצְרַפְתִּים] as one refines silver,
> > and test them [וּבְחַנְתִּים] as gold is tested.
> They will call on my name,
> > and I will answer them.
> I will say, "They are my people";
> > and they will say, "The Lord is our God."

In Zechariah's vision the refining and testing process will only be conducted after the selection of a remnant where the subject is the Lord himself. While the description of the two-thirds who are "cut off" does not employ the metaphor of smelting, the effect is the same. Those for whom there is no chance of purification are removed. For the one-third who remain and are brought to the fire, two distinct stages are collapsed into a single metaphor through the use of synonymous parallelism: the silver will be refined (צָרַף) and the gold will be tested (בָּחַן). This refining and testing by means of fire is anticipated in the opening verse of the chapter where Zechariah prophesies that "a fountain shall be opened for the house of David and the inhabitants of Jerusalem, to cleanse them from sin and impurity" (13:1). In light of this parallel between the cleansing by means of water and fire, Meyers and Meyers note a possible allusion to Num 31:23, which speaks of purification by means of water and by fire. Indeed, they note that Zech 13:9a uses some of the same vocabulary as Num 31:23: both passages employ the verb בא (bring, or put) and the noun אֵשׁ (fire) preceded by the preposition בְּ (in, into).[86] The purpose of the refining and testing is thus apparently one of purification.

Even though this third has been spared destruction, they are not spared some form of punitive judgment. As Meyers and Meyers interpret the refining and testing metaphor, "the hardships of the remnant in the land are viewed as the mechanism that will rid them of their flaws. Just as the technique of silver refining involves subjecting the impure metal to intense fire to remove the dross, so will the struggles of those left in Yehud to survive in a war-torn land dominated by a distant superpower constitute a purifying experience."[87] Whether or not this purification is successful is never explicitly stated, though a positive answer is implied in the following verses where the people call on the name of the Lord and more clearly in the promise that "he will answer them" (v. 9b)

We conclude with Mal 3:1–4:

> See, I am sending my messenger to prepare the way before me [הִנְנִי שֹׁלֵחַ מַלְאָכִי וּפִנָּה־ דֶרֶךְ לְפָנָי], and the Lord whom you seek will suddenly come to his temple. The messenger of the covenant, in whom you delight—indeed, he is coming, says the Lord of hosts. But who can endure the day of his coming, and who can stand when he appears? For he is like a refiner's fire [כְּאֵשׁ מְצָרֵף] and like fullers' soap [וּכְבֹרִית מְכַבְּסִים]; he will sit as a refiner and purifier of silver [מְצָרֵף וּמְטַהֵר כֶּסֶף], and he will purify [וְטִהַר] the descendants of Levi and refine them like gold and silver [וְזִקַּק אֹתָם כַּזָּהָב], until they present offerings to the Lord in righteousness. Then the offering of Judah and Jerusalem will be pleasing to the Lord as in the days of old and as in former years.

It is fascinating that in one short book, indeed, in the span of one chapter, the prophet Malachi employs the motif of the oven that consumes stubble as a metaphor for the destruction of evildoers (MT 4:1; LXX 3:19) and the

---

86. Meyers and Meyers, *Zechariah 9–14*, 393.
87. Ibid., 405.

motif of the כְּאֵשׁ מְצָרֵף "refiner's fire" as a metaphor for refining and purification of the elect (3:1–4). The rhetorical question "who can endure the day of his coming, and who can stand when he appears?" is reminiscent of Isa 33:14b: "Who among us can live with the devouring fire? Who among us can live with everlasting flames?" However, whereas Isaiah answered with a general reference to the righteous who could endure the fire, in Malachi those who will experience the fire that attends the coming day as a purifying and refining power are very narrowly identified—they are the priesthood, the descendants of Levi in the temple, purified in preparation for the coming of the Lord. Significantly the refiner's fire is paired with the image of וּכְבֹרִית מְכַבְּסִים "fullers' soap," underlining the purificatory function of the fire.

Strikingly, whereas the subject in Mal 4:1, where the fire functions punitively, is presumably the Lord, here in 3:1–4 it is the messenger of the Lord who is likened to a refining fire and is the subject of the verbs זָקַק "purify" and צָרַף "refine." The precise identity of this messenger is debated. He may be a human figure such as a prophet, Elijah, or Malachi himself, whose name מַלְאָכִי literally means "my messenger."[88] Beth Glaizer-McDonald notes a striking parallel between Mal 3:1 (הִנְנִי שֹׁלֵחַ מַלְאָכִי וּפִנָּה־דֶרֶךְ לְפָנָי) and Exod 23:20 (הִנֵּה אָנֹכִי שֹׁלֵחַ מַלְאָךְ לְפָנֶיךָ לִשְׁמָרְךָ בַּדָּרֶךְ), suggesting that the messenger in Malachi is an angel like the one who lead the Israelites through the desert.[89] Regardless of the precise identity of the messenger, or whether he is human or angelic, his role suggests that an intermediary figure could occasionally function as the bearer of the Lord's judgment by fire. Further, whereas several of the texts we have thus far considered employ the motif of the refining fire metaphorically, the present text is unquestionably eschatological and may envision a literal fire that attends the coming of the day of the Lord.

Conclusion

With this we conclude our discussion of the functions of fire in the Hebrew Bible. In addition to observing that fire features in theophanies and functions punitively as a weapon of the Lord both within history and in the eschatological future, we have seen that there is a significant, if frequently overlooked, prophetic tradition that employs the motif of a refining and testing fire, according to which fire functions to purify those towards whom it is directed. Given the significance of these last two functions of fire for our study, it may prove fruitful to ask when and why fire began to play and

---

88. See the discussion in Malchow, "Messenger of the Covenant."
89. Glazier-McDonald, *Malachi*, 130.

eschatological and soteriological function in Jewish thought. Fire first gains its prominence in eschatology and soteriology among the prophets. The pre-exilic prophet First-Isaiah (Isa 1–39) introduces fire as an instrument of future judgment (5:24; 29:5–6; 30:27–28; 33:10–16) and explicitly as a means of judgment on the Day of the Lord (34:8–10). Similarly, the call narrative in Isa 1:25–26 is the earliest text to attribute a purificatory function to fire in the cleansing of human beings, here the cleansing of Isaiah's lips in preparation for his prophetic ministry and not in an eschatological context. We should draw special attention to this purificatory function of fire. In some cases the refiner's fire is applied to the inhabitants of Jerusalem as a whole (Jer 6:27–30; Ezek 22:18–22; 24:1–13) while in other cases it is more limited in scope (Mal 3:1–4), but generally speaking it is the elect or righteous who are the object of the fire. Admittedly, Jeremiah and Ezekiel only use the motif to show that the hoped for purification failed. The irony with which it is used in these cases, however, underscores the fact that such a failure was contrary to the expected purification with which the motif is typically associated. Notably, while fire functions as a means of purification in the thinking of pre-exilic (Isa 1:25–26; Jer 6:27–30) and exilic (Ezek 22:18–22; 24:1–13) prophets in non-eschatological contexts, it is only among the post-exilic prophets (Zech 13:8–9; Mal 3:1–4) that it becomes a means of *eschatological* purification explicitly on the Day of the Lord.[90] This may suggest that while Jewish prophets such as Isaiah, Jeremiah, and Ezekiel had already prepared the way for the appropriation of this eschatological function of fire, the prominent role given fire in Persian eschatology was a significant factor in Israel's appropriation of this motif.[91] It may be significant that in Malachi, which was composed in the Persian period, fire functions both to purify the elect (3:1–4) and to destroy the wicked (4:1), just as it does in Zoroastrian literature. Having considered several of the key Hebrew Bible texts in which fire plays a significant role, we now turn to the literature of Second Temple Judaism, where the eschatological and soteriological functions of fire attain greater prominence.

---

90. In Isa 34:8–10, fire is a means of judgment but not purification on the Day of the Lord.

91. Persian influence on this particular text is likewise suggested by Whitley, *Prophetic Achievement*, 220. More generally, see Oesterley and Robinson, *Hebrew Religion*, 391–93. This perspective coheres with the general conclusion of Hultgård ("Origins of Apocalypticism," 80), who judges, "the encounter with Iranian religion produced the necessary stimulus for the full development of ideas that were slowly under way within Judaism."

CHAPTER 3

Judgment by Fire in Second Temple Apocalyptic Literature

Introduction

IN CHAPTER 2 WE considered a wide range of texts in the Hebrew Bible in which fire features prominently and functions in multiple ways in order to establish the variability of this motif. In this chapter, which focuses on the apocalyptic literature of Second Temple Judaism, we narrow our focus to the fire of judgment, whether that be judgment within history or eschatological judgment at the end or beyond the realm of history, and we do so with a particular interest in the potential purificatory function of fire or, even if it has no explicit purgative function, the possibility that the righteous may be able to endure or be spared the testing fire of judgment. In the interest of narrowing the field to a manageable scope, we shall consider only the literature of Second Temple Judaism that can be classified as apocalyptic, and only those texts where fire plays a prominent role in scenes of future or eschatological judgment. As a consequence of this, the texts that comprise *First Enoch* and the Dead Sea Scrolls will receive the most attention, while other texts, such as *Jubilees*, for instance, are excluded due to that fact that in such texts the fire of judgment plays only a minimal role and its functions are not dealt with at length.[1] In some cases, as we shall see, fire maintains its negative function as a purely destructive force regardless of those with whom it comes into contact; in others fire assumes a more discerning as-

---

1. Fire is mentioned thirty-three times in *Jubilees*; however, most of these occurrences refer to sacrificial fire, fire as a penalty imposed by human judges, or to theophanic fire. Only *Jub.* 9:15, which makes a brief reference to a judgment by sword and fire on the Day of Judgment, speaks explicitly of a future judgment by fire.

pect, destroying the wicked while preserving the righteous; and in others still, the purificatory function of fire is highlighted and those who come into the presence of fire are left cleansed by it.

The literature to be discussed includes *1 Enoch*, the relevant texts discovered at Qumran, and various other apocalyptic texts of the Second Temple period. While we cannot offer an exhaustive survey of the relevant literature, the selected texts represent the major voices of Second Temple Jewish apocalyptic thought and will furnish us with a fairly thorough perspective on the variety of beliefs regarding the eschatological functions of fire in Second Temple Judaism. Our approach shall be roughly chronological, insofar as that is possible, although all texts belonging to a certain corpus (e.g., *1 Enoch*, Qumran, the *Sibylline Oracles*) shall be treated together. We begin with one of the earliest Second Temple apocalypses, *1 Enoch*.

First Enoch

The book now known as *1 Enoch* is in fact a composite work composed of five booklets, all five of which, with the exception of *The Parables of Enoch* (or *The Similitudes of Enoch*), were discovered among the Scrolls unearthed at Qumran, where they were apparently quite influential. These different sections were composed over a rather expansive period of time; the earliest sections, *The Book of the Watchers* (*1 En.* 1–36) and *The Astronomical Book* (*1 En.* 72–82), were written as early as the third century BCE whereas *The Similitudes* have been dated to the early first century CE.[2] As some of the earliest examples of the apocalyptic literature of Second Temple Judaism, the texts that make up *1 Enoch* present a welcome starting point for our survey of literature from that period. Throughout the composite work, fire plays a substantial role in eschatological judgment, frequently appearing as a feature in the depictions of places of final punishment. We shall take a chronological approach to the texts that now make up *1 Enoch* with an eye to the development and or preservation of any traditions relating to the eschatological functions of fire.[3]

While the full text of *1 Enoch* exists only in late Ethiopic translations, the earliest of which may date to the fifteenth (or possibly fourteenth)

---

2. For a dating of these texts representative of mainstream Enoch scholarship, see the brief overview in Nickelsburg, *Jewish Literature*, 43–53.

3. Fire also occurs in *1 Enoch* in a theophanic context. Enoch receives visions of the house of God, which is composed of fire (14:9–22). Chapter 23 also discusses the fire of the luminaries of heaven. However, in this chapter we shall focus only on the fire of eschatological judgment.

century CE, early fragments survive in Aramaic, and expansive sections of the text are extant in Greek. There also exists an eighth-century Latin fragment of *1 En.* 106:1–18.[4] We shall engage the Greek text where relevant.[5]

The Book of the Watchers (*1 Enoch* 1–36)

One of the earliest sections of *1 Enoch* is widely known as *The Book of the Watchers*, for it is primarily concerned with the angels (or watchers) who sinned with the daughters of humans in Gen 6:1–4. *The Book of Watchers* is itself likely composite in origin, though it is not possible to tease apart the originally distinct strands of tradition. The earliest fragments of the text, including parts of chapters 1–12, were discovered at Qumran (4QEna), while the book of *Jubilees*, written in the mid second century, appears to have had knowledge of the text.[6] Further, evidence from chapters 85–90, which contain allusions to the Maccabean wars and which demonstrate familiarity with *1 En.* 1–36, suggests that *The Book of Watchers* had become influential before death of Judas Maccabeus in 160 BCE.[7] These fragments and allusions allow us to date *The Book of Watchers* at least prior to 175 and possibly as early as the late third century BCE.[8]

According to *1 Enoch*, angels, referred to as watchers, are to blame for the introduction of the great wickedness into the world that required God to send the deluge that wiped out all of humanity, save for Noah and his family. Thus, a significant theme in *The Book of the Watchers* is the judgment of these fallen angels. Azaz'el is among the angels, and in 10:6 God sends Raphael to bind him and throw him into a dark hole in the middle of the desert in Dudael and to cover him with rocks so that "on the day of the great judgment, he will be led away to the burning conflagration [τὸν ἐνπυρισμόν]."[9] In having Raphael cover him with rocks, the "author is implying the imagery of death, burial, and a resurrection to judgment."[10] "The day of the great

---

   4. Isaac in Charlesworth, *OTP* 1:6.
   5. For the Greek text see Black, *Book of Enoch*.
   6. VanderKam, "Enoch Traditions in Jubilees," 235.
   7. Nickelsburg, *Jewish Literature*, 46.
   8. Ibid.

   9. Unless otherwise noted, all translations of *1 Enoch* are from Nickelsburg and VanderKam, *1 Enoch*. I have also consulted the following translations and critical editions: Charles, *Book of Enoch*; Knibb, *Ethopic Book of Enoch*; Charlesworth, *OTP*; Black, *Bok of Enoch*; Stuckenbruck, *1 Enoch 91–108*. The name Dudael may derive from the Hebrew דודא אל "cauldron of God," which would be fitting given the fate of Azaz'el who is buried there. See Dillmann, *Das Buch Henoch*, 100; Knibb, *Book of Enoch*, 1:87.

   10. Nickelsburg, *1 Enoch 1*, 221.

judgment" is a phrase frequently invoked in *1 Enoch*, and fire is commonly associated with it. It is on that day that Azaz'el will be punished with fire. The fiery judgment is spelled out in further detail in 10:13–14, where a similar fate is reserved for Semyaza and the Watchers who are buried in the ground under rocks until the day of judgment, when they will be thrown into an abyss of fire. The angels will be cast into the abyss of fire where they will be imprisoned and have a painful and fiery punishment inflicted upon them. The duration of their punishment is somewhat confused, for it is said that it will last αἰῶνος "forever" (v. 13), and yet the author also states that ἀφανισθῇ "they will be destroyed" (v. 14),[11] implying a point at which they may be annihilated and their punishment terminated.

The fullest description of the abyss of fire is found in 21:3–10, where Enoch visits the abyss of fire, where he sees "seven of the stars of heaven, bound and cast in it together, like great mountains, and burning in fire [ἐν πυρὶ καιομένους]" (v. 3). Uriel explains, "These are the stars of heaven that transgressed the command of the Lord; they have been bound here until ten thousand years are fulfilled—the time of their sins" (v. 6). Enoch then goes on to describe "another place, more terrible than this one" where there is "a great fire burning and flaming [πῦρ μέγα ἐκεῖ καιόμενον καὶ φλεγόμενον]" (v. 7) and an abyss "full of great pillars of fire [στύλων πυρὸς μεγάλου]" of inestimable size. Uriel explains to him, "This place is a prison for the angels. Here they will be confined forever" (10). The seven stars burning with fire are presumably angels, and are to be identified with the fallen watchers. This reading is supported by the description that they were ὁμοίους ὄρεσιν μεγάλοις "like great mountains," which indicates their immense size, for the angels were believed to be enormous creatures, and even their offspring with human women were believed to be giants. The phrase ἐν πυρὶ καιομένους "burning in fire" clearly indicates the means of their punishment.

Enoch proceeds to an even more disturbing place where he witnesses a great fire and immense pillars of fire spouting from a large fissure. The scope of what he witnesses is beyond words and evokes great fear in him. Noting Enoch's fear, the angel explains that what he is witnessing is the place where the angels, again presumably the watchers, are to be imprisoned forever. R. H. Charles suggests that the place described in verses 1–6 where the angels are imprisoned for ten million years is a "place of preliminary punishment" whereas the more terrifying place described in verses 7–10 is "the final place of punishment for the fallen angels."[12] While the former period

---

11. So also the translation of Charles, *Book of Enoch*, 25. Isaac's translation in Charlesworth, *OTP* has "they will burn and die."

12. Charles, *Book of Enoch*, 44–45.

of imprisonment was only for a set period of time, albeit an unimaginably protracted one, the latter punishment is said to last forever. Presumably the fire plays some role in the punishment, but its precise function is not spelled out in any detail.

In chapters 17–19 beings ὡς πῦρ φλέγον "like a flaming fire" take Enoch on a tour of heaven, earth and Sheol, during which he sees many fiery phenomena, including a fiery sword, a river of fire, a mountain of fire, and a deep pit with pillars of fire:

> And I saw the place of the luminaries and the treasuries of the thunders, and to the depths of the ether, where the bow of fire [τόξον πυρός] and the arrows and their quivers (were) and the sword of fire and all the lightnings [ἀστραπὰς πάσας]. And they led me away to the living waters and to the fire of the west [πυρὸς δύσεως], which provides all the sunsets. And I came to the river of fire [ποταμοῦ πυρός], in which fire flows down like water [τὸ πῦρ ὡς ὕδωρ] and discharges into the great sea of the west. (17:3–5)

It is probable that those who are "like a flaming fire" are some sort of angelic figures, possibly the seraphim whose name derives from שָׂרַף and means "ones who burn."[13] The fiery sword (absent from the Greek text) is a weapon of God's judgment (see Deut 32:41) as are the lightning bolts (ἀστραπάς), which he shoots from his bow of fire (τόξον πυρός). The significance of the fire in the west is unclear. Blau, citing *B. Bat.* 84a, which says that the sun is red in the evening because it passes the gate of Gehenna and red in the morning because it passes the roses of the Garden of Eden, suggests that the fiery hue in the west is Gehenna;[14] however, as Charles points out, in *1 Enoch* Gehenna is not in the west.[15]

One of the more arresting features of this passage is the ποταμὸς πυρός "river of fire," in which the fire flows like water (τὸ πῦρ ὡς ὕδωρ). *1 Enoch* 14:19 describes rivers of fire flowing out from the divine throne, just as LXX Dan 7:10 has a ποταμὸς πυρός issuing from before the Ancient of Days; however, the river of fire envisioned here appears to be something quite different. It has much more in common with Pyriphlegethon, the river of fire located in the underworld in Greek mythology.[16] In Enoch's vision it is a permanent feature of the landscape, not a singular element in the divine judgment. As Kelley Bautch indicates, "[a]lthough biblical traditions

---

13. Nickelsburg, *1 Enoch 1*, 281; Bautch, *Geography of 1 Enoch*, 44–49.
14. Cited in Charles, *Book of Enoch*, 39.
15. Ibid.
16. Nickelsburg, *1 Enoch 1*, 283; Bautch, *Geography of 1 Enoch*, 82–83.

certainly know of a river of fire as a description of God's anger or as a part of an end time catastrophe, it is not given any geographical significance."[17] Moreover, unlike the river of fire found in biblical traditions that have to do with divine wrath or judgment, *1 En.* 17:5 ascribes no function to the river of fire. Perhaps the author attributed some function to the river of fire, as he presumably did to the fiery sword and the bow of fire with its fiery arrows of lightning, but if he did so, he left it to his readers to infer what it was. The river's location in the great darkness where no flesh walks points to the netherworld and confirms that the author probably has in mind Pyriphlegethon. If so, the author may have implicitly attributed to it the function of punishing sinners in the underworld.[18]

Shortly after his vision of the fiery river, Enoch is taken on a tour to the west to "a place that was burning night and day" where he sees seven mountains of stone. Describing his vision he writes:

> And I saw a burning fire [πῦρ καιόμενον]. And beyond these mountains is a place, the edge of the great earth; there the heavens come to an end. And I saw a great chasm among pillars of heavenly fire. And I saw in it pillars of fire [τοὺς στύλους τοῦ πυρός] descending; and they were immeasurable toward the depth and toward the height. Beyond this chasm I saw a place where there was neither firmament of heaven above, nor firmly founded earth beneath it. Neither was there water upon it, nor bird; but the place was desolate and fearful. There I saw seven stars like great burning mountains [ὄρη μεγάλα καιόμενα]. To me, when I inquired about them, the angel said, "This place is the end of heaven and earth; this has become a prison for the stars and the hosts of heaven. The stars that are rolling over in the fire [οἱ κυλιόμενοι ἐν τῷ πυρί], these are they that transgressed the command of the Lord in the beginning of their rising, for they did not come out in their appointed times. And he was angry with them and bound them until the time of the consummation of their sins—ten thousand years." (18:9–16)[19]

The initial sight of πῦρ καιόμενον "a burning fire" is probably related to the likeness of the throne of God and thus serves a theophanic function (see 24:1).[20] The notion of a mountain serving as the throne of God draws upon

---

17. Bautch, *Geography of 1 Enoch*, 83.
18. Seneca, *Phaedra* 1226 associates Pyriphlegethon with punishment in Hades.
19. Trans., Nickelsburg and VanderKam, *1 Enoch*, 39.
20. Nickelsburg, *1 Enoch 1*, 286.

a long biblical tradition (see Isa 14:13; Ezek 28:14; 16).[21] It may also recall the fire that descended upon Mount Sinai.[22] The theophanic fire, however, is not the only fire present in Enoch's vision of the seven mountains. There are also a deep pit with τοὺς στύλους τοῦ πυρός "pillars of fire," seven stars that are likened to seven ὄρη μεγάλα καιόμενα "great burning mountains." Moreover, stars are rolled ἐν τῷ πυρί "in the fire" for their punishment. The pillars of fire are open to multiple interpretations. First, it is possible that the descending pillars of fire are simply part of the architecture of the immense pit and that their immeasurable depth indicates that the pit is likewise immeasurably deep. However, Nickelsburg points to the possibility that they could be manifestations of the fire of God's punishment. In support of this reading he cites Gen 19:24, Ps 11:6, and Ezek 38:22, none of which speaks of pillars of fire, but only of the fire of judgment raining from heaven. Surely this connection is tendentious at best. Nickelsburg appears to realize this, for he prefers an alternative reading. The fiery description of the stars suggests to him that "the pillars are themselves the angels."[23] This last reading is supported by 19:1 which indicates that the angels "stand in many different appearances," one of which may be pillars of fire. Further, in Exod 13:21 the LORD who goes before the Hebrews in a pillar of fire (LXX: ἐν στύλῳ πυρός) may be identified with the Angel of the LORD, as in the burning bush in Exod 3:2-4. Thus, the suggestion that the descending pillars of fire represent the luminous angels descending into the abyss is probably correct.

The seven stars that burn with fire, in keeping with our previous exegesis, may be identified with the fallen watchers.[24] More interesting is the detail that they are κυλιόμενοι ἐν τῷ πυρί "rolling over in the fire." That the place is identified as a prison, and that it is mentioned directly in connection with God's fury over their transgression of the divine commandments, indicates that this is a reference to their punishment. The reason for their punishment is that they did not keep their course and arrive punctually. They failed to keep the commandments laid out before them by God. Thus, they are bound, and their punishment is to be administered for ten thousand years, indicating that their punishment will come to an end at some time, but not in the foreseeable future. However, whether or not that end results in their rehabilitation or their destruction is not clear in this text.

---

21. Charles, *Book of Enoch*, 41.
22. Bautch, *Geography of 1 Enoch*, 128.
23. Nickelsburg, *1 Enoch 1*, 287.
24. Bautch, *Geography of 1 Enoch*, 145. Nickelsburg, however, identifies the pillars of fire with the watchers, *1 Enoch 1*, 276.

Given that in 10:13–14 the end of the watchers' punishment is their death, that may be the case here as well.

The above passages from *The Book of Watchers* demonstrate that while fire is occasionally used in theophanies or to describe the appearance of angelic beings, its predominant function is one of punishment for the fallen angels. The fire is administered at the great judgment and is most often located in the abyss or the pit that serves as the prison for the angels who transgressed God's commandments. Most of these texts indicate that the fire is purely punitive and will result in the destruction of those subjected to it. The only text in *The Book of Watchers* in which fire played no discernible function was in Enoch's description of the river of fire in the underworld. However, given its close relationship to Pyriphlegethon, which functions to punish sinners in the underworld, it may be appropriate to attribute a punitive function to Enoch's river of fire as well, especially given the frequency with which the *Book of Watchers* elsewhere associates fire with punishment.

Apocalypse of Weeks (*1 Enoch* 93:1–10; 91:11–17)

While it is not the earliest section of *1 Enoch*, the Apocalypse of Weeks is a likely candidate for "[t]he oldest surviving historical apocalypse in Jewish literature."[25] Discovery of an Aramaic manuscript at Qumran (4Q212) indicates that the Ethiopic text reversed the order and that the text now located in 93:1–10 originally preceded that found in 91:11–17.[26] The text contains no clear references to historical events or persons and is therefore very difficult to date with much precision, but an allusion in *Jub.* 4:18 and fragments from Qumran suggest a date of composition prior to the first century BCE.[27] The *Apocalypse of Weeks* outlines the history of the world from its creation until its end and divides it into ten periods that are referred to as weeks. The text is, unfortunately not extant in Greek. The only mention of fire occurs in the sixth week. For context, one must be aware that in the fifth week a house and a kingdom were built, for the house is subjected to fire in the sixth week:

> And after this <there will arise a sixth week, and> all who live in it will become blind, and the hearts of all will stray from wisdom; and in it a man will ascend. And at its conclusion, the temple of the kingdom will be burned with fire, and in it the whole race of the chosen root will be dispersed. (93:8)

---

25. VanderKam, *Introduction to Early Judaism*, 103.
26. See Ibid.
27. See VanderKam, "Enoch Traditions in Jubliees."

The house of the kingdom[28] referred to in this reiteration of Israel's history is the First Temple, which was burned with fire by the Babylonians in 586/587 BCE (see 89:66). The reference is to a literal event in Israel's history just as the mention of the dispersion of those who were in the house refers to Israel's literal exile. Despite the fact that the Babylonians were responsible for the conflagration "[t]he divine source of punishment is implied by the passive verbs, both here ('will be burned') and in the next phrase ('will be dispersed')."[29] Thus the historical burning of the temple is understood in terms of divine judgment by fire.

The reason for the burning of the temple appears to be the blindness of the sheep and their forgetfulness of wisdom. Despite the gifts of the Torah, the Ark of the Covenant, and the temple cult, all of which are recounted in weeks three through five of the apocalypse, Israel's blindness will cause them to go astray through their idolatry and improper temple observances.[30] The corruption of the people is obliquely reflected in the phrase "a man will ascend." This obscure phrase probably points to Elijah, of whom 2 Kings 2:11 says "and he went up [ויעל]," and it is possible that here in the Apocalypse of Weeks "the stress falls less on the manner of Elijah's departure than on his escape from the evil generation."[31] Destruction of the temple by fire is thus depicted as a just and fitting punishment for the corruption of those portrayed as blind sheep.

The Animal Apocalypse (*1 Enoch* 85–90)

Another very old section of *1 Enoch*, is *The Book of Dreams*, most of which is taken up by *The Animal Apocalypse* (85–90), in which biblical figures are represented by bulls, cows, and sheep while their opponents are depicted as unclean or wild animals. On the basis of internal evidence, the text can be dated to the Maccabean Revolt.[32] The lamb who is murdered in 90:8 is a likely reference to the high priest Onias III, and the great horn that grows on one of the sheep in 90:9 is almost certainly an allusion to Judas Maccabeus.[33] As in *The Book of the Watchers*, the angels who sinned with human women

---

28. Knibb (*Ethiopic Book of Enoch*, 224) translates this as "the house of sovereignty."
29. Stuckenbruck, *1 Enoch 91–108*, 117.
30. Ibid., 113.
31. Ibid., 116.
32. Tiller, *Commentary on the Animal Apocalypse*, 61–82; Collins, *Apocalyptic Imagination*, 67.
33. Collins, *Apocalyptic Imagination*, 69; Nickelsburg, *Jewish Literature*, 85.

play a significant role, and they are once again depicted as stars.³⁴ The only passage in *The Book of Dreams* in which the fire of judgment appears is in 90:24-27, which recalls several of the scenes discussed above in the *Book of the Watchers*:

> And judgment was exacted first on the stars, and they were judged and found to be sinners. And they went to the place of judgment, and they threw them into an abyss; and it was full of fire, and it was burning and was full of pillars of fire. And those seventy shepherds were judged and found to be sinners, and they were thrown into that fiery abyss. And I saw at that time an abyss like it was opened in the middle of the earth, which was full of fire. And they brought those blinded sheep, and they were all judged and found to be sinners. And they were thrown into that fiery abyss, and they burned. And that abyss was to the south of that house. And I saw those sheep burning and their bones burning.³⁵

This scene is explicitly identified as one of judgment. Among those judged and found guilty are the stars, the seventy shepherds, and the blinded sheep. We have already identified the stars with the fallen watchers. The seventy shepherds are also angelic beings, though they are not to be confused with the watchers.³⁶ Lastly, the blind sheep represent the Israelites who went astray. They correlate with those who were blindfolded in the temple in 93:8.

Two abysses are described. Noting this distinction, Patrick Tiller states, "[t]he stars (fallen Watchers) share their place of torment with the angelic shepherds. The blinded sheep (wicked Israelites) have a separate but equal abyss, to be identified with Gehenna (90.26)."³⁷ The location of the second abyss is described in relation to the house from which Enoch observes the judgment. It is located to the south of the house, indicating that it lies to the south of Jerusalem,³⁸ which is the precise location of the Valley of Hinnom.³⁹ The description of the punishment of the sheep is accorded

---

34. On the Book of Dreams see VanderKam, *Introduction to Early Judaism*, 105–6. On the *Animal Apocalypse* in particular see Nickelsburg, *Jewish Literature*, 83–86.

35. This section is extant only in Ethiopic.

36. Nickelsburg, *1 Enoch 1*, 404.

37. Tiller, *Commentary on the Animal Apocalypse*, 371.

38. Isaac's translation (in Charlesworth, *OTP* 1:71) is more literal in stating that the abyss is on the right side of the house. See Knibb, *Ethiopic Book of Enoch*, 215; Charles, *Book of Enoch*, 212. Behind the Ethiopic text no doubt stands the Hebrew term תֵּימָן, which means both right and south. Thus, either translation allows an identification of the abyss with the Valley of Hinnom.

39. Nickelsburg, *1 Enoch 1*, 404.

a longer description and more detail and is therefore probably of greater significance to the author. This is fitting given the identification of the blind sheep with Israelites. One of the most striking details of the entire passage is the mention of the burning bones of the sheep. This arresting image may have simply been "intended to make the description more graphic"[40] or may indicate "the intensity of the heat."[41] Even more likely, however, that their bones are consumed indicates their complete destruction.[42] Thus complete destruction by fire, once a fate reserved only for the fallen angels responsible for introducing sin into creation, is envisioned as the fate for those Israelites who have gone astray. In the historical context of the Maccabean revolt, the fate of the fallen angels and wicked Israelites who failed to keep the commandments of God may be read as a warning to Hellenized Jews choosing to abandon Jewish religious practices. This would be especially true under the oppressive rule of Antiochus Epiphanes, who forbade Jews from keeping certain laws, such as keeping the Sabbath, performing circumcision, and studying the Torah. The *Animal Apocalypse* may be warning readers that whereas it is only earthly fire that may be in store for those who violate Seleucid laws (see 4 Macc 5:32; 6:24; 7:12; 9:19, 22; 10:14; 11:19, 26; 14:9–10; 15:15; 18:12–20), eschatological fire awaits those who violate the Lord's commandments (see also 4 Macc 9:9; 12:12).

The Epistle of Enoch (*1 Enoch* 91–107 [108])

First Enoch 91:9 finds its literary context in Enoch's *Exhortation*, which is frequently included as part of the *Epistle*. Most likely, however, the *Exhortation* was a later expansion added to bridge the *Apocalypse of Weeks* and the *Epistle*.[43] In the *Exhortation*, Enoch shares with his children a description of the final judgment. While the textual Enoch addresses his words to his children, the author is himself addressing the implied readers, to whom Enoch refers as "a future generation" (92:1).[44] Scholars have dated those implied readers to the late Hasmonean period.[45] As we shall see, the

---

40. Tiller, *Commentary on the Animal Apocalypse*, 372.

41. Nickelsburg, *1 Enoch 1*, 404.

42. See our discussion of Isa 33:12 above, where the phrase "the peoples will be as if burned to lime" suggests that their bones are burned, thus indicating complete destruction. See Amos 2:1b: "he burned to lime the bones of the king of Edom."

43. Stuckenbruck, *1 Enoch 91–108*, 154.

44. Nickelsburg, *Jewish Literature*, 110.

45. Nickelsburg (*1 Enoch 1*, 8) notes the "explicit appeal for the righteous to stand fast," which fits the proposed historical context.

Exhortation directs strong words against those who live sumptuously, and the author may have in view "the excesses of Alexander Janneus or perhaps John Hyrcanus."[46] At 91:9 states the following:

> And all the idols of the nations will be given up, and the tower(s) will be burned with fire. They will be removed from all the earth, and they will be thrown into the fiery judgment, and they will be destroyed in fierce, everlasting judgment.[47]

One element of the final judgment, according to Enoch, is the destruction of all idols. As Loren Stuckenbruck observes, the eschatological destruction of idols is a widespread motif (see *T. Mos.* 10:7; Tob 14:6; 4Q198 1.13; Wis 14:11).[48] The destruction of the idols may reflect the expectation of the eschatological conversion of the Gentiles.[49] It may also be the case that the tower(s) destroyed by fire are pagan temples.[50] However, Stuckenbruck avers that "[t]he term *māxfad* ('tower') occurs frequently in the *Animal Apocalypse*, where it refers either to the 'heavenly temple' (87:3) or to the temple in Jerusalem (89:50, 54, 56, 66–67, 73)."[51] On the basis of this observation and the negative view of the temple held in the *Apocalypse of Weeks* (see 89:54, 56, 66, 73), Stuckenbruck suggests that the author of the *Exhortation* envisioned the eschatological burning of the Second Temple and that "it expresses the author's genuine anticipation of divine punishment against the Temple and its establishment."[52]

The second sentence indicates that "they" will be thrown into the fire and destroyed; however, there is some dispute regarding to whom the pronoun refers. While Charles believes that the pronoun has "the heathen" as its antecedent,[53] Nickelsburg suggests that the idols and their temples are in view.[54] In keeping with the positive expectation of the conversion of the nations, Nickelsburg's view seems preferable. However, he has read the refer-

---

46. Nickelsburg, *Jewish Literature*, 114. Vanderkam (*Introduction to Early Judaism*, 120), however, suggests a date closer to 170 BCE.

47. Extant only in Ethiopic.

48. Stuckenbruck, *1 Enoch 91–108*, 179.

49. Nickelsburg, *1 Enoch 1*, 413.

50. Isaac (in Charlesworth, *OTP*, 1:72) suggests that the reference may be alternatively to "palaces" or "castles." Charles (*Book of Enoch*, 227) has "temples;" Knibb (*Ethiopic Book of Enoch*, 218) has "towers."

51. Stuckenbruck, *1 Enoch 91–108*, 179.

52. Ibid., 180. See *Barn.* 16:5: "And it shall be in the last days that the Lord shall hand over the sheep of the pasture and the sheepfold and their tower to destruction."

53. Charles, *Book of Enoch*, 227; Knibb, *Ethiopic Book of Enoch*, 2:218.

54. Nickelsburg, *Jewish Literature*, 413.

ence to the "tower(s)" as referring to pagan temples whereas Stuckenbruck's reading, which is more grounded in the passage's broader literary context, would indicate that it is not pagan temples, but the Jewish temple that is cast into the fire and destroyed. Thus a similar eschatological fate is expected for the second Jewish temple as that which the first temple faced, the burning of which is alluded to in 93:8.

Similarly warning of an impending fiery punishment is 98:3b, in which Enoch, addressing the wise and righteous, promises judgment on their oppressors who horde their wealth and abuse their power: "Thus they will perish, together with all their possessions, and all their splendor and honor; and for dishonor and slaughter and the great destitution, their spirits will be cast into the fiery furnace" (τὰ πνεύματα ὑμῶν εἰς τὴν κάμινον τοῦ πυρὸς ἐμβληθήσεται).[55]

Echoing this exhortation to the righteous is 100:7–9, which is one among a series of woes addressed to the wicked themselves:

> Woe to you, unrighteous, when you oppress the righteous on a day of hard anguish, and burn them in fire [φθλάξητε αὐτοὺς ἐν πυρί]; for you will be recompensed according to your deeds. Woe to you, hard of heart, who lie awake to devise evil; fear will overtake you, and there will be no one to help you. Woe unto you, all you sinners, because of the words of your mouth and the works of your hands, for you have strayed from the holy deeds; in the heat of a blazing fire you will burn.[56]

In connection with 93:8b and 100:7–9, the *Epistle* offers one more related description of the fate of sinners. Concluding the series of woes pronounced upon the wicked is 103:5–8:

> Woe to you, dead sinners. When you die in your sinful wealth, those who are like you say about you, "Blessed are the sinners all the days they have seen. And now they have died with goods and wealth, and affliction and murder they have not seen in their life; they have died in splendor, and judgment was not executed on them in their life." Know that down to Sheol [εἰς ᾅδου] they will lead your souls; and there they will be in great distress, and in darkness [ἐν σκότει] and in a snare and in a flaming fire [ἐν φλογί]. Into great judgment your souls will enter, and the great judgment will be for all the generations of eternity [ἐν πάσας ταῖς γενεαῖς τοῦ αἰῶνος]. Woe to you, you will have no peace.[57]

---

55. Nickelsburg and VanderKam, *1 Enoch*, 149.
56. Translated in ibid., 155.
57. Trans., Ibid., 159–60. The final line of this section is missing from the extant

These three passages, the first comforting the righteous and the latter two condemning the wicked, are all variations on the same theme. The first observation to be made regarding these texts is that all bespeak a reversal of fortunes and warn of a judgment with fire. Those who prosper through their wickedness and oppression and who appear to enjoy blessings in their earthly lives without recompense for their impiety and malice will face a "great judgment" in Sheol (literally: to Hades εἰς ᾅδου), where they will be cast into a "fiery furnace" (τὴν κάμινον τοῦ πυρός) and burned "in the heat of a blazing fire." Typical of many depictions of Gehenna is the paradoxical juxtaposition of darkness (ἐν σκότει) and fire (ἐν φλογί) (see 1QS II.8; IV.13; Matt 8:12; 22:13; 25:30; 2 Pet 2:17; Jude 14), which adds to the fear and distress experienced by those imprisoned there. Notably, 98:3b and 103:8 explicitly state that it is the spirits (τὰ πνεύματα) of the wicked, as distinct from their bodies, that will be punished. It is therefore in a disembodied afterlife that their punishment is enacted on the day of the final judgment. This is something new that we have not yet encountered in the Enochic corpus. The belief in a disembodied afterlife reflects the historical context of the Hasmonean period with its influx of Hellenistic notions of the afterlife, including an immortal soul.[58]

The "fiery furnace" motif is an interesting one, recalling several biblical passages. The κάμινον τοῦ πυρός of *1 En.* 98:3 is reminiscent of the κάμινον τοῦ πυρός of LXX Daniel 3. However, whereas Daniel's righteous friends are preserved bodily in the fiery furnace, the spirits (τὰ πνεύματα) of the oppressors of the righteous in Enoch are apparently destroyed by the fire.[59] The fate of Shadrach, Meshach, and Abednego is, of course, an exception to the rule. Thus, as a motif of eschatological judgment *1 En.* 98:3 has more in common with Matt 13:42, 50, which also condemns the wicked to the κάμινον τοῦ πυρός.[60] The punishment envisioned lasts "for all the generations of eternity," and thus there will be no end.

Yet another tradition in the *Epistle* envisions a judgment of fire, but not in Sheol. In 102:1–3, the judgment by fire appears to take place upon the earth in the last days:

> Then, when he hurls against you the flood of the fire of your burning [τὸν κλύδωνα τοῦ πυρὸς καύσεως ὑμῶν], where will you flee and be saved? And when he utters his voice against you with

---

Greek fragments. Knibb (1978: 235) has "you will burn in blazing flames of fire."

58. See Segal, *Life after Death*, 367–8.

59. On death by burning in the ancient Near East, see Beaulieu, "Babylonian Background."

60. See Stuckenbruck, *1 Enoch 91–108*, 336.

a mighty sound, will you not be shaken and frightened? The heavens and all the luminaries will be shaken with great fear; and all the earth will be shaken and will tremble and be thrown into confusion. All the angels will fulfill what was commanded them; and all the sons of earth [οἱ υἱοὶ τῆς γῆς] will seek to hide themselves from the presence of the Great Glory, and they will be shaken and tremble. And you, sinners [ἁμαρτωλοί], will be cursed forever; you will have no peace.

Nickelsburg notes that similarities between this passage and the Sinai theophany in Exod 19:16–18 exist: both mention fire, the voice, and trembling. It is thus possible that "the author has in mind a kind of repetition of the Sinai theophany."[61] Alongside the theophanic elements, however, are themes of eschatological judgment. We encounter the shaking of the luminaries and of the earth, reminiscent of the expectation that the stars will fall from heaven and the earth will tremble as part of the eschatological scenario. Also explicitly tied to the theme of judgment is the woe on the sinners, that they "will be cursed forever" and "will have no peace."

Most striking, however, is the depiction of the fire as a flood (τὸν κλύδωνα τοῦ πυρὸς καύσεως ὑμῶν).[62] Nickelsburg observes that this passage may bear some relation to 1QXI.28–36, noting that in addition to the fiery torrent, both passages mention "God's voice (34), the heavenly entourage as executors of judgment (35–36||102:3), the quaking of the cosmos (35||102:2), and the finality of the judgment (36||102:3)."[63] The flood of fire is here employed as a weapon of God, an instrument of his judgment directed at a specific target. The object of his judgment is the unrighteous, but it is unclear whether those referred to as οἱ υἱοὶ τῆς γῆς "the sons of earth" are to be identified with or distinguished from ἁμαρτωλοί "the sinners." Some have equated the two categories.[64] However, based on the positive outcome attributed to the "sons of the earth" in 100:5, Stuckenbruck surmises that "the author thinks of the frightened 'children of the earth' as those who do not at the moment belong to his own community (more narrowly defined) but who may eventually understand the Enochic message and recognize their wrongdoing, while 'the sinners' are those who will never at all come to

---

61. Nickelsburg, *1 Enoch 1*, 509.

62. In his translation of the Ethiopic, Knibb (*Ethiopic Book of Enoch*, 237) has "he brings a fierce fire upon you." There is no reference to the flood. Charles (*Book of Enoch*, 253) has "grevious fire."

63. Ibid. See our treatment of this text from the *Thanksgiving Hymns* on pages 97–99 below.

64. Ibid.

an admission of what they have done."[65] Thus the righteous are presumably spared from or preserved through the flood of fire, the "sons of the earth" may be tried or tested by it, their fate being contingent upon their perseverance, and the sinners destroyed by it.

Finally, the *Admonition*, which is extant only in the Ethiopic tradition, is—like the *Exhortation*—a later expansion, which could have been added anytime between the second century BCE and the fourth century CE. Most plausible, however, is a date sometime in the mid-to-late-first century CE. The *Admonition* is therefore not too late to be treated here. On the one hand, the *Admonition* functions as "a summarizing and interpretive conclusion to the corpus."[66] It is clearly drawing upon traditions and motifs employed elsewhere in *1 Enoch*. On the other hand, it does contain some new ideas that are lacking elsewhere in *1 Enoch*. Most notable is the introduction of the notion of immediate punishment of the wicked in the afterlife, which elsewhere in *1 Enoch* is postponed until the final judgment.[67] Thus the *Admonition* clearly reflects a later and more developed eschatology than that present elsewhere in the book, yet not necessarily one that postdates the New Testament period. It may therefore be of some relevance in illuminating our study of the function of fire in New Testament eschatology.[68]

In 108:2–6 Enoch begins his exhortation to the righteous with a description of the fate of the wicked:

> You who have observed it will wait for these days until an end is made of those who do evil and the power of wrongdoers comes to an end. And you, wait until sin passes away; for their names will be erased from the book of life and from the books of the holy ones, and their seed will be destroyed forever, and their spirits will be killed, and they will cry out and lament in a deserted place that is invisible and burn in fire. For there will be no earth. And I saw there (something) like a cloud which was unfathomable, since on account of its depth I could not look upon it, and I saw a flame of fire which was burning brightly, and (something) like brightly shining mountains were turning over and shaking from one side to the other. And I asked one of the holy angels who was with me and said to him, "What is it that is bright? For there is no heaven, but only a fiery flame that burns, and the sound of weeping and crying out and groaning

---

65. Stuckenbruck, *1 Enoch 91–108*, 487.
66. Nickelsburg, *1 Enoch 1*, 552.
67. Stuckenbruck, *1 Enoch 91–108*, 692.
68. See in particular the excursus on the parallels between *1 En.* 108 and "Hell in the Gospel Traditions" in Nickelsburg, *1 Enoch 1*, 556.

and powerful anguish." And he said to me, "This place which you see—there the spirits of the sinners and godless will be led, and (the spirits of) those who do evil and of those who alter everything that the Lord has done through the mouth of the prophets (about) all the things which will happen."[69]

Stuckenbruck observes that the "[t]he combination of three elements—burning fire (21:3; see 18:15), mountains (18:13; 21:3), and the 'turning over', or revolving, motion (18:15)—establish the indebtedness of the text to the patriarch's cosmic journeys in the *Book of Watchers*."[70] We should not be surprised, therefore, to find similar functions attributed to the fire in this passage as those discovered in the *Book of Watchers*. In this passage it is apparent that the fire is punitive—note their "weeping and crying out and groaning and powerful anguish"—and it is reserved only for "the spirits of the sinners and the godless." The specific identification of "those who alter everything that the Lord has done [said] through the mouth of the prophets (about) all the things which will happen" may serve as a warning against anyone wishing to revise the content of Enoch's vision (see 99:2; 104:10, 11). As elsewhere in the *Epistle*, it is the disembodied souls or spirits that are tormented by fire, and the punishment is administered in some post-mortem place of punishment. The fire is not merely punitive, but appears to bring about, or at least contribute to, complete destruction: "their names will be erased from the book of life and from the books of the holy ones, and their seed will be destroyed forever, and their spirits will be killed." The fire contributes to their annihilation. While Enoch's reference to audible cries and groans would seem to indicate that the punishment of sinners has already been initiated, its context in an apocalyptic vision of the future and the future tense which is used throughout would suggest that it is a vision of a future state not yet inaugurated.

Interestingly, the author of this text states that in these days "there will be no earth" (v. 3) and "there is no heaven" (v. 5). It may be that this is meant to recall the primordial chaos when "the earth was without form and void [תֹהוּ וָבֹהוּ]"[71] However, that "a fiery flame that burns" remains in the absence of earth and heaven may indicate that the Stoic concept of the cosmic conflagration stands in the background. If so, it provides an interesting antecedent to 2 Peter 3, where heaven and earth are dissolved in fire to

---

69. Trans., Stuckenbruck, *1 Enoch 91–108*, 697–98, 704, 709. This section is extant only in Ethioptic.

70. Ibid., 707.

71. Ibid., 704.

be replaced by a new heaven and a new earth. Here, however, the imagery is closely linked with the place of punishment for sinners.

In nearly all of the passages in the *Epistle of Enoch* discussed above, fire plays a role in post-mortem eschatological judgment and functions punitively in some place of torment in the underworld, whether that is Hades, Sheol, or Gehenna. Those against whom it is directed are specifically identified as the wicked and impious, and in certain cases it is specified that it is their disembodied souls that will be punished. The major exception to this generalization that we noted was in 102:1–3, where a flood of fire surges upon the earth as an instrument of God's judgment in the last days. In that sole text in the *Epistle of Enoch* we observed the possibility that the righteous might be spared from or preserved through the eschatological judgment by fire, whereas those identified as the "sons of the earth" might be tested by it and face differing fates based on their faithfulness. Regarding those identified as sinners, however, the flood of fire maintains its destructive function.

The Parables of Enoch (*1 Enoch* 37–71)

The *Book of Parables* is undoubtedly the most controversial text in the Enochic corpus and can be listed among the most disputed in all of Second Temple literature, owing primarily to its use of the Son of Man title and difficulties in dating the text. The *Book of Parables* is especially tantalizing for those doing Gospel and historical Jesus research because it may provide evidence for a pre-Christian use of "Son of Man" as a messianic title. At least that was the view championed by R. H. Charles, who dated the text to the late Hasmonean period (94–79 BCE).[72] Perceptions of the *Book of Parables* changed significantly with the discovery of the Dead Sea Scrolls, for when the Qumran documents were unearthed, despite the many Aramaic fragments of other portions of *1 Enoch*, no portion of the *Book of Parables* was accounted for. Its absence from Qumran led J. T. Milik to posit an alternative and very late date. He famously argued that it was a Christian document influenced by the eschatology of the later *Sibylline Oracles* and written sometime after 270 CE.[73] More recent work on the *Book of Parables* has questioned the underpinnings of Milik's proposal, noting especially that his proposal is predicated on an argument from silence. The absence of the *Book of Parables* from Qumran need not indicate that its composition postdates the Dead Sea Scrolls, for many early Jewish documents are not represented in the Qumran library. Moreover, all evidence indicates that

72. Charles, *Book of Enoch*, 67.
73. Milik and Black, *Books of Enoch*, 89–98.

the composition of *1 Enoch* was rather fluid. Most scholars working on the *Book of Parables* today tend to date it sometime between the turn of the millennium and the mid to late first century CE.[74] The relevance of the *Book of Parables* can therefore be maintained, for it either antedates Jesus and the New Testament writings, or provides us with non-Christian Jewish traditions contemporaneous with and parallel to some of the earliest Christian documents. The texts in which fire plays a prominent role in the *Book of Parables* can help therefore in providing a fuller context for our study of the eschatological functions of fire in the New Testament.

The first pertinent text in the *Book of Parables* is at 48:9, which describes the judgment of the powerful kings and landowners:

> And into the hands of my chosen ones I shall throw them. As straw in the fire and as lead in the water, thus they will burn before the face of the holy, and they will sink before the face of the righteous; and no trace of them will be found.[75]

The first-person speaker in this passage is unclear. In the context it is Enoch who is recounting his vision; however, the content—"my chosen ones" ... "I shall throw"—suggests that the speaker is God. Nonetheless, it is significant that God turns the abusive landowners and kings over to the hands of the righteous elect ones for their judgment. That the kings and landowners are likened to grass indicates their wickedness, drawing upon the motif of useless stubble consumed by fire (Exod 15:7, 10; Isa 5:24; Obad 18; Mal 4:1). Black suggests that the fire in question is the fire of Gehenna.[76] The destruction of the wicked by fire before the faces of the holy ones does indeed recall Isa 66:24, where the righteous go out to look upon the burning of the sinners' bodies, which is sometimes associated with Gehenna (e.g., Mark 9:44, 48). Strikingly, however, "[a]t 27.2, 3, 90.26–27 the torments of the wicked appear to be an ever present spectacle for the righteous, whereas in the Parables it appears to be a temporary spectacle only; the wicked are to vanish forever in Gehenna from the sight of the righteous."[77] Further on in

---

74. For some of the most recent trends in dating the *Book of Parables* see the lively discussion by the contributors to the 2005 Camaldoli Seminar on the Parables of Enoch compiled in Boccaccini, *Enoch and the Messiah Son of Man*, 415–96. The essay "Enoch in Sheol: Updating the Dating of the Book of Parables" by David Suter (415–43) gives a particularly helpful summary of the current state of research.

75. The entire Book of Parables is absent not only from the extant Scrolls found at Qumran but also from the Greek fragments.

76. Black, Vanderkam, and Neugebauer, *Book of Enoch*, 211.

77. Ibid.

54:1 there is a reference to "a valley, deep and burning with fire" where kings are thrown, which is probably also an allusion to Gehenna.[78]

Another judgment with fire, the judgment at the resurrection of the dead when those in Sheol and Gehenna will rise, is described in much greater detail in 52:6–7:

> These mountains that your eyes saw—the mountain of iron, and the mountain of copper, and the mountain of silver, and the mountain of gold, and the mountain of soft metal, and the mountain of lead—all these will be before the Chosen One like wax before the fire, and like the water that comes down from above upon these mountains, and they will be weak before his feet. And in those days none will save himself either by gold or silver, and none will be able to flee.

The melting of the mountains at the coming of the Chosen One is a striking detail, for it borrows theophanic imagery from Ps 97:5, Nah 1:5, and Mic 1:4[79] and applies it not to God, but to "the Chosen One."[80] The function of the fire, in addition to its theophanic role, is unclear, but it may be inferred that it plays a role in the judgment. The line "no one shall be able to escape" is probably a reference to eschatological judgment, which all must face. And the connection with fire is enforced by the edict that "no one shall be saved either by gold or silver," for these are among the elements melted by the fiery manifestation of "the Elect One." The desire to escape, moreover, suggests a punitive function.

What is perhaps the most fascinating passage concerning fire in the *Book of Parables* is found in chapter 67, wherein Noah describes a vision that has been revealed to him. In reference to the punishment of the Watchers, Noah recounts the following:

> And he will confine those angels who showed iniquity in that burning valley that my great-grandfather Enoch had shown me previously in the West by the mountains of gold and silver and iron and soft metal and tin. And I saw that valley in which there was a great disturbance and troubling of waters. And when all this happened from that fiery molten metal and the troubling of (the waters) in that place, the smell of sulphur was generated, and it mixed with those waters;[81] and the valley of those angels

78. See Ibid., 219.
79. Charles, *Book of Enoch*, 103.
80. See Black, Vanderkam, and Neugebauer, *Book of Enoch*, 216.
81. Knibb (*Ethiopic Book of Enoch*, 157) has "it was associated with those waters" instead of "it mixed with those waters." In his reading, therefore, the molten metal does

who had led astray burned beneath that ground. And through the valleys of that (area) rivers of fire issue, where those angels will be judged who led astray those who dwell on the earth. (67:4–7)

Two valleys are juxtaposed: one that burns with fire and one that is filled with turbulent waters. Just prior to this passage, Noah alludes to "angels of punishment, who are ready to go forth and let loose all the power of the water that is beneath the earth, that it might be for the judgment and destruction of all who reside and dwell on the earth" (66:1). Like the waters of the flood, the turbulent waters Noah describes here also originate from under the earth and are likewise waters of judgment. Thus the same waters are probably in view. The fire from the one valley produces a flow of molten metal that is united with the waters and flows underground, much like the Greek subterranean river Pyriphlegethon. In what follows, the mixing of the molten metal and the water is so complete that "those waters will be changed and become a fire that burns forever" (v. 13b). It is stated explicitly that the fire burns punitively. There can be no question, then, as to the function. As already noted, the water with which the fire is joined is identified with the water of the great flood, the judgment scene without equal in the Jewish tradition. Moreover, the fire is directed against the angels, specifically the Watchers, and for the authors of *1 Enoch* the flood was a direct result of the sin of the Watchers. These links with Noah's flood underscore the destructive role of the fire.

Although it is too fragmentary to be treated in any detail below in our discussion of the Dead Sea Scrolls, a fragment from the *Book of Giants* discovered at Qumran offers a suggestive parallel to Noah's vision above. In 4Q530 II.6–12 two of the giants dream dreams that portend their fates. It is the first dream that may be of relevance here. Describing his dream, the giant named 'Ohyah explains:

> II. 6 ... [... in] my dream I have seen in this night: [...] 7 [... ga]rdeners and they were watering 8 [...] numerous [roo]ts issued from their trunk 9 [עקרהן] [...] I watched until tongues of fire from 10 [...] all the water and the fire burned in all 11 [...] ... [...] 12 [...] Here is the end of the dream.[82]

Despite the poor state of this text, much of significance can be gleaned from it in connection with our passage from the *Book of Parables*. While Milik

---

not mix with the waters, but the sulpherous smell is associated with them.

82. Trans., García Martínez and Tigchelaar, *Dead Sea Scrolls*, 2:1063.

proposed that the gardeners may be guardian angels,[83] in his analysis of this passage, Loren Stuckenbruck observes that the work of these supposed angels ends in destruction, and so he questions whether the "gardeners" ought to be viewed in such a positive light. Noting that the word עקרהן, which he renders "their rootage," ends in a feminine suffix, Stuckenbruck suggests that the large shoots that originate from this feminine source refer to the giants who were born from human women, and he concludes that "[t]he 'watering' activity is hence a metaphor for impregnation and the 'gardeners' represent the Watchers."[84] The impending destruction envisioned in lines 9 and 10 is thus a reference to the great flood of judgment. Notably, fire has been imported into the description of the flood, thus providing an interesting parallel to the above passage from the *Book of Parables* in which fire likewise mingles with the water of Noah's flood.

These passages from the *Book of Parables* provide us with some new examples of eschatological motifs that we will encounter again elsewhere in the literature of Second Temple Judaism as well as some motifs we have already observed. In the first case we have the striking case of the wicked being handed over to "the elect ones" to be punished by fire, thus providing a possible instance of divine representatives administering the fiery judgment instead of God doing so. We also encountered here the notion of a second judgment by fire after the resurrection. And lastly we have observed a fascinating instance in which descriptions of the great flood, from which Noah and his family were saved, combine imagery of water and fire, resulting in a flood of fire comparable to that we observed in the *Epistle of Enoch* 102:1–3. In every case we have considered in the *Book of Parables*, fire functions punitively and plays no apparent purgative function. Having considered this wide assortment of texts from *1 Enoch*, we are now in a good position to discuss the literature of the Essenes at Qumran, for whom the books that now make up *1 Enoch*, with the exception of the *Book of Parables*, were obviously authoritative texts.

The Dead Sea Scrolls

When expounding upon the beliefs of the Essenes, Hippolytus of Rome writes, "they affirm that there will be both a judgment and a conflagration of the universe, and that the wicked will be eternally punished" (*Adv. Haer.* IX. 22).[85] The so-called "Essene hypothesis" identifies the community(ies)

---

83. Milik and Black, *Books of Enoch*, 304.
84. Stuckenbruck, *Book of Giants*, 114.
85. ANF V:137.

responsible for the composition and preservation of the literature found at Qumran with the Essenes, based on similarities between the descriptions of the Essenes by Josephus, Philo, and Pliny and the Scrolls themselves. While the Essene hypothesis has some detractors and may be in need of some qualification, it remains the most plausible theory regarding the identity of the sectarians responsible for writing and/or preserving the Dead Sea Scrolls. The above quotation from Hippolytus—while it is found in a much more extensive description that bears some relation to Josephus' account and is probably to some extent dependent upon it—indicates that the Essenes held to the belief that the final judgment would be accompanied by fire.[86] As we shall see, fire plays a prominent role in the eschatology of the Scrolls. Hippolytus' description of the Essenes thus overlaps with what we know of the Dead Sea sect on this point. However, we are now in a position to offer a more nuanced discussion of the eschatological functions of fire in the belief system of the sectarians who lived there than that which Hippolytus was able to present.

Our treatment of the Dead Sea Scrolls will focus primarily on the major sectarian texts. Some texts, particularly those found in Cave 4, are far too fragmented for their full significance to be felt. We can note, however, that many of these fragments associate fire with judgment and thus confirm that the prominence of fire in the eschatology of the Essenes was widespread, but their fragmentary nature allows us to say nothing more. A number of texts that deal with fire do so in the context of theophanies, particularly in descriptions of the "throne-chariot," while others have to do with halakhah on kindling fire or sacrifice. In the interest of a fuller discussion of the eschatological functions of fire in the Dead Sea Scrolls, such texts will receive no further treatment than their brief mention here. With those few caveats aside, let us consider the eschatological functions of fire in the Dead Sea Scrolls.

The Damascus Document

The *Damascus Document* has been reconstructed from two medieval manuscripts discovered in the Cairo Genizah (storeroom) and fragments discovered at Qumran from ten manuscripts of the work. Certain aspects

---

86. For a discussion of Hippolytus' description of the Essenes and an evaluation of its relationship to Josephus's account see M. Black, "The Account of the Essenes in Hippolytus and Josephus," in Davies and Daube, *Background of the New Testament*, 172-75. Black, who suggests that Hippolytus may have been aware of some traditions about the Essenes that went unreported by Josephus, observes, "At one point at least the new scrolls agree with Hippolytus' account; there is little doubt that the Zadokites believed in a final 'conflagration of everything'" (175).

of this document, for instance the legislations concerning women and children, conflict with the typical assumption that the Qumran community was composed entirely of celibate males, a notion derived from the *Rule of the Community*.[87] Josephus refers to some Essenes who marry and some who do not (*War* 2.160–61), while according to the *Damascus Document* itself some of the sectarians live "in perfect holiness" (VII.5) while others "live in the camps according to the rule of the law" taking wives and having children (VII.6–7). The *Damascus Document* may therefore indicate some level of diversity within the sect. Regardless, the text was apparently of some importance at Qumran, given the numerous fragments discovered there. The fragments assist us in dating the document, for they can be dated by means of paleography to several time periods within the Community's existence, ranging from the Hasmonean period (4Q266) to the late Hasmonean or early Herodian period (4Q271 and 5Q12) to the Herodian period (4Q267, 4Q269, and 4Q272). Some fragments may even date to the early first century (4Q268 and 4Q279). An allusion to the book of *Jubilees* in CD-A XVI.2–4 suggests a date of composition sometime after the Maccabean War, the period to which most scholars date *Jubilees*, suggesting a date sometime after 160 BCE.[88] And the reference to the death of the Teacher of Righteousness in CD-B XX.13–15 as a past event suggests a date of final composition around 100 BCE.[89]

The document itself consists of two major sections: an admonition, which recounts Israel's history from the exile in Babylon to the author's period, emphasizing the survival of a remnant who come to form a righteous community, and the laws, which consists of legal regulations, including purity rules, priestly requirements, dietary laws and the like, concluding with a section on the rules of the community.[90] In the introduction preceding the admonition, the author describes the fate of those who turn away from the covenant:

> [S]trength and power and a great anger with flames of fire [בלהבי אש] by the <hand> of all the angels of destruction [מלאכי חבל] against those turning aside from the path and abominating the precept, without there being for them either a remnant or survivor. (CD-A II.5b-6)[91]

---

87. On other tensions between these two texts, see Charlesworth, *Dead Sea Scrolls*, 2:6–7.

88. Ibid., 3:1–3.

89. Nickelsburg, *Jewish Literature*, 123.

90. Ibid., 122–23.

91. Trans., García Martínez and Tigchelaar, *Dead Sea Scrolls*, 1:553. I rely primarily

According to this passage, judgment by fire is reserved apparently not only for enemies of the community, but even for those who stumble and fall away from the teachings of the community and fail to maintain the sect's standards of righteousness. Strikingly, the fire of judgment is delivered by the מלאכי חבל "angels of destruction." This is significant, for it suggests an eschatological scenario such as the last judgment, in which angelic mediators may be the bearers of fiery judgment.[92] Nothing seems to indicate that anything other than a literal fire is envisaged. The statement that there will be neither remnant nor survivor also makes clear that their destruction will be complete.

Another text that mentions fire, though not in an explicitly eschatological context is CD-A V.13:

> They are all igniters of fire, kindlers of blazes [קדחי אש ומבערי זיקות]; webs of a spider are their webs, and their eggs are vipers' eggs. Whoever comes close to them will not be unpunished.[93]

As Ben Zion Wacholder notes, this text paraphrases Isa 50:11: "But all of you are kindlers of fire, lighters of firebrands [קֹדְחֵי אֵשׁ מְאַזְּרֵי זִיקוֹת]. Walk in the flame of your fire, and among the brands that you have kindled!" and 59:5: "They hatch adders' eggs, and weave the spider's web; whoever eats their eggs dies, and the crushed egg hatches out a viper."[94] Quite unlike the text above, which envisions angels administering a fiery judgment, this text is apparently metaphorical and uses a poetic figure to teach that not only will one have to face the consequences of his or her own wicked deeds but of those with whom he or she associates as well. It is a bit of conventional wisdom exhorting those in the community to be careful with whom they associate and contributes little to our understanding of the fire of eschatological judgment.

In the *Damascus Document*, then, the motif of fire appears to play a relatively fluid role. In some cases it is applied literally as an instrument of eschatological judgment whereas in other instances it is most likely a metaphorical way of speaking of the consequences of associating with those of questionable morals. Significantly, as we saw in the *Parables of Enoch*, the bearers of the fiery judgment may be divine mediators. In the former text

---

on the translation of García Martínez and Tigchelaar. I have also consulted Charlesworth, *Dead Sea Scrolls*; Vermes, *Complete Dead Sea Scrolls*; Wise, Abegg, and Cook, *Dead Sea Scrolls*. In most instances the translations do not differ in substance. I shall note only where they do.

92 See Gregg, *Historical Jesus*, 55–56.
93. García Martínez and Tigchelaar, *Dead Sea Scrolls*, 1:557.
94. Wacholder, *New Damascus Document*, 210.

we witnessed "the elect ones" administering the judgment by fire, and in the latter we observed that God's angels are the bearers of God's fiery wrath.

1QPesher Habakkuk

Among the numerous pesharim discovered at Qumran, the Habakkuk Pesher, which comments on Habakkuk 1–2, is the best preserved and most frequently discussed.[95] While the script of the Habakkuk Pesher suggests it was written by two different scribes—the second scribe beginning at XII.13 on through the conclusion—both scripts are Herodian, and the text may be dated to the second half of the first century BCE.[96] Its relevance to New Testament scholars stems largely from its comments on Hab 2:24, "But the righteous shall live by his faith," a verse that is also central to Paul's theology (e.g., Gal 3:11; Rom 1:17). More significant for our interest in the eschatological functions of fire in the New Testament, however, is 1Qp Hab IX.9–X.13, which employs two extended metaphors of building, first with reference to a house and second to a city, each of which concludes with the threat of punishment by fire of the spiritual leader responsible for the construction.

> IX *Hab 2:9–11* Woe to anyone putting ill-gotten gains in his house, placing 13 his perch high up to escape the power of evil! You have planned things that will bring disgrace 14 to your house, destroying many nations and sinning against your [so]ul. For 15 the sto[ne] will shout from the wall, the wooden beam will an[swer . . .] 16 [The interpretation of the wor]d concerns the pr[iest] who [. . .]
>
> X 1 for its stones to be by oppression and the beam of its wood by pillage. And what 2 it says: *Hab 2:10* <<Destroying many nations and sinning against your soul>>. *Blank* 3 Its interpretation: it is the house of judgment [בית המשפט], for God will give 4 his judgment among many nations and from there will lead him to punishment. 5 And in their midst he will proclaim him guilty and will punish him with sulphurous fire [ובאש גופרית]. *Hab 2:12–13*

---

95. Pesher (plural: pesharim) is a transliteration of the Hebrew word meaning "interpretation." In genre a pesher is something akin to an ancient commentary. The Habakkuk Pesher, a continuous verse by verse commentary, is one of fifteen to eighteen pesharim discovered at Qumran. On the pesharim in general see Horgan, *Pesharim*, esp. 1–9, 229–59. On the Habakkuk Pesher itself see Harris, *Qumran Commentary*; Brownlee *Midrash Pesher of Habakkuk*. For introductory comments see VanderKam, *Introduction to Early Judaism*, 46–48; Nickelsburg, *Jewish Literature*, 128–30.

96. Charlesworth, *Dead Sea Scrolls*, 6B: 157.

Woe 6 to him who builds a city with blood and founds a town on wickedness! Does 7 this not stem from YHWH of Hosts that the nations wear themselves out for fire 8 and the peoples are exhausted for nothing? *Blank* 9 The interpretation of the word concerns the Spreader of the Lie, who has misdirected many, 10 building a useless city with blood and erecting a community with deceit 11 for his own glory, wearing out many by useless work and teaching them 12 a[c]ts of deceit, so that their labours are for nothing; so that 13 those who derided and insulted God's chosen will go to the punishment of fire [למשפטי אש].[97]

The first half of this text speaks of the construction of a building that is referred to as בית המשפט. The phrase has been interpreted variously as "the house of judgment," indicating a courtroom or "divine tribunal" in which the judgment of "the Wicked Priest" will take place,[98] or as "the house of damnation," and thus "the prison of the wicked in the netherworld,"[99] or as "the condemned house," which will itself be subjected to future eschatological judgment.[100] That the context speaks of a house that will be disgraced because it has been built by means of oppression and pillage suggests a possible reference to the literal temple in Jerusalem, especially given its association with the Wicked Priest, and thus tips the balance in favor of the last translation.[101] Strikingly, however, the description of the literal house takes on a metaphorical slant. The building supplies—wooden beams and stones—which were acquired through oppression and pillage, cry out and testify against the builder, "the Wicked Priest," whom God will judge and punish באש גופרית "with sulphurous fire" (see Gen 19:24).

The second half extends the building metaphor, but now widens the scope to include an entire city, perhaps Jerusalem, and makes specific reference to the foundations of the city. Now, however, it is "the Spreader of the Lie" who is directing the building. Strikingly the building of the city becomes an explicit metaphor for "erecting a community." The Spreader of the Lie meets a fate identical to that of the Wicked Priest—משפטי אש "the

---

97. García Martínez and Tigchelaar, *Dead Sea Scrolls*, 1:19.

98. Dupont-Sommer, *Essene Writings*, 265. See Charlesworth, *Dead Sea Scrolls*, 6B 179.

99. Brownlee, *Midrash Pesher of Habakkuk*, 161–62.

100. Vermès, *Complete Dead Sea Scrolls*, 514.

101. There is widespread agreement that the title הכהן הרשע (the Wicked Priest) plays on the title הכהן הראש (the high priest). Regarding his precise identity, however, there have been many proposals. Most often he is identified as Jonathan or Simon Maccabeus; see Ibid., 18. However, an alternative theory, called the Groningen Hypothesis, posits a series of six different Wicked Priests; see García Martínez and van der Woude, "A 'Groningen' Hypothesis."

punishment of fire." In fact, some have understood the Wicked Priest and the Spreader of the Lie to be one and the same.[102] While this is contested, if the identification is correct, it is probable that we ought to read these two images, the construction of a condemned house and the founding of the useless city, as an instance of synonymous parallelism. Thus, while it may be a literal reference to the corruption the sectarians believed reigned among the Jerusalem priesthood in charge of the temple precincts, the condemned house also points to the larger community, the builder (i.e. spiritual leader) of which will be judged by God and must face the punishment of fire. The motif of building on a foundation and the judgment of the builder by fire may be of relevance to our later discussion of 1 Cor 3:10–15.[103]

Fire clearly functions negatively here. This is apparent first from the fact that the fire is referred to as a punishment directed at "those who derided and insulted God's chosen." Specifically, it is directed at "the Wicked Priest" and "the Spreader of the Lie," the enemies of the Dead Sea sect, who, according to the sect were deserving of the most extreme punishment imaginable. Whether or not these two sobriquets ought to be associated with the same individual, there is no question that they point to enemies of the sect, and we can very safely assume that the fire is understood negatively. This is made even clearer by the use of the phrase באש גופרית "sulphurous fire," which draws on the description of the destruction of Sodom and Gomorrah, one of the quintessential scenes of judgment in Jewish tradition, which envisaged the total destruction of those consumed by the fire.[104] The destruction of the Wicked Priest and the Spreader of the Lie, and his community (or their communities) is complete.

Thanksgiving Hymns

Perhaps the most significant texts for our discussion of the Scrolls are the collection designated the *Thanksgiving Hymns*. While earlier scholarship tended to view all of the Hymns as compositions of the Teacher of Righteousness, they are now commonly divided into two categories, those

---

102. Vermès, *Complete Dead Sea Scrolls*, 60–61.

103. Some of the potential parallels between 1 Cor 3:9–15 and 1Qp Hab IX.9–X.13 are briefly noted in Eisenman, *James the Just*, 55. Eisenman's overarching thesis regarding the identification of James with the Teacher of Righteousness, however, is so speculative and controversial that some of his more plausible suggestions have been overlooked.

104. See pages 39–52 above.

attributed to the Teacher and those attributed to the community.¹⁰⁵ Based on linguistic and conceptual similarities to other sectarian Scrolls, the Hymns are widely believed to belong to the main corpus of texts authored by the community, and paleographical analysis of a collection of the Hymns found in Cave 4 (4QHb) reveals that those particular Scrolls were written just after 100 BCE, suggesting an original date of composition sometime in the late second century BCE.¹⁰⁶ Fire plays a considerable role in several of the *Thanksgiving Hymns*, and the passages mentioning fire in the Hymns, particularly 1QHa XI.29-37 (formerly 1QH III.1-18), have garnered far more scholarly attention than any other of the sectarian texts mentioning fire. Given the poetic nature of these texts, it is not surprising to find several cases in which fire is used metaphorically to describe the weapons of warfare (X.26),¹⁰⁷ the melting of one's heart (XII.33), or one's inward passions (XVI.30; 33). And, very interestingly, 1QHa VI.3-4 speaks metaphorically of "the men of truth" and the "poor in spirit" who are "refined by poverty and purified in the crucible." Much more characteristic of the *Thanksgiving Hymns*, however, is the appearance of fire in an eschatological context with which we are concerned. It is interesting indeed that three of the five texts we shall examine below juxtapose the imagery of fire and water. It is with those texts and their common motifs that we shall begin.

The most well-known and widely-discussed passage that mentions fire in the *Thanksgiving Hymns* is 1QHa XI.27b-37:

> 27 ... When the measuring line falls upon judgment, and the lot of anger 28 on the forsaken and the outpouring of wrath against the hypocrites, and the period of anger against any Belial, and the ropes of death enclose with no escape, 29 then the torrents of Belial [נחלי בליעל] will overflow all the high banks like a devouring fire [כאש אוכלת] in all their watering channels (?), destroying every tree, green 30 or dry [כול עץ לח ויבש], from their canals. It roams with flames of fire until none of those who drink are left.¹⁰⁸ It consumes [תאוכל] the foundations of clay 31 and the tract of dry land; the bases of the mountains does he

---

105. Nickelsburg, *Jewish Lilterature*, 133-37. Those considered herein are generally regarded as belonging to the latter group.

106. For an excellent discussion of the dating of the *Hodayot*, see Puech, "Hodayot," 366.

107. It is not uncommon to find fire or flame as a metaphor for the sword or the spear in the Hebrew Bible (see Nah 3:3; Job 39:23; Isa 39:23). See the comments of Holm-Nielsen, *Hodayot*, 42, n. 17.

108. Vermes (*Complete Dead Sea Scrolls*, 262) has "unto the end of their courses it shall scourge with flames of fire."

> burn [ותאוכל] and converts the roots of flint rock into streams of lava. It consumes right to the great deep. 32 The torrents of Belial [נחלי בליעל] break into Abaddon [לאבדון]. The schemers of the deep howl at the din of those extracting mud. The earth 33 cries out at the calamity which overtakes the world, and all its schemers scream, and all who are upon it go crazy, 34 and melt away in the great calamity. For God will thunder with the roar of his strength, and his holy residence echoes with the truth of 35 his glory, and the host of the heavens adds to their noise, [and] the eternal foundations melt and shake, and the battle of heavenly heroes 36 roams unceas[ingly] over the earth, [un]til the determined eternal unparalleled destruction [כלה ונחרצה לעד ואפס]. *Blank* 37 *Blank* I give you thanks, Lord, for you are a massive rampart for me 38 [. . . al]l destroyers and all [. . .] you hide me from the turbulent calamities . . . 40 [. . .] iron [ba]rs. Not shall enter [. . .] 41 [. . .] around it lest . . . [. . .]¹⁰⁹

This text envisions a time of upheaval, referred to variously as "the outpouring of wrath" and "the period of anger," that accompanies or precedes the final judgment. Perhaps the most striking feature of this period of tribulation is the נחלי בליעל "torrents of Belial," which are described as being כאש אוכלת "like a devouring fire." As observed by Svend Holm-Nielsen, "the imagery is really a combination of fire and water, as is shown by the comparative particle in כאש."¹¹⁰ Thus, what is being described is "the river of fire," a common motif in apocalyptic literature (see Dan 7:10; *1 En.* 17:5; 67:13).¹¹¹ Paralleling the image of the river of fire, the rocks forming the bases of the mountains are melted into streams of molten lava. The function of the river of fire is clearly negative and the breadth of its destruction is complete, as is indicated by the fact that the waters destroy כול עץ לח ויבש "every tree, green or dry" and even reach down to the depths of Sheol.¹¹² According to Helmer Ringgren, "[t]here is no doubt that the last lines of this psalm describe the final decision, God's intervention and the final battle through which the evil are annihilated."¹¹³ Indeed, the text speaks of the eschatological events that are to transpire as resulting in the כלה ונחרצה לעד ואפס "determined eternal unparalleled destruction." And John Joseph Collins notes that although it is

---

109. García Martínez and Tigchelaar, *Dead Sea Scrolls*, 1:167.
110. Holm-Nielsen, *Hodayot*, 72, n. 43.
111. See discussion in Ibid., 71, n. 39.
112. Holm-Nielsen (*Hodayot*, 73, n. 47) notes that "Lines 29 and 32 together may adversatively signify a totality: no one and nothing escapes the destruction, see Amos 9:1–4."
113. Ringgren, *Faith of Qumran*, 157.

the only text in the Dead Sea corpus to do so, this passage "possibly speaks of the destruction of the world." However, despite the totality of the destruction of the rivers of fire, which consume every green tree and every dry tree and reach down to Abaddon, lines 37–41—fragmented though they may be—indicate that the psalmist expects to be delivered from the impending judgment. The phrases "you are a massive rampart for me" and "you hide me from turbulent calamities" suggest protection in the face of extreme danger, and indicate that the psalmist and his community expect divine intervention on their behalf.

One issue of continued debate revolves around the question of Iranian influence. As Ringgren notes, "[e]ver since this psalm was first published it has been assumed that at least the description of the stream of fire which annihilates the foundations of the earth belongs to the eschatological context, and that this indicates that under the influence of Iranian ideas the Qumran congregation expected their destruction of the world by fire." Ringgren, however, thinks Iranian influence is unlikely based on the fact that the phrases "cords of death encompass" and וְנַחֲלֵי בְלִיַּעַל "torrents of Belial" occur in Ps 18:4 (MT 18:5), that the devouring fire motif is common among the prophets (see Amos 1:4; 7:4), and the phrases עֵץ לָח "green tree" and עֵץ יָבֵשׁ "dry tree" is used in Ezek 17:24.[114] The biblical phrases and allusions do not necessarily rule out Iranian influence, however. Indeed, specific motifs such as the river of fire and the metal or stone melting in the foundations of the mountains, which are present in both the Scrolls and Zoroastrian literature, may support it (see *1 En.* 52:6 above and 1QHa IV.13–14 below).[115] The notion that at the end everyone would pass through a fiery river of molten metal is anticipated in the Zoroastrian literature we have discussed above.[116] The Zoroastrian material is significant, for if it has influenced 1QHa XI.27b–37, it supports our reading that the righteous are preserved despite the broad scope of the destructive rivers of fire. The parallels are not exact, however, for in the Zoroastrian texts the stream of molten metal proves purificatory for the upright whereas in the Hodayot the righteous are simply spared the destruction. More, whereas the Zoroastrian texts have the gods melting the mountains, thus indicating a divine source for the streams of molten metal, the phrase "torrents of Belial" in the *Thanksgiving Hymn* indicates a darker, demonic source. Thus we cannot speak in terms of the wholesale adoption of eschatological doctrines, merely in terms of the borrowing of select motifs.

---

114. Ibid., 158.
115. Mansoor, *Thanksgiving Hymns*, 120. Winston, "Iranian Component."
116. See pages 2, 3, and 7 above.

Nonetheless, both traditions maintain that those subjected to the fire will face different outcomes based upon their moral standing.

A bit more subtle, yet no less striking is 1QHa XIV.16b–19a, which speaks metaphorically of the righteous and holy ones of the community as a flowering shoot that grows into an eternal planting that covers the whole world with its shade. The streams that water the planting are described as follows:

> 16 ... All the streams of Eden [נהרות עדן] [will water] its [bra]-n[ch]es and they will be [seas without] 17 limits; and its forest will be over the whole world, endless, and as deep as to Sheol [ועד שאול] [its roots.] The source of light [מעין אור] [will] be an eternal spring 18 [למקור עולם], inexhaustible, in its shining flames [בשביבי נוגהו] all the son[s of injustice] [בנ(י)י עולה] will burn [and it will be turned] into a fire [לאש] that singes all the men of 19 guilt [אנשי אשמה] until destruction [עד כלה].[117]

What is noteworthy in this passage is how the positive image of the נהרות עדן "streams of Eden" quickly gives way to the negative image of fire and flame that burn and destroy the [בנ]י עולה "sons of injustice" and the אנשי אשמה "men of guilt." Obviously, we have here another juxtaposition of water and fire. More interesting in this passage, however, are the functions of these elements. There is no apparent change of imagery between "the streams of Eden," which cover the entire earth, mentioned in line 16 and the למקור עולם "eternal spring,"[118] which is characterized by the אור "light" it provides for the righteous. It is therefore possible to read the two as referring to the same source of water. If this identification is correct, it is remarkable that the very בשביבי נוגהו "shining flames" that are a מעין אור "source of light" for the righteous are turned into a raging אש "fire" that consumes and destroys the wicked.[119] The stream of light, which provides life to the righteous, becomes a river of fire, which consumes the wicked עד כלה "until destruction." Notably, like "the torrents of Belial," which break לאבדון "into Abbadon" (i.e. the depths of hell) in 1QHa XI.32, the streams of Eden are here said to go down ועד שאול "to Sheol," which suggests a conceptual link between the two texts.

The third passage in the *Thanksgiving Hymns* that pairs fire and water is 1QHa XVI.11–20. The passage begins with a description of the righteous depicted as a plantation of trees drawing on streams of living water, similar to the imagery in the above passage, and continues:

---

117. García Martínez and Tigchelaar, *Dead Sea Scrolls*, 1:175.

118. Vemes (*Complete Dead Sea Scrolls*, 272) has "an eternal ever-flowing fountain."

119. See Isa 10:17: "The light of Israel will become a fire, and his Holy One a flame; and it will burn and devour his thorns and briers in one day."

11 ... But you, [O G]od, you protect its fruit with the mystery of powerful heroes 12 and spirits of holiness, so that the flame of the searing fire [ולהט אש] [will] not [reach] the spring of life [מעין חיים(ב)], nor with the everlasting trees 13 will it drink the waters of holiness [מי קודש], nor produce its fruit with [...] of the clouds. For it sees, but does not know, 14 notices, but does not believe, in the spring of life, and gives eternal [...] But I had become the mockery of the raging torrents [זאי נהרות(ב)ל] for 15 they threw their mire over me. *Blank* 16 But you, my God, you have placed in my mouth as it were an early rain for all [...] and a spring of living water; 17 the skies will not fail to open, they will not stop, but will become a torrent overflowing in[to ...] of water and into the seas, without e[nd.] 18 They will swell suddenly from secret hiding-places, [...] they will become waters of [... for every tree,] 19 green and dry [... (לכול עץ לח ויבש)], a marsh for every animal. [...] like lead in powerful waters, [...] 20 ... of fire and dry up. But the plantation of fruit [...] eternal, for the glorious garden and will bear [fruit always.]

Unlike the above passages where fire and water are combined into one image, here there seems to be an opposition between the two elements. The [במעין חיי[ם] "spring of life" and מי קודש "waters of holiness" are protected from the ולהט אש "searing fire," and so the plantation of trees is likewise spared.[120] However, the water imagery takes a negative turn from the spring of life and the waters of holiness to the ל[ב]זאי נהרות "raging torrents" that hurl mire over the righteous.

Again the phrase [לכול עץ לח ויבש] "every tree, green and dry" is employed, but this time in reference to the trees that are overcome by the waters. The text mentioning fire in line 20 is unfortunately too fragmented to decipher. However, it may indicate that after the flood of water will come the scorching punishment of fire that will consume the green tree and the dry tree as that is the standard use of the phrase. This is, however, only speculation. Surprisingly, although the psalmist laments that he had been mocked by the torrents in line 14, and despite the fact that lines 18 and 19 envision the destruction of every tree, green and dry, the end of line 20 indicates that the righteous plantation is eternal and will continue to bear fruit forever, suggesting escape from the torrents of destruction and the raging fires. Thus the seemingly all-encompassing destruction, indicated by the reference to the green and dry trees, appears to exclude the fruit-bearing trees, namely

---

120. According to Vermes (*Complete Dead Sea Scrolls*, 279), "the whirling flame of fire" protects the waters so that "No [man shall approach] the well-spring of life or drink the waters of holiness."

the righteous. As in 1QHa XI.27b–37, the righteous are preserved from the destruction of both fire and water while the wicked are destroyed.

Very briefly, we note that in 1QHa IV.13–14 we once again encounter the motif of the foundations of mountains being burned:

> [Even though you burn] the foundations of mountains and fire [sears] the base of Sheol [בשאול תחתיה], those who … […] in your regulations. You [protect] [שמר(תה)] the ones who serve you loyally, [so that] their posterity is before you all the days.[121]

Although there is no mention of the rocks or metal being melted, we do see that here again the fire reaches down [בשאול] to Sheol, indicating the wide scope of the fiery judgment. However, those who serve God are preserved [שמר(תה)] "you protect" from its destructive force.

Finally, we turn to 1QHa XIII.15b–19:

> 15 … And to show your greatness /through me/ before the sons of Adam, you did wonders 16 with the poor [באביון], you placed him [like g]old in the cruci[ble] [במצר(ף כז(הב] to be worked by fire, and like purified silver in the furnace [ככסף מזוקק בכור] of the smiths to be refined [נופחים לטהר] seven times. 17 The wicked of the nations hustle me with their trials, and the whole day they crush my soul. *Blank* 18 But you, my God, have changed [תשיב] [my soul] the storm to a calm and have freed the soul of the poor [אביון] like […] prey from the power of 19 lions. *Blank*[122]

Line 16 draws on the imagery in Mal 3:2, where the LORD will come as "a refiner's fire" (כְּאֵשׁ מְצָרֵף) and v. 3, which states: "he will sit as a refiner and purifier of silver [מְצָרֵף וּמְטַהֵר כֶּסֶף], and he will purify [וְטִהַר] the descendants of Levi and refine them like gold and silver [וְזִקַּק אֹתָם כַּזָּהָב וְכַכָּסֶף]." Whereas in Malachi the Levites in the temple will be purified so as to be fit to offer sacrifices in the eschatological temple, in the *Hodayot*, it is the אביון "poor," the common name applied to the upright in the sect, who are tested and tried by the fires. In connection with the allusion to Mal 3:3, it may be significant that prominent among the Qumran community numbered descendants of the Zaddokite priesthood, and thus the priesthood at Qumran may have applied to themselves the expectation of eschatological purification by fire that Malachi applies to the Levites.[123]

---

121. García Martínez and Tigchelaar, *Dead Sea Scrolls*, 1:149.

122. Ibid., 1:173.

123. On the central role of the priesthood at Qumran see Kugler, "Priesthood at Qumran."

The fire clearly does not function punitively; it is explicitly identified with purification and refinement in the crucible (במצר[ף כז[הב) or furnace (ככסף מזוקק בכור).[124] Closely associated with this is the mention of a trial at the hands of the nations, which may include an element of suffering.[125] This link probably indicates that the motif of testing in the crucible and purification and refinement through fire is being used metaphorically for the period of earthly trial being endured by the righteous at the hands of the wicked. Line 18 mentions that the trial results in a change in the soul of the righteous. While the first few lines of this passage do speak of a purifying and refining function of the fire, it is not at all clear that it is this cleansing that is referred to when the psalmist speaks of "change" (תשיב). More likely, as the phrases "the storm to a calm" and "like [. . .] prey from the power of lions" indicate, the change in mind is conceptualized in terms of deliverance from the period of trial and testing.

Our discussion of the *Thanksgiving Psalms* reveals that the psalmists employed the motif of eschatological fire in multiple ways. While the crucible motif was employed metaphorically to refer to the testing and trials that the righteous must endure, the majority of texts appear to speak of a literal fire to accompany the eschatological ordeal. It is significant, moreover, that more than half of these texts pair the image of fire with that of water, often suggesting that the two elements become one, forming a river or flood of fire. Many of the psalms indicate the vast scope of the eschatological fire, indicating that it will consume the green tree and the dry tree alike and that it will even reach down into Abaddon or Sheol. Despite the broad scope of these eschatological fires, whereas the righteous may be tested in the crucible of suffering in those passages that speak metaphorically of fire, they are apparently spared the deluges of fire that are envisioned and will escape their destructive torrents.

1QS Community Rule

The *Community Rule*, also known as the *Manual of Discipline*, is one of the most interesting of the scrolls for considering the sociology of the inhabitants of Qumran, and it is this document that has contributed most to our ability to identify, with at least some degree of probability, the sectarians with the Essenes described by Josephus, Philo, and Pliny the Elder. One

---

124. Vermes (*Complete Dead Sea Scrolls*, 268) translates these terms as "the melting-pot" and "the melting-pot of the smelters."

125. Holm-Nielsen, *Hodayot*, 96, n. 42; Mansoor, *Thanksgiving Hymns*, 134, nn. 7–8.

does not expect to find in them, however, much about the community's eschatological expectations, or more precisely their beliefs concerning the eschatological functions of fire. Surprisingly, two texts do make specific reference to the fire of eschatological judgment, albeit in passing, and two other passages, while not explicitly using the word fire, refer to the related notion of testing in a crucible.

One passage that refers explicitly to fire is 1QS II.7–8, which states that the Levites will curse the Lot of Belial, saying:

> 7 ... Accursed are you, without mercy, according to the darkness of your deeds, and sentenced 8 to the gloom of everlasting fire [באפלת אש עולמים]. May God not be merciful [לוא יחונכה] when you entreat him. May he not forgive by purifying your iniquities [ולוא יסלח לכפר עווניך].[126]

"The Lot of Belial" is, like "the sons of darkness," one of many frequently employed sobriquets in the Dead Sea Scrolls used in reference to the enemies of the sect. According to this text, these enemies will be judged according to their wickedness and sentenced באפלת אש עולמים "to the gloom of everlasting fire." Clearly, this text has in mind the fire of punishment in Sheol or Gehenna. The phrase "everlasting fire" recalls the language used by Jude 7 (πυρὸς αἰωνίου) to describe the fire that fell on Sodom and Gomorrah, and may therefore be drawing on a similar tradition. That the fire is everlasting need not indicate that the punishment is so.[127] To be sure, the fire is undoubtedly expected to have a negative effect, but it need not be understood as an eternal punishment.

What follows is intriguing. After sentencing his enemies to Sheol or Gehenna, the author includes in his condemnation his further desire that God לוא יחונכה "not be merciful" and ולוא יסלח לכפר עווניך "not forgive by purifying your iniquities."[128] Strikingly, in condemning the Lot of Belial to the fire of judgment the author felt the need to express his wish that their iniquities *not* be purified and forgiven. Implicit in this is at least the possibility of purification, and given the appearance of the words אש "fire" and לכפר "purifying" in such close proximity, it is difficult not to assume a link between the two.[129] To be clear, I do not wish to suggest that this text

---

126. García Martínez and Tigchelaar, *Dead Sea Scrolls*, 1:73.

127. Leaney, *Rule of Qumran*, 131.

128. Vermes (*Complete Dead Sea Scrolls*, 99) has "May God not heed when you call on Him, nor pardon you by blotting out your sin!" See Charlesworth (*Dead Sea Scrolls*, 1:11): "May God not be compassionate unto you when you cry out. May he not forgive (you) by covering over your iniquity."

129. While a more literal translation of כפר might be "to cover, to atone" (s.v. BDB,

indicates that the author believed "the Lot of Belial" would be purified by the fires of Gehenna. Rather, this text explicitly rules that out; however, the very fact that the author felt the need to exclude explicitly the possibility of purification indicates that even reference to the everlasting fire of Gehenna could provoke thoughts of purification.

Similarly making an explicit reference to fire, although without expressing any concern that it might prove purificatory, 1QS IV.11–14 warns:

> 11 ... And the visitation 12 of all those who walk in it [the spirit of deceit] will be for an abundance of afflictions at the hands of all the angels of destruction [כול מלאכי חבל], for eternal damnation [לשחת עולמים] by the scorching wrath of the God of revenges, for permanent terror and shame 13 without end with the humiliation of destruction by the fire of the dark regions [כלה באש מחשכים]. And all the ages of their generations (they shall spend) in bitter weeping and harsh evils in the abysses of darkness until 14 their destruction, without there being a remnant or a survivor for them.[130]

This passage, which falls in the section known as the *Treatise on the Two Spirits*, takes up the fate of "those who walk in the spirit of deceit" in contrast to the rewards that will be received by "the sons of truth."[131] Whereas at the last judgment the sons of truth will receive "healing, plentiful peace in a long life, fruitful offspring with all everlasting blessings, eternal enjoyment with endless life, and a crown of glory with majestic raiment in eternal light" (IV.6b–7), those who walk in the spirit of deceit will face לשחת עולמים "eternal damnation" which the מלאכי חבל "angels of destruction" have in store for them.

While the eschatological outcome awaiting those who walk in the spirit of deceit is indisputably negative, Paul Garnet remarks, "it is not clear whether the ultimate destiny of the wicked is annihilation or eternal punishment."[132] Indeed, there is some ambiguity here. On the one hand, angels of destruction, perhaps after the final eschatological battle, will subject them to the eternal punishment by fire. On the other hand, כלה באש מחשכים "destruction by the fire of the dark regions" seems to promise an-

---

497), "to purify" is a contextually appropriate translation.

130. García Martínez and Tigchelaar, *Dead Sea Scrolls*, 1:77.

131. Nickelsburg notes that the treatise on the Two Spirits reflects the same sort of thinking underlying the Deuteronomistic teaching on the Two Ways; however, whereas the latter promises blessings and curses in the present life, the former looks to eschatological rewards and retributions.

132. Garnet, *Salvation and Atonement*, 114.

nihilation of the wicked in Sheol. It is perhaps possible, however, that the phrase "eternal damnation" should not be taken too literally and that it may point to a protracted period of punishment that concludes with the destruction of the wicked at the final judgment. This appears to be the notion expressed in the final sentence quoted above. That there will be no remnant or survivor underscores the completeness of the destruction.

In the same column of the Rule we find a passage that, although it does not mention fire, requires our attention at this juncture. In describing the end, the author writes:

> 20 . . . Then God will refine [יברר], with his truth, all man's deeds, and will purify [וזקק] for himself the structure of man, ripping out all spirit of injustice from the innermost part 21 of his flesh, and cleansing him with the spirit of holiness [ולטהרו ברוח קודש] from every wicked deeds [sic]. He will sprinkle over him the spirit of truth like lustral water [כמי נדה] (in order to cleanse him) from all the abhorrences of deceit and (from) the defilement 22 of the unclean spirit, in order to instruct the upright ones with knowledge of the Most High, and to make understand the wisdom of the sons of heaven to those of perfect behaviour. (IV.20b–22a)[133]

Notably, shortly after the previous passage on eschatological fire, this passage speaks of רוח קודש "the spirit of holiness" and מי נדה "lustral water."[134] This concentrated occurrence of the words fire, spirit, and water in column four has caught the attention of many commenting on John the Baptist's proclamation in Matt 3:11//Luke 3:16 in which he juxtaposes his own water baptism with the baptism in Holy Spirit and fire that will be administered by the Coming One.[135] Whereas reference to fire in the previous text was explicitly negative and appears to presume annihilation for the wicked, this passage speaks of the purification and refinement of the upright with a spirit of truth and a spirit of holiness that is likened to purification in "lustral waters." The potential that these passages have for illuminating Matt 3:11/Luke 3:16 is manifest, and we shall return to them in our discussion of John the Baptist in chapter 6.

Although they make no explicit mention of fire, two additional passages must be examined briefly before we conclude our discussion of the

---

133. García Martínez and Tigchelaar, *Dead Sea Scrolls*, 1:79.

134. Vermes (*Complete Dead Sea Scrolls*, 103) translates this as "like purifying waters" and Charlesworth (*Dead Sea Scrolls*, 1:19) has "like waters of purification."

135. See Dunn, "Spirit-and-Fire Baptism," 87.

Rule of the Community. As part of the introduction, which speaks of initiation into the community, 1QS I.16–18 states:

> And all who enter in the Rule of the Community shall establish a covenant before God in order to carry out all that he commanded and in order not to stray from following him out of any fear, dread, or testing [ומצרף] (that might occur) during the dominion of Belial.[136]

The word מצרף, which is here translated as "testing," can mean "furnace" or "refiner's fire" as it does in Prov 17:3 and 27:21.[137] As Leaney notes, this notion "recalls Mal. 3.2 f. where the fire is again a refiner's fire and is also to purify the righteous."[138] While the testing no doubt points to some form of suffering that the believer must endure to prove him- or herself faithful, it does not appear to refer to a literal fire. Rather the fire functions as a metaphor for the intensity of suffering, much as it is used in 1 Pet 4:12. Similarly, 1QS VIII.4 speaks of the righteous "undergoing trials" (מצרף) in order to atone for sin. The word may be translated as above, and must again be taken metaphorically.

Thus, for the *Rule of the Community*, fire is generally reserved as a fate for the wicked as a punitive judgment in Sheol. We did note, however, that the author felt compelled to rule out the possibility of purification in 1QS II.8, suggesting at least an awareness of the purificatory function of fire. Yet it is clear that where the wicked are concerned, the author of the *Rule of the Community* could only imagine their annihilation. For the righteous, however, he envisioned refinement, not through fire, despite the fact that the purification is mentioned in close proximity to fire, but through truth and a spirit of holiness. Lastly, as did the author of the *Thanksgiving Hymns*, the author of the *Rule of the Community* uses the metaphor of a crucible for the testing of the righteous in their present situation.

The War Scroll

One final Scroll requires our attention. The *War Scroll* anticipates an eschatological war that is to take place between the Sons of Light and the Sons of Darkness in which God will ultimately intervene to win the cause of his chosen. The dating of the text is complicated by the fact that textual

---

136. García Martínez and Tigchelaar, *Dead Sea Scrolls*, 1:71.

137. Leaney, *Rule of Qumran*, 126, citing Flusser, "Baptism of John." On the term מצרף see especially Sander, "ΠΥΡΩΣΙΣ."

138. Leaney, *Rule of Qumran*, 126.

evidence reveals that there were at least two different recensions of the *War Scroll* in circulation.[139] While the composite nature of the text does not allow any certainty regarding the dating of the Scroll in its entirety, the evidence does allow some general parameters. An allusion to Dan 11:40—12:3 in 1QM I suggests a *terminus a quo* of 160 BCE while paleographical dating indicates a *terminus a quem* of 110 BCE.[140] More, the "Table of Nations" listed in 1QM I.1–2a "may be said to reflect a Maccabean or immediately post-Maccabean situation."[141]

The references to fire in the *War Scroll* are few in number. Speaking of the battle between the "anointed ones" and "the troops of Belial," 1QM XI.10–11 states, "The stricken of spirit [ונכאי רוח] you shall set aflame, like a torch of fire [כלפיד אש] in straw, devouring wickedness, without ceasing until, the sin has been consumed."[142] The ונכאי רוח "stricken of spirit" probably refers to "the poor, those you saved," mentioned in 1QM XI.9.[143] At the very least it can be identified as a favorable designation, for it is used in the Hebrew Bible to denote the humility of God's elect (see Isa 66:2: וּנְכֵה־רוּחַ). They are likened to a "torch of fire," which most clearly recalls Zech 12:6: "On that day I will make the clans of Judah like a blazing pot on a pile of wood, like a flaming torch [וּכְלַפִּיד אֵשׁ] among sheaves; and they shall devour to the right and to the left all the surrounding peoples, while Jerusalem shall again be inhabited in its place, in Jerusalem."[144] However, the imagery is familiar from Gen 15:17 as well, where the torch signifies the divine presence.[145] Whatever text may stand in the background, "the torch of fire" is generally a positive motif. The torch, however, is said to consume the wicked, who are likened to grass. No question is left as to the function of the fire, which consumes and destroys wickedness and sin, which are attributed to "the troops of Belial." While the fire may be intended as a metaphor for the weapons of war, given the eschatological nature of the *War Scroll*, we cannot rule out the expectation of literal fire accompanying the more mundane weaponry.

The only other explicit reference to fire in the *War Scroll* is in the second half of a fragmentary line, concluding, כאש עברתו באלילי מצרים "like the fire of his wrath against the idols of Egypt" (XIV.1). Here fire functions

---

139. Charlesworth, *Dead Sea Scrolls*, 2:83.
140. Ibid., 2:84.
141. See Davies, *IQM*, 59–60.
142. García Martínez and Tigchelaar, *Dead Sea Scrolls*, 1:131.
143. On the title "the poor" see Davies, *IQM*, 98–99.
144. See Yadin, *Scroll of The War*, 312.
145. See discussion in chapter 2 above.

as a metaphor for something, but the fragmentary nature of the line precludes us from identifying exactly what it is. Nonetheless, the connotation is clearly negative, for the fire is a metaphor of wrath.[146]

Although it does not mention fire explicitly, the references to "burning," "the crucible," and Nadab and Abihu in 1QM XVI.15–XVII.3; 8–9 require our attention:

> Col. XVI 15 And starting to speak he will say: <<[Go]d [has risen, and] the h[ea]rt of his people he has tested in the crucible [יבחן במצרף], and not [. . .] your slain, for from ancient times you heard 16 the mysteries of God. [You then, be strong, stand in the breach, and do not fear . . .] their [. . .] . . . 17 . . .

> Col. XVII. 1 and he has placed their success in the burning [ושם שלומם בדלק] [. . .] those tested in the crucible [בחוני מצרף]; he has whetted her weapons of war and they shall not be blunted until [every] wicked [people is destroyed.] 2 And you, remember the trial [of Nadab and] Ab[i]hu, sons of Aaron; by judging them God showed his holiness to the eyes [of all the people. And Eleazar] 3 and Itmar he confirmed for the covenant of an everlasting [priesthood].

> 8 [. . .] And you, sons of his covenant, 9 be strong in God's crucible [במצרף אל] until he shakes his hand and finishes his testings [מצרפיו], his mysteries concerning your existence.[147]

This passage constitutes a speech by the Chief Priest, which spans from XVI.15b–XVII.9, in which he exhorts his hearers to be fearless in the face of the impending battle.[148] Holding this passage together are three references to מצרף "the crucible," the first and last of which function as an inclusio—the first in XVI.15b, and the last in XVII.9. As we have already seen repeatedly, "the crucible" is frequently used as a metaphor for testing, and it is so used here, specifically with reference to testing "the heart" (XVI.15).[149] According to Davies, it functions "as an explanation for the casualties suffered."[150] In light of this inclusio, we may conclude that the whole passage has in mind a period of testing.

---

146. On the idols of Egypt see Isa 19:1; Ezek 30:13. See Yadin, *Scroll of The War*, 325.

147. García Martínez and Tigchelaar, *Dead Sea Scrolls*, 1:139, 141.

148. Davies, *IQM*, 79–80.

149. See Prov. 17:3: "The crucible is for silver, and the furnace is for gold, but the LORD tests the heart."

150. Davies, *IQM*, 80.

Within this inclusio, a few lines stand out as worthy of a close examination. First, the opening line, ושם שלומם בדלק "he has placed their success in the burning" is quite remarkable, for it may attribute a positive function to the fire of testing.[151] Indeed, the word here translated "success" is related to the word שָׁלוֹם, which means "peace" or even "salvation." Unfortunately, the line is broken following this phrase, and it is impossible to ascertain its full context. As it stands, it is open to at least two possible readings: it is possible that the successful outcome is contingent upon the burning process; that is, the success is *because of* the burning. Alternatively, the text envisions at the very least the possibility of a positive outcome *despite* the fiery test. Second, we have an allusion to the infamous story of Nadab and Abihu, which we discussed in chapter 2.[152] It is probable that Yigael Yadin is correct in suggesting that the story of Nadab and Abihu is recollected in order "to prove that God deals justice even to the most select."[153] They function as a foil, warning those being tested in the crucible of suffering that not all of God's chosen will endure the period of testing. On the other hand, the reference to Eleazar and Itmar holds out the promise of a remnant and a renewed covenant between God and his people (XVII.2–3; see Num 3:4). Some will fare better than others in the crucible of suffering, and the passage functions to exhort the suffering to continue in their endurance so that they are found worthy. Lastly, despite the reference to the literal fires that consumed Nadab and Abihu, the crucible motif probably continues to function as a metaphor for earthly sufferings.

As evidenced by the numerous texts considered above, fire played a preeminent role in the eschatological conceptions of those responsible for the composition of the Dead Sea Scrolls, thus confirming Hippolytus's assertion that the Essenes believed in a conflagration of the universe and a judgment by fire. As we have seen, the beliefs of those whose literature was found at Qumran were quite varied. Several texts testify to the widespread belief that the wicked would face a post-mortem punitive judgment by fire, resulting in their annihilation. A number of others indicate belief in a flood of fire upon the earth that would destroy the impious, but by all appearances leave the righteous unharmed. Notably, a significant percentage of

---

151. Charlesworth (*Dead Sea Scrolls*, 133) has "He will set peace for them in the burning." Vermes (*Complete Dead Sea Scrolls*, 181) suggests the alternative "He will pay their reward with burning" with the implication that it is the wicked who are burned at the hand of those tested in the crucible. It is more probable, however, that the burning should be identified with the testing in the crucible itself. The use of שָׁלוֹם connote judgment would be exceptional.

152. See pages 43–48 above.

153. Yadin, *Scroll of the War*, 339.

these texts combine fire and water into a single instrument of divine judgment. And a handful of passages employ the metaphor of being tested in the crucible in reference to the sufferings of the righteous elect in this world.

The Septuagint

The Septuagint (LXX), the Greek translation of the Hebrew Bible, was composed over the course of over two centuries and by translators in multiple geographical regions. According to the *Letter of Aristeas*, the translation of the Pentateuch was carried out by seventy-two Jewish translators, six from each of the twelve tribes of Israel, who were sent to the library of Alexandria around 250 BCE. Although several elements of this tradition are certainly legendary,[154] the general region and time period of composition are probably accurate. While it is impossible to trace its precise history, and although the work was carried out by numerous translators in disparate regions, it is reasonable to surmise that the rest of the Hebrew Bible was translated into Greek by the end of the first century BCE.[155] As is any translation, the Septuagint is an interpretive work that reflects the interests and concerns of its authors.[156] These particular authors are themselves witnesses to the diversity of Second Temple Judaism.[157] It is fitting, therefore, that in this chapter we should call attention to the motif of judgment by fire in the Septuagint as it bears on our discussion of the fiery ordeal in New Testament eschatology. However, as we have already discussed this theme in our chapter on the Hebrew Bible, we shall only treat those passages where the Greek texts differ significantly from the Hebrew and where those differences contribute to our understanding of the motif of judgment by fire in Second Temple Judaism. This means we will not consider every text discussed in our section on judgment by fire in the Hebrew Bible. It also means, however, that there may be some passages included here that were not considered above, for in some instances judgment by fire appears in the Greek text where it was absent from the Hebrew.

---

154. Note especially the later tradition that each of the translators independently arrived at identical translations (see *b. Meg.* 9a–9b).

155. On the history of the translation and its complexity, see Jobes and Silva, *Invitation to the Septuagint*, 29–37.

156. For a discussion of the translators of the LXX as "storytellers" in their own right, see Beck, *Translators as Storytellers*, esp. 1–11.

157. As Nickelsburg (*Jewish Literature*, 192) notes, "the collection constitutes the largest corpus of Jewish writings from the Greco-Roman period."

First, let us consider how the Greek translation of Isa 1:25b differs from the Hebrew of that verse:[158]

MT: וְאֶצְרֹף כַּבֹּר סִיגָיִךְ וְאָסִירָה כָּל־בְּדִילָיִךְ

"I will smelt away your dross as with lye and remove all your alloy."

LXX: πυρώσω σε εἰς καθαρόν τοὺς δὲ ἀπειθοῦντας ἀπολέσω καὶ ἀφελῶ πάντας ἀνόμους ἀπὸ σοῦ καὶ πάντας ὑπερηφάνους ταπεινώσω.

"I will burn you to bring about purity. But the disobedient I will destroy, and I will remove from you all the lawless and humble all who are arrogant."[159]

There are several significant interpretive moves made by the Greek translator. First we note the LXX's use of the word πυρώσω where the MT has וְאֶצְרֹף. While the use of fire is implicit in the verb צָרַף "to refine, test" the verb πυρόομαι "to set on fire, to burn," with its obvious relationship to the Greek noun πῦρ "fire," gives fire a more central and explicit role in the process. Indeed, in the MT the focus is on the function of בֹּר "lye" in the process of smelting, not on the fire. Second the LXX states explicitly that the end result is purification. While this may be implicit in the MT, which alludes to the removal of alloy resulting in a pure metal, the Greek translator's choice of the adjective καθαρός "clean, pure" makes the final goal of purification explicit. Lastly, according to the LXX the burning plays a dual function: while some will be purified by the burning, others will be destroyed. This final clause regarding the destruction of the disobedient, the removal of the lawless, and the humbling of the arrogant is entirely absent from the MT. Thus we encounter in the LXX of Isaiah a greater emphasis on the dual function of fire; it is both purificatory and destructive.

Another text where we encounter a subtle though significant change in the LXX is Jer 6:27–30. In our discussion of the purificatory function of fire in the Hebrew Bible, we noted that this is one of two texts where the expectation of the purification of Israel is thwarted.[160] Here I present the English translation of the LXX:

---

158. For our discussion of the Hebrew text see pages 62–63 above.

159. Because it is the translation best suited for comparison with the NRSV, all translations of the LXX, unless otherwise noted, are from Pietersma and Wright, *NETS*, here 826.

160. See page 63–64 above. The other text is Ezek 24:12.

I have given you as a tester among tested peoples, and you will know me when I test their way. All are noncompliant, going about with slanders; bronze and iron, all of them have been corrupted. Bellows failed from a fire; lead failed; in vain does a silversmith coin silver; their wickedness did not melt. Call them "rejected silver," because the Lord has rejected them.[161]

The only noteworthy difference between the LXX and the MT is in the opening verse, where we encounter a minor change that carries larger implications:

MT: בָּחוֹן נְתַתִּיךָ בְעַמִּי מִבְצָר וְתֵדַע וּבָחַנְתָּ אֶת־דַּרְכָּם

I have made you a tester and a refiner *among my people* so that you may know and test their ways.[162]

LXX: δοκιμαστὴν δέδωκά σε ἐν λαοῖς δεδοκιμασμένοις καὶ γνώσῃ με ἐν τῷ δοκιμάσαι με τὴν ὁδὸν αὐτῶν

I have given you as a tester *among tested peoples*, and you will know me when I test their way.[163]

Significantly, whereas the MT states that Jeremiah is to be a tester בְעַמִּי "among my people," the LXX has ἐν λαοῖς δεδοκιμασμένοις "among tested peoples." The intentional removal of the possessive pronoun "my" indicates that the translator may not wish the party in question to be identified with Israel. Supporting this possibility, the plural λαοῖς "peoples" has prompted some English translators to render this word as "nations."[164] If this translation is correct, it eases the tension created by the failure of the refining process. For the LXX translator, it is apparently of less concern that the gentile nations are not purified by the refining fire of the LORD and are deemed "rejected silver" than if this were true of God's own people.

Similarly, the failure of the purification process in the allegory of the pot in Ezek 24:3–13 is somewhat mitigated in the LXX. Both the MT and the LXX liken Jerusalem to a filthy pot that the LORD attempts to cleanse with fire. Yet at the close of the allegory in vv. 12–13 the LXX parts ways with the MT, which is decidedly more pessimistic:

---

161. Pietersma and Wright, *NETS*, 888.
162. Italics added.
163. Pietersma and Wright, *NETS*, 888, italics added.
164. See Brenton, *Septuagint*, 911.

MT: תְּאֻנִים הֶלְאָת וְלֹא־תֵצֵא מִמֶּנָּה רַבַּת חֶלְאָתָהּ בְּאֵשׁ חֶלְאָתָהּ
בְּטֻמְאָתֵךְ זִמָּה יַעַן טִהַרְתִּיךְ וְלֹא טָהַרְתְּ מִטֻּמְאָתֵךְ לֹא תִטְהֲרִי־עוֹד עַד־
הֲנִיחִי אֶת־חֲמָתִי בָּךְ

In vain I have wearied myself; its thick rust does not depart. To the fire with its rust! Yet, when I cleansed you in your filthy lewdness, you did not become clean from your filth; you shall not again be cleansed until I have satisfied my fury upon you.

LXX: καὶ οὐ μὴ ἐξέλθῃ ἐξ αὐτῆς πολὺς ὁ ἰὸς αὐτῆς καταισχυνθήσεται ὁ ἰὸς αὐτῆς ἀνθ᾽ ὧν ἐμιαίνου σύ καὶ τί ἐὰν μὴ καθαρισθῇς ἔτι ἕως οὗ ἐμπλήσω τὸν θυμόν μου.

and much of her rust may not come out of her, and her rust shall be completely shamed, because you were defiling yourself. And what if you are no longer cleansed until I will sate my fury?[165]

Whereas the Hebrew text forthrightly claims that the process of purification is done תְּאֻנִים "in vain," the Greek translator has sought to reduce the pessimism by removing this statement. Further, the LXX employs the subjunctive mood to indicate the provisional nature of two negative pronouncements.[166] When the Greek text states that the rust οὐ μὴ ἐξέλθῃ ἐξ αὐτῆς "may not come out of her," the subjunctive of ἐξέρχομαι suggests that this outcome is not a foregone conclusion.[167] There may still be hope of purification. Further, in v. 13 the translator shifts the final clause from the indicative לֹא תִטְהֲרִי "you shall not be cleansed" to the conditional τί ἐὰν μὴ καθαρισθῇς "what if you are no longer cleansed," here employing the subjunctive of καθαρίζω. While the LXX certainly retains some of the pessimism of the MT, the translator has taken clear steps to reduce the tension created by the failure of the fire to cleanse the pot. Thus we see that the translators of both LXX Jer 6:27–30 and LXX Ezek 24:12–13, each in their own way, express some discomfort with the conclusion of the refining process in the Hebrew of those texts, suggesting that they were more at home with the expectation that the testing fire would succeed in the process of purification.

Two additional texts not discussed above in our section on the Hebrew Bible demand our attention here, for they employ the word πύρωσις, which

---

165. Pietersma and Wright, *NETS*, 965.

166. On the many uses of the subjunctive mood see Wallace, *Greek Grammar*, 461–80.

167. Admittedly, however, the dominant sense of the subjunctive mood is to indicate that the outcome is "uncertain but probable." See Ibid., 461.

will be of importance in our discussion of 1 Peter. I first wish to draw attention to the Septuagint's handling of Amos 4:9a and 10b:

MT: הִכֵּיתִי אֶתְכֶם בַּשִּׁדָּפוֹן וּבַיֵּרָקוֹן . . . וָאַעֲלֶה בְאֹשׁ מַחֲנֵיכֶם וּבְאַפְּכֶם

I struck you down with blight and mildew . . . and I made the stench of your camp go up in your nostrils.

LXX: ἐπάταξα ὑμᾶς ἐν πυρώσει καὶ ἐν ἰκτέρῳ . . . καὶ ἀνήγαγον ἐν πυρὶ τὰς παρεμβολὰς ὑμῶν ἐν τῇ ὀργῇ μου.

I struck you with fever [lit. "burning"] and jaundice . . . and I brought up your camps with fire in your wrath.[168]

The alterations here are significant. First, the Greek translator, has translated בַּשִּׁדָּפוֹן "with blight" as ἐν πυρώσει "with burning." While this may refer to "human diseases,"[169] πύρωσις is a curious word choice, for elsewhere in the Bible it refers to "a burning ordeal" (see Prov 27:21; 1 Pet 4:12).[170] In determining the Greek translator's meaning, therefore, we should take our cue from the other references to fire in the immediate context. In LXX Amos 4:2 the translator introduces a reference to ἔμπυροι λοιμοί "fiery destroyers" who "shall cast those with you into boiling cauldrons." This clause is entirely absent from the MT.[171] Glenny argues that these "fiery destroyers" are "agents of judgment and destruction, who are possibly angelic, and are sent by God to judge oppressors; thus the judgment here is potentially eschatological (final judgment?), but it could also be temporal."[172] Further, we encounter another intrusion of fire into the Greek text through an apparent misreading of v. 10b. Working with the unpointed Hebrew, the translator seems to have mistaken the word בְאֹשׁ "stench" for the preposition בְּ and the noun אֵשׁ "fire," thus rendering it as ἐν πυρί. This error may have influenced the translator's earlier choice of the word πυρώσει. Further, the initial confusion may itself have been precipitated by the following verse which refers to Sodom and Gomorrah and makes explicit reference to fire: "you were like a firebrand snatched from the fire" (Amos 4:11b). Clearly the author of LXX Amos 4 had a keen interest in judgment by fire, and since the word πύρωσις here appears in the context of references to fire as a means of divine judgment, it too probably adverts not simply to fevers or "human diseases" but

---

168. Pietersma and Wright, *NETS*, 792.
169. So Kellerman, "יָרָק," 366.
170. BDAG s.v. πύρωσις 2.
171. The MT has וְאַחֲרִיתְכֶן בְּסִירוֹת דּוּגָה "even the last of you with fishhooks."
172. Glenny, *Finding Meaning in the Text*, 210. On angels as dispensers of divine fiery judgment see pages 28, 38, 40-43, 67, 92-93, 105 above.

to some sort of fiery trial. As the text concludes, however, it is a fiery trial from which God delivers.¹⁷³

Whereas LXX Amos employs πύρωσις in a prophetic, perhaps even eschatological, context, LXX Prov 27:21 does so in the context of wisdom literature. In both the MT and the LXX the proverb is non-eschatological:

MT: מַצְרֵף לַכֶּסֶף וְכוּר לַזָּהָב וְאִישׁ לְפִי מַהֲלָלוֹ

> The crucible is for silver, and the furnace is for gold, so a person is tested by being praised.

LXX: δοκίμιον ἀργύρῳ καὶ χρυσῷ πύρωσις ἀνὴρ δὲ δοκιμάζεται διὰ στόματος ἐγκωμιαζόντων αὐτόν.

> Burning is a test for silver and gold, but a man is tested by the mouth of them who praise him.¹⁷⁴

The point is simply this: a person's response to praise from others will reveal his or her true character, just as the smelting process reveals the true quality of precious metals.¹⁷⁵ What is significant for our purposes, though, is that the LXX translator uses the word πύρωσις to translate the Hebrew word מַצְרֵף, which we have already encountered in our discussion of the Dead Sea Scrolls.¹⁷⁶ There it took on an explicitly eschatological hue, so while the context in Proverbs is clearly not eschatological, the association of πύρωσις with מַצְרֵף prepares us for its use in eschatological contexts. This association may also support our suggestion that Amos employs the term πύρωσις to refer to a judgment by fire.

Lastly, we must consider Theodotian's translation of Dan 11:35. Theodotian was a second-century CE convert to Judaism whose translation of the book Daniel eventually replaced the "Old Greek" of that text. Yet some evidence suggests he may have been working with older Greek translations. For instance, as Jobes and Silva indicate, while in most cases the author of Hebrews appears to be dependent on the "Old Greek," Heb 11:33 cites a rendering identical to Theodotian's translation of Dan 6:23.¹⁷⁷ Thus, while Theodotian's own translation is slightly later than the composition of the majority of the books of the New Testament, it presents us with some

---

173. We shall return to the LXX of Amos 4:11b (ὡς δαλὸς ἐξεσπασμένος ἐκ πυρός) in our discussion of 1 Cor 3:15 on pages 201 and 209 below.
174. Pietersma and Wright, *NETS*, 645.
175. Perdue, *Proverbs*, 228.
176. On which, see especially Sander, "ΠΥΡΩΣΙΣ."
177. Jobes and Silva, *Invitation to the Septuagint*, 41–42.

traditions that appear to have antedated those writings. Here is the Hebrew compared with Theodotian's translation (TH) of Dan 11:35:

MT: וּמִן־הַמַּשְׂכִּילִים יִכָּשְׁלוּ לִצְרוֹף בָּהֶם וּלְבָרֵר וְלַלְבֵּן עַד־עֵת קֵץ כִּי־עוֹד לַמּוֹעֵד

> Some of the wise shall fall, so that they may be refined, purified, and cleansed, until the time of the end, for there is still an interval until the time appointed.

TH: καὶ ἀπὸ τῶν συνιέντων ἀσθενήσουσιν τοῦ πυρῶσαι αὐτοὺς καὶ τοῦ ἐκλέξασθαι καὶ τοῦ ἀποκαλυφθῆναι ἕως καιροῦ πέρας ὅτι ἔτι εἰς καιρόν.

> And some of them that understand shall fall, to try them as with fire, and to test them, and that they may be manifested at the time of the end, for the matter is yet for a set time.[178]

This passage speaks of the tribulations that the wise will endure during the interval before the end of time, just prior to the great tribulation (Dan 12:1). Our chief concern here is not to offer an exegesis of this section of Daniel, but simply to observe the Greek translator's introduction of fire into the text. Significantly, where the MT states that the wise לִצְרוֹף . . . וּלְבָרֵר וְלַלְבֵּן "may be refined, purified, and cleansed," the text preserved by Theodotian makes explicit that this period is meant πυρῶσαι αὐτοὺς "to try them as with fire."[179] Notably, it was no great leap for the translator to attribute to fire the qualities of refining, purifying, and cleansing, even with reference to human beings. In Theodotian's text it is a fiery trial that will test and refine the righteous in the last days, here in the days leading up to the great tribulation. As a result, they will "be manifested [ἀποκαλυφθῆναι] at the time of the end."[180] The presence of fire in this text fits the general trend we have been observing in the translational strategy of the LXX, where fire appears to gain a more prominent role in scenes of eschatological purification. Indeed, in the preceding pages we have observed two interrelated tendencies: where the MT presents a scene of purification and refining through the metaphor of smelting or during a period of testing, the LXX makes explicit that fire plays some prominent role in this process; and where the purification process fails in the MT, the LXX takes observable steps to mitigate the tension created by this failure. The texts surveyed above may therefore suggest that

---

178. Brenton, *Septuagint*, 1068–9.
179. The "Old Greek" has καθαρίσαι ἑαυτοὺς here.
180. The fire that tests and manifests in the days leading to the end of time presents intriguing parallels to the similar functions of fire in 1 Cor 3:10–15.

the fiery ordeal, and especially the purificatory function of fire, takes on a greater significance in the Second Temple Period.

The *Sibylline Oracles*

The *Sibylline Oracles*, which is composed of a prologue followed by twelve books (numbered 1–8 and 11–14), is, like *1 Enoch*, a composite work written over a period of several centuries. Its composition history, however, is much vaster than that of *1 Enoch*, spanning from the second century BCE to the seventh century CE.[181] While in its final redaction the *Sibylline Oracles* is a Christian work, it incorporates earlier Jewish texts that bear no traces of Christian influence and thus may be of early origin. In form the work is modeled upon the Greek oracles uttered by aged female prophets known as sibyls, who, like the Jewish prophets, often foretold the doom of cities and empires, though sometimes offered hope.[182] It is possible that the Oracles' Greek literary form was employed with a gentile readership in mind. However, despite their genre and potential gentile audience, the oracles draw deeply from the well of Jewish eschatological thought, and it is with these traces of early, Jewish apocalyptic thought that we are concerned. In what follows, we shall consider the earliest strands of the Jewish material with hopes of identifying pre-Christian beliefs regarding the eschatological functions of fire in these apocalyptic texts.

The Third Book

It is widely agreed that the earliest oracle preserved in the final redaction of the *Sibylline Oracles* can be found in the third book. While even this earliest layer is not devoid of Christian redaction (see v. 776), most of the text can be dated to the period ranging from the middle of the second century to the end of the first century BCE and is of Egyptian provenance.[183] The Oracles are often dated on the basis of their references to several very specific political events. For instance, *Sib. Or.* 3.77 alludes to the reign of Cleopatra, but anticipates that the end of time will occur during her administration, suggesting that at least this section was written after her ascension but prior to her demise. The

---

181. On the date and provenance of each individual book of the *Sibylline Oracles*, see Collins's introductions before each book section in Charlesworth, *OTP*, 1:317–427.

182. For introductory matters on the genre, content, and purpose of the *Sibylline Oracles*, see Collins, *Sibylline Oracles*, 1–19; VanderKam, *Introduction to Early Judaism*, 107–10; Nickelsburg, *Jewish Literature*, 193–96.

183. See Collins in Charlesworth, *OTP*, 1:360.

latest traditions may thus be dated to c. 30 BCE.[184] The main corpus, which comprises the earliest stratum, includes verses 97–349 and 489–829, and so we shall treat the relevant passages from that section first.

In a passage that speaks of the period of the Babylon exile, the sibyl prophesies, "the heavenly God will send a king and will judge each man in blood and the gleam of fire [κρινεῖ δ' ἄνδρα ἕκαστον ἐν αἵματι καὶ πυρὸς αὐγῇ]" (286–87).[185] Most discussion of this passage centers on the identity of the king who is sent by God, some identifying him as an eschatological king and others, noting that the following lines speak of the kings of the Persians who bring aid to the Jews to rebuild the temple, identifying him as Cyrus.[186] The context of exile and restoration prophesied in the oracle has led most to favor the latter interpretation. What is more difficult is the question of who ἄνδρα ἕκαστον "each man" is who will be judged in ἐν αἵματι καὶ πυρὸς αὐγῇ "blood and the gleam of fire." If the identification of the king with Cyrus is correct, it may be a reference to the Babylonian armies that he vanquished. The text may also allow that the outcome of this judgment will be positive for the Jews, who are identified as "a certain royal tribe whose race will never stumble," and that the judgment by blood and fire finds them righteous, though here we enter into the realm of conjecture.

Oracles of destruction, which are uttered on specific kingdoms, sometimes invoke a judgment of fire. The woe against Crete, for instance, warns, "To you will come affliction and fearful, eternal destruction. The whole earth will again see you smoking and fire will not leave you forever, but you will burn" (504–7).[187] Here, the oracle, which may draw upon biblical pronouncements of destruction on Crete (see Ezek 25:16; 30:5; Zeph 2:5–6), employs the imagery of fire to denote complete and eternal destruction.[188] Elsewhere, however, the fire appears to play a more selective role. In a later section which describes a period of judgment that affects the "whole earth" and "all mortals," the sibyl proclaims, "the one who created heaven and earth will set down much lamented fire on the earth. One third of all mankind will survive" (542–44).[189] There is some dispute as to whether the passage is meant to refer literally to all of humankind or whether it applies

---

184. Nickelsburg, *Jewish Literature*, 194.
185. Trans., Collins in Charlesworth, *OTP*, 1:368.
186. See Collins, *Sibylline Oracles*; Buitenwerf, *Book III of the Sibylline Oracles*, 23.
187. Trans., Collins in Charlesworth, *OTP*, 1:373.
188. Buitenwerf, *Book III of the Sibylline Oracles*, 248.
189. One is reminded of the similar notions in Ezek 5:2, where Ezekiel's sign act suggests the destruction of Israel in measurements of one third, and in Rev 8:9–10; 9:13, where one third of creation, including humankind, is destroyed.

universalizing apocalyptic language to the Greeks.[190] In either case, what is striking is that one third will survive, indicating that the fire, whether it is directed solely towards the Greeks or more broadly towards all of humankind, plays a selective role, destroying some while preserving others. The evidence is too scanty, however, to determine whether this selection is arbitrary or intentional.[191]

Later, when the third oracle takes up the topic of the judgment of those who would destroy the Jewish temple, fire plays a central role, as one might expect. The judgment is described as follows: "And God will speak, with a great voice, to the entire ignorant empty-minded people, and judgment will come upon them from the great God, and all [πάντα] will perish at the hand of the Immortal. Fiery swords [ῥομφαῖαι πύρινοι] will fall from heaven on the earth. Torches [λαμπάδες], great gleams, will come shining into the midst of men" (669-74).[192] According to the analysis of Buitenwerf, the ῥομφαῖαι πύρινοι and λαμπάδες are probably meant as harbingers of the impending divine judgment, for he notes, "in lines 798-799, the appearance of swords in the heavens at night is mentioned among the signs preceding the end. Falling swords are also mentioned as portents elsewhere" (see Isa 34:5; Josephus, *J.W.* 6.288-289).[193] Further on the sibyl states, "God will judge all men by war and sword and fire and torrential rain [πυρὶ καὶ ὑετῷ τε κατακλύζοντι]. There will also be brimstone [θεῖον] from heaven, and stone and much grievous hail" (689-92).[194] All of this results in the death and destruction of many. For the elect, however, there will be protection: "But the sons of the great God will all live peacefully around the Temple, rejoicing in these things which the Creator, just judge and sole ruler, will give. For he alone will shield them, standing by them magnificently as if he had a wall of blazing fire around about [ὡσεὶ τεῖχος ἔχων πυρὸς αἰθομένοιο]" (702-6). Drawing on an image from Zech 2:5 in which the Lord promises to be "a wall of fire all around [Jerusalem]" (LXX Zech 2:9: τεῖχος πυρὸς κυκλόθεν), the sibyl similarly promises the Jews divine protection from the destructive forces that come to bear upon their enemies. It is quite interesting that the

---

190. Despite the apparently universalizing language, Buitenwerf (*Book III of the Sibylline Oracles*, 254) argues that the focus does not in fact shift from Greece here, but "that the author concludes his warnings to Greece by applying universal, apocalyptic language.... The passage should be interpreted, therefore, as part of the description of the ruin of Greece."

191. See, by contrast, the intentional preservation of the righteous in verses 702-6 below.

192. Trans., Collins in Charlesworth, *OTP*, 1:377.

193. Buitenwerf, *Book III of the Sibylline Oracles*, 278.

194. Trans., Collins in Charlesworth, *OTP*, 1:377.

element of fire which serves to protect the Jews is one of the elements that functions to destroy their wicked oppressors.

The remaining texts we have to consider from Book 3 are found in verses 1–96, which as a unit can be dated to the period of Cleopatra shortly after the Battle of Actium on 2 September 31 BCE.[195] The first passage we shall consider is found in verses 46–62.

> But when Rome will also rule over Egypt
> Guiding it toward a single goal, then indeed the most great kingdom
> of the immortal king will become manifest over men.
> For a holy prince will come to gain sway over the scepters of the earth
> forever, as time presses on.
> Then also implacable wrath will fall upon Latin men [Λατίνων ... ἀνδρῶν].
> Three will destroy Rome with piteous fate.
> All men [πάντες δ' ἄνθρωποι] will perish in their own dwellings
> when the fiery cataract flows from heaven [οὐρανόθεν πύρινος ῥεύσῃ καταράκτης].
> Alas, wretched one, when will that day come,
> and the judgment of the great king immortal God?
> Yet, just for the present, be founded, cities, and all
> be embellished with temples and stadia, markets and golden
> silver and stone statues so that you may come to the bitter day [πικρὸν ἦμαρ].
> For it will come, when the smell of brimstone [θείου] spreads
> among all men [πᾶσιν ἐν ἀνθρώποισιν]. But I will tell all in turn,
> in how many cities mortals will endure evil.[196]

Here we have a prophecy of the awaited Day of Judgment on which a πύρινος ... καταράκτης "fiery cataract" will proceed οὐρανόθεν "from heaven." Some confusion persists regarding precisely who will face the judgment by fire. While line 51 indicates that the wrath will fall upon Λατίνων ... ἀνδρῶν "Latin men," which may lead one to suspect that Rome is the sole object of the fiery judgment, just two lines later it is stated in no uncertain terms that the destructive cataract of fire will consume πᾶσιν ἐν ἀνθρώποισιν "all men." And again in lines 60–61 there is the prediction that the smell of

---

195. See Collins in Charlesworth, *OTP*, 1:360. There are no significant discrepancies between this and the translation of Terry, *Sibylline Oracles*, 57–58.
196. Trans., Collins in Charlesworth, *OTP*, 1:363.

θείου "brimstone" will spread πᾶσιν ἐν ἀνθρώποισιν "among all men." Yet in the closing lines of 61 and 62, where the sibyl promises that she is yet to pronounce "in how many cities mortals will endure evil," hints that not all cities will face the fiery destruction.

While the scope of the fire is unclear, it is evident that the fire plays a destructive role. It is a weapon of God's wrath that will cause "all men" to "perish in their own homes." Moreover, the day on which it will be unleashed is regarded as a πικρὸν ἦμαρ "bitter day." The smell of brimstone mentioned in line 60 may also recall the fire and brimstone (LXX: θείου) that rained down on Sodom in Gen 19:24. Given its wholly negative aspect, then, we may presume that the scope of the fire is limited, but clarity regarding this question is of no apparent concern to the sibyl. As Collins observes, "The difference then between the later and earlier stages of the sibyllina is a transfer of emphasis from the blessings of God's judgment to its destructive aspect."[197]

If confusion regarding the scope of the fire prevails in lines 47–62, it is quite the opposite in lines 80–92, which lie near the close of this section of Book 3. Here we encounter an oracle wherein fire functions unambiguously as an element in the complete destruction of the world.[198]

> ... then all the elements [στοιχεῖα] of the universe
> will be bereft, when God who dwells in the sky
> rolls up the heaven as a scroll is rolled,
> and the whole variegated vault of heaven falls
> on the wondrous earth and ocean. An undying cataract
> of raging fire will flow [ῥεύσει δὲ πυρὸς μαλεροῦ καταράκτης],
>   and burn earth, burn sea [φλέξει δὲ γαῖαν, φλέξει δὲ θάλασσαν],
> and melt the heavenly vault [πόλον οὐράνιον] and days and creation [κτίσιν] itself
> into one and separate them into clear air.[199]
> There will no longer be twinkling spheres of luminaries,
> no night, no dawn, no numerous days of care,
> no spring, no summer, no winter, no autumn.
> And then indeed the judgment of the great God
> will come into the midst of the great world, when all these
>   things happen.[200]

---

197. Collins, *Sibylline Oracles*, 100.

198. See van der Horst, *Hellenism—Judaism—Christianity*, 281–82.

199. Terry (*Sibylline Oracles*, 59) has "... and melt / Creation itself together and pick out / What is pure."

200. Trans., Collins in Charlesworth, *OTP*, 1:364. Compare *Sib. Or.* 2.196–213.

The emphasis is on the cosmic scope of the judgment, and what is described is nothing short of the "destruction of the entire cosmos."²⁰¹ The stream of fire that flows from heaven functions systematically to undo all of creation. Whereas in Gen 1 God creates the heavens (οὐρανόν; v. 1), the earth (γῆν; v. 1), the seas (θαλάσσας; v. 10) and all of creation over the course of six days, here we see the complete reversal of that act of creation when the heavenly vault (πόλον οὐράνιον), earth (γαῖαν), sea (θάλασσαν), and all of creation (κτίσιν) are melted in the cosmic fire. The imagery draws on biblical apocalyptic motifs of the stream of fire and the heavens being rolled up like a scroll, but the eschatological conflagration that consumes the entire world draws ultimately on the Stoic doctrine of *ekpyrosis*, according to which fire results in the return of all things to a primal substance.²⁰² Nothing remains, neither stars nor seasons nor days; all is turned into rarified air.

It is interesting to note that in contrast to some of the earlier oracles, this passage shows no interest in differentiating the lot of the righteous from that of the wicked. To be sure, the emphasis is on the melting of the elements, and none of humanity, neither the upright nor the unjust, is given any mention. We are left to presume that all of humankind faces the same fate as the cosmos, to be consumed by the fiery cataract and dissolved into nothing. If there is a future individual resurrection and judgment, it is not mentioned.²⁰³

The Fourth Book

The Fourth Book of the *Sibylline Oracles* has a long composition history, beginning with the original oracle, which was probably composed shortly after Alexander the Great's reign, and concluding with the final redaction sometime around the year 80 CE.²⁰⁴ A few lines make reference to fire as a place of punishment for those who are found unrighteous at the great judgment, when God will judge "impious and pious at once, then he will also send the impious down into the gloom in fire" (42–43).²⁰⁵ And elsewhere a description is given of the fiery destruction of specific cities: "When the flame of Aetna belches forth a stream of great fire it will burn all miserable

201. Collins, *Sibylline Oracles*, 100.
202. See ibid; van der Horst, *Hellenism—Judaism—Christianity*, 271–77. Note the reference to στοιχεῖα, a significant term in Stoic cosmology, in line 47. See our discussion on the use of this term in 2 Pet 3:10, 12 on pages 236–39 below.
203. See, however, *Sib. Or.* 4.159–92 below.
204. See Collins in Charlesworth, *OTP*, 1:381–82.
205. Ibid., 1:385.

Sicily, and the great city Croton will fall into the great stream" (80–82).[206] This passage is peculiar for the fact that the stream of fire originates in Aetna and not from the divine throne or, more generally, from heaven.

Certainly the most remarkable appearance of fire in the Fourth Book comes in a lengthy passage in lines 159–92, which describe the cosmic destruction of the world by fire:

> ... God is no longer benign
> but gnashing his teeth in wrath and destroying the entire
> race of men at once by a great conflagration [ἐμπρεσμοῦ μεγάλοιο].
> Ah, wretched mortals, change these things, and do not
> lead the great God to all sorts of anger, but abandon
> daggers and groanings, murders and outrages,
> and wash your whole bodies in perennial rivers [ἐν ποταμοῖς ... ἀενάοισιν].
> Stretch out your hands to heaven and ask forgiveness
> for your previous deeds and make propitiation
> for bitter impiety with words of praise;
> God will grant repentance and will not destroy.
> He will stop his wrath again if you all
> practice honorable piety in your hearts.
> But if you do not obey me, evil-minded ones, but love
> impiety, and receive all these things with evil ears,
> there will be fire throughout the whole world [κόσμον ὅλον], and a very great sign
> with sword and trumpet at the rising of the sun.
> The whole world [κόσμος ἅπας] will hear a bellowing noise and mighty sound
> [μύκημα καὶ ὄμβριμον ἦχον].
> He will burn the whole earth [χθόνα πᾶσαν], and will destroy the whole race of men
> [ὀλέσει γένεος ἀνδρῶν]
> and all cities and rivers at once, and the sea.
> He will destroy everything by fire, and it will be smoking dust
> [κόνις ... αἰθαλόεσσα].
> But when everything is already dusty ashes [τέφρη σποδόεσσα],
> and God puts to sleep the unspeakable fire [πῦρ ... ἄσπετον],
> even as he kindled it,

---

206. Ibid., 1:386.

> God himself will again fashion the bones and ashes [ὀστέα καὶ
> σποδινήν] of men
> and he will raise up [στήσει] mortals again as they were before
> [ὡς πάρος ἦσαν].
> And then there will be a judgment over which God himself will
> preside,
> judging the world again. As many as sinned by impiety,
> these will a mound of earth cover,
> and broad Tartarus and the repulsive recesses of Gehenna.
> But as many as are pious, they will live on earth again
> when God gives spirit and life and favor
> to these pious ones. Then they will all see themselves
> beholding the delightful and pleasant light of the sun.
> Oh most blessed, whatever man will live to that time.[207]

This passage is noteworthy for several reasons. First, many have seen a parallel to John the Baptist's call to a baptism of repentance in the sibyl's exhortation to abandon evil practices and to "wash your whole bodies in perennial rivers [ἐν ποταμοῖς . . . ἀενάοισιν]" and ask forgiveness in order to be spared the coming judgment by fire.[208] Indeed, the focus on ablutions is one of the primary reasons that scholars have argued in favor of a Syrian or Palestinian provenance for this text given the emphasis both John the Baptist and the Essenes placed on such practices.[209]

Secondly, fire plays a central role in the cosmic destruction in which God's πῦρ . . . ἄσπετον "unspeakable fire" will consume the χθόνα πᾶσαν "all the earth" and the ὀλέσει γένεος ἀνδρῶν "the whole race of men," the fate ascribed to those who fail to repent. As in 3.80–92, what is envisioned here is the ἐμπρεσμοῦ μεγάλοιο "great conflagration" that devours all of the elements, again clearly drawing on the Stoic notion of *ekpyrosis*. The cosmic scope of the fire is emphasized by the repetition of the phrases κόσμον ὅλον and κόσμος ἅπας "the whole world" along with the assertion that the burning of all of the elements leaves a world of nothing but κόνις . . . αἰθαλόεσσα "smoking dust" and τέφρη σποδόεσσα "dusty ashes." In this regard it is slightly different from 3. 80–92, where the cosmic conflagration turns all elements into one primal element, air.

---

207. Ibid., 1:388–89.

208. Webb, *John the Baptizer*, 120–21; Collins, *Cosmology and Eschatology*, 221–22; Taylor, *The Immerser*, 91.

209. See Thomas, *Le mouvement baptiste*, 47–50; Collins in Charlesworth, *OTP*, 1:382.

Third, and perhaps most substantial, this passage combines the Stoic doctrine of *ekpyrosis* with the Jewish eschatological belief in individual resurrection. From the ashen ruins of the burnt creation God στήσει "will raise" all of humanity from their ὀστέα καὶ σποδινήν "bones and ashes" and restore them to ὡς πάρος ἦσαν "as they were before" for the day of individual judgment. Whereas the fire of the cosmic conflagration showed no distinction between the wicked and the righteous, at the final judgment God will send the impious to Gehenna, where they may again encounter a fiery judgment and will reward the pious with joyful and agreeable lives on the renewed earth. The fiery destruction may not be discerning, but the God who passes judgment at the resurrection is. The combination of Jewish and Hellenistic eschatological expectations is remarkable. Indeed, as Collins observes, "[w]hile the idea here may be the traditional Jewish resurrection for judgment or its rewards, the manner in which it is expressed suggests a rapprochement with the myth of the eternal return."[210] In this text we see one of the most syncretistic understandings of the functions of fire in Jewish eschatology.

In both the third and fourth books of the *Sibylline Oracles*, we have seen that fire appears in several oracles prophesying the destruction of individual cities. Most intriguing in these texts, however, are those passages that adopt the Stoic notion of *ekpyrosis* and depict a cosmic conflagration that consumes all the elements in creation. Whereas the conflagration in the Third Book concludes with all turning into the single primal element of air with no further explanation, the Fourth Book envisions a post-conflagration resurrection and recreation of both the righteous and the wicked for the final judgment. It appears that in both of these texts the pious and impious alike are consumed in the initial conflagration. But in the latter text, after the last judgment, the upright are restored to a happy earthly life while the unjust are condemned to Gehenna, where they will presumably undergo further fiery punishments. It does not appear that the fires of the cosmic conflagration have any purifying effect on the righteous, but the righteous do receive their reward on the Day of Judgment.

Fourth Ezra

Chapters 3–14 of the Christian work *2 Esdras*, which is found in the Apocrypha or Deuterocanonical Literature, contain a Jewish apocalypse commonly referred to as *4 Ezra* or the *Ezra Apocalypse*. The final form of this text is well aware of the fall of the Jerusalem temple, and is in fact thought to be a theodicy attempting to make sense of the events of 70 CE. It has thus

---

210. Collins, *Sibylline Oracles*, 103.

been dated to sometime around the end of the first century CE.²¹¹ More precisely, based on specific references in the eagle vision, Michael Stone has suggested a date sometime near the end of the reign of Domitian (81–96 CE).²¹² Despite the fact that the completed form of this fascinating text postdates the composition of much of the New Testament, it may still contain some old traditions that have been collected and edited by a later redactor. Two passages in *4 Ezra* may, therefore, be of significance to our study of fire in the New Testament.

The first text is *4 Ezra* 7:6–11, which is the second of two analogies spoken to Ezra by an angel that are meant to illumine how one is to enter the world to come:

> There is a city built and set on a plain, and it is full of all good things; but the entrance to it is narrow and set in a precipitous place, so that there is fire on the right hand and deep water on the left; and there is only one path lying between them, that is, between the fire and the water [*ignis et aqua*], so that only one man can walk upon that path.²¹³ If now that city is given to a man for an inheritance, how will the heir receive his inheritance unless he passes through the danger set before him?²¹⁴

The motifs present here are quite widespread. Oesterley notes that the reference to the righteous passing through *ignis et aqua* "fire and water" may be an allusion to Ps 66:12b: "we went through fire and through water; yet you have brought us out to a spacious place."²¹⁵ Stone observes several other parallels ranging from Wisdom and Rabbinic literature to Iranian Eschatology.²¹⁶ The prevalence of the motif suggests that it antedates its present context in *4 Ezra* and that the ideas implicit in it may have been known to the authors of the New Testament.

The text is explicitly allegorical and despite the obvious eschatological concern with entering into the future life, the explanation of the allegory makes clear that neither the fire nor the water is literal. After Ezra replies that one cannot enter the city without passing through the narrow and treacherous path, the angel explains that because of Adam's transgression,

---

211. See Nickelsburg, *Jewish Literature*, 270.

212. Stone, *Fourth Ezra*, 10.

213. Box (*Ezra-Apocalypse*, 101) has "that it can contain only one man's footstep at once."

214. Trans. Metzger in Charlesworth, *OTP*, 1:537. I have also consulted the translation of Box, *Ezra-Apocalypse*.

215. Oesterley, *II Esdras*, 64.

216. See Stone, *Fourth Ezra*, 197.

"unless the living pass through the difficult and vain experiences, they can never receive those things that have been reserved for them" (v. 14). Whereas other texts we have previously observed have combined the images of fire and water, here they are distinct from one another. Nonetheless, as the explanation of the allegory makes clear, both are images of danger that threaten those who wish to walk on the narrow way. Few other details are given in the explanation of the allegory, and many questions are left unanswered, but presumably the righteous will succeed while the wicked will fail in traversing the path set before them.[217]

Also relevant to our discussion is the Sixth Vision in chapter 13, which contains a vision of the eschatological destruction of the wicked at the hands of God's Messiah. Although the final form of *4 Ezra* was probably completed in the Common Era, this particular vision may predate the destruction of the Jewish temple in 70 CE and thus fall within our purview.[218] In the vision, Ezra sees "something like the figure of a man"—to whom divine attributes such as pre-existence are assigned—rising out of the sea and then standing upon a mountain to face his enemies. Fire plays an exceptional role in the judgment of his opponents. Ezra the seer describes it as follows:

> And behold, when he saw the onrush of the approaching multitude, he neither lifted his hand nor held a spear or any weapon of war; but I saw only how he sent forth from his mouth as it were a stream of fire [*fluctum ignis*], and from his lips a flaming breath [*spiritum flammae*], and from his tongue he shot forth a storm of sparks [*scintillas tempestatis*]. All these were mingled together, the stream of fire and the flaming breath and the great storm, and fell on the onrushing multitude which was prepared to fight, and burned them all up, so that suddenly nothing was seen of the innumerable multitude but only the dust of ashes and the smell of smoke. When I saw it, I was amazed. (9–11)[219]

As in the *Book of Parables*, theophanic language is here applied to the messianic figure, for "whenever his voice issued from his mouth, all who heard his voice melted as wax melts when it feels the fire" (v. 4; see Mic 1:4). Very noteworthy are several links with Daniel, especially from chapter 7; both texts are dream reports made by seers; in both visions the winds stir up the sea (*4 Ezra* 13:2; Dan 7:1); Ezra's "something like the figure of

---

217. If so, it is reminiscent of the Chinavat Bridge in Zoroastrian eschatology. See Duchesne-Guillemin *Hymns of Zarathustra*, 145, 360; Stone, *Fourth Ezra*, 197.

218. Box, *Ezra-Apocalypse*, 286.

219. Trans., Metzger in Charlesworth, *OTP*, 1:551–52. There are no substantial differences in the translation of Box (*Ezra-Apocalypse*, 287–88).

a man" (v. 3) is reminiscent of Daniel's "one like a human being" (7:13); both figures ride on clouds (*4 Ezra* 13:3; Daniel 7); where the man in *4 Ezra* 13:6-7 carves out a mountain on which to stand, the source of which cannot be identified, Dan 2:34 and 45 speak of a stone cut not by human hands; and just as a fiery stream proceeds from the mouth of the messianic figure in *4 Ezra* 13:10-11 to consume his opponents, in Dan 7:10 a stream of fire issues from under the throne to destroy the enemies of God. Dependence of the *Ezra Apocalypse* on the book of Daniel is quite evident. The fire that issues from the mouth of the man may also recall Isa 11:4, which in describing the ideal Davidic king states, "he shall strike the earth with the rod of his mouth, and with the breath of his lips he shall kill the wicked" (see *1 En.* 102:1; Rev 19:15).[220]

The function of the fire in this passage is unquestionably punitive. It is employed by a messianic figure against his enemies, and it annihilates the multitude, leaving only ashes and dust. It is significant that it is within the power of a messianic figure or mediator figure to deliver the fiery punishment. Also striking is the image of the *fluctum ignis* "stream of fire," for like many other texts already discussed it combines fire and water imagery. A new element here, however, is the *spiritum flammae* "flaming breath," which some have found significant for interpreting John the Baptist's proclamation of the Coming One's baptism in πνεύματι (spirit or breath) and fire.[221] In all of the above respects it also provides a significant parallel to Christ's confrontation with his eschatological adversary in 2 Thess 2:8: "And then the lawless one will be revealed, and the Lord Jesus will slay him with the breath of his mouth [πνεύματι τοῦ στόματος] and destroy him by his appearing and his coming." Significantly, in the interpretation of the vision that is given by the angel in 13:25-55 the fire is interpreted not literally but as a metaphor for the law (v. 39). It is possible, however, that the allegorical interpretation, which is secondary, turns into a metaphor that which was intended literally in the original apocalyptic vision.

In these two passages fire plays decidedly different roles. In the former it, along with water, is a metaphor for the hardships that the righteous must endure before they may receive their reward in the world to come. The righteous may pass through it successfully while presumably the wicked will succumb to the flames or the waters. The fire is a dangerous obstacle, but not an insurmountable one. In the latter text, however, it functions to destroy the wicked, either as a literal fire, as perhaps in the original apocalyptic vision, or as a metaphor for the law, as in the allegorical interpretation. It is

---

220. See Stone, *Fourth Ezra*, 386.
221. Dunn, "Spirit-and-Fire Baptism."

solely and inescapably punitive, and it is directed only towards the wicked. Also significant in the latter text is the role of the messianic figure as the deliverer of God's fiery wrath.

Conclusion

It is not our intention here to summarize each of the texts surveyed above, but to draw together some of the strands from those discussions. The first observation to be made is that whenever fire is mentioned in reference to the post-mortem state of the wicked in the Sheol or Gehenna, it functions punitively and appears to result in their annihilation. When the eschatological judgment by fire appears on earth, it is typically directed against the wicked in the form of a localized fire, though in some texts it manifests itself in the form of a cosmic conflagration. In some cases the righteous are preserved through or spared the destructive flames of the cosmic fire while in others they appear to be consumed along with the wicked. Third, it is very common for these texts where the fire appears on earth to juxtapose fire and water and to combine them into a river or flood of fire drawing on the river of fire in Daniel 7 or drawing on the story of Noah's flood as typological of the last judgment while importing fire into it. Fourth, whenever the motif of the crucible is employed, the fire is understood as a metaphor for the trials and sufferings of the righteous, which function to test and purify them. Fifth, it is not uncommon for an angelic, messianic, or human intermediary to play the role of administering the fiery judgment. Sixth, although a less well attested motif, a handful of texts appear to predict a fiery destruction of the second Jewish temple or recall the destruction of the first temple. And seventh, as we noted especially in the LXX, there is a tendency to introduce fire into texts that speak of purification and at the same time to qualify those passages where purification by fire is done in vain. With this context for understanding the roles of fire in the eschatological thought world of Second Temple Judaism, we are now in a position to proceed to our consideration of the fiery ordeal in the New Testament.

CHAPTER 4

John the Baptist and the Baptism in Fire

Introduction

WE BEGIN OUR DISCUSSION of fire in the Gospel traditions where Mark begins his gospel—with John the Baptist. John the Baptist's message is dominated by the motif of fire. He preaches that "every tree that does not bear good fruit will be cut down and thrown into the fire" (Matt 3:10//Luke 3:9) and that the Coming One will separate the wheat from the chaff and "the chaff he will burn with unquenchable fire" (Matt 3:18//Luke 3:17). These passages assume that the fire plays a destructive function; it is the non-fruit-bearing trees and useless chaff that are "thrown into" the fire, which is described as "unquenchable." Clearly this is a reference to the fire of Gehenna, into which individuals are thrown and which is likewise described as unquenchable. However, far and away the most intriguing logion attributed to John the Baptist is his proclamation of the Coming One's baptism by fire and Holy Spirit (Matt 3:11//Luke 3:16; see Mark 1:8; John 1:33; Acts 1:5; 11:16). Indeed, when commenting upon this verse in his *Philosophical Dictionary*, the French Enlightenment writer Voltaire lamented, "[t]hese words, He will baptize with fire, have never been explained."[1] Whether or not Voltaire's judgment that the Baptist's words have remained a perpetual enigma is correct, there certainly exists a lack of clarity surrounding this text. This perplexity cannot, however, be ascribed to a dearth of exegetical effort, for many have attempted to elucidate the meaning of this saying.

---

1. Voltaire, *Philosophical Dictionary*, s.v. Baptism.

Matthew 3:11//Luke 3:16

While modern critical scholars overwhelmingly agree that some form of this saying can be traced back to the historical John the Baptist, three questions have dominated the discussion of his pronouncement, all of which have some bearing on our present investigation. First, disparate proposals have been offered in an attempt to discern the identity of John's expected Coming One. The most plausible candidates are as follows: the Coming One was God;[2] he was Elijah *redivivus*;[3] he was the Son of Man;[4] or he was the Messiah.[5] Second, the similar sayings in Mark 1:8; John 1:33; Acts 1:5; 11:18, which speak of a baptism with the Holy Spirit alone, raise questions about the earliest form of the saying. Did John originally speak of a baptism with both the Holy Spirit and fire?[6] Did he proclaim a baptism with the Holy Spirit alone?[7] Or, as some modern critics contend, did he anticipate a baptism with fire only?[8] Third, precisely who would receive the baptism in fire and what effect it would have on its recipients are both subjects of debate. Some believe that while the righteous would receive a gracious baptism with the Holy Spirit, the baptism with fire would be reserved as punishment for the wicked.[9] Likewise, those who regard the reference to baptism in the Holy Spirit to be secondary typically hold that the Coming One's baptism in fire would be solely destructive.[10] Others have maintained that the baptism with the Holy Spirit (or breath) and fire is a single baptism (hendyadis), which would be required of all and would serve the dual function of refining the repentant while simultaneously punishing the unrepentant.[11] Of primary importance for our study is this last question regarding the function of the baptism in fire. Tangentially related, however, are the first two, for the identity of the Coming One who administers the fiery baptism may

---

2. See Bretscher, "Whose Sandals?"; Hughes, "John the Baptist"; Reiser, *Jesus and Judgment*, 181–86.

3. Robinson, "Elijah, John and Jesus," 30. See also Brown, "John the Baptist," 132–40.

4. Gnilka, *Jesus of Nazareth*, 74–75; Becker, *Jesus of Nazareth*, 46–47.

5. Manson, *Sayings of Jesus*, 41; Bultmann, *History of the Synoptic Tradition*, 246.

6. Dunn, "Spirit-and-Fire Baptism"; Webb, *John the Baptizer*, 275. Others interpret πνεῦμα not as "spirit," but as "wind"; see, for instance, Barrett, *Holy Spirit*, 126; Best, "Spirit-Baptism."

7. See Ellis, *Gospel of Luke*, 89; Meier, *Marginal Jew*, 2:36.

8. Manson, *Sayings of Jesus*, 41; Bultmann, *History of the Synoptic Tradition*, 246; Becker, *Jesus of Nazareth*, 45; Collins, *Mark*, 146.

9. Scobie, *John the Baptist*, 71; Lang, "πῦρ," 943.

10. Reiser, *Jesus and Judgment*, 185; Becker, *Jesus of Nazareth*, 34.

11. Dunn, "Spirit-and-Fire Baptism," 86; Davies and Allison, *Matthew*, 1:316.

have some bearing on its function, and it is of obvious importance to know whether John prophesied a baptism in fire at all. We shall therefore briefly treat the first two questions before addressing the function of the Coming One's baptism in fire.

The Identity of the Coming One

When considering the main candidates for the Coming One, we can observe an obvious division between God, on the one hand, and the human or intermediary figures, on the other. Given our primary concern with the function of the baptism in fire, in the present study we must limit the question to whether John expected the Coming One to be God himself or some other intermediary figure, whether human, angelic, or otherwise. The primary arguments in favor of the Coming One being identified with God are, firstly, that it is typically God who comes with fire on the day of judgment (see Joel 2:30; Isa 66:15f.; Ezek 38:22; 39:6; Mal 3:19). Secondly, if the traditions that associate John the Baptist with Elijah are authentic (see Mark 9:11-13)—there is some debate and even tension between Gospel traditions (see John 1:21) over this—it is possible that John saw himself in Mal 4:5 as "a precursor of none save of God only in his coming to judge the world."[12] Indeed, by all accounts it appears that in Malachi's prophecy, the Lord whom Elijah precedes is God, for there is no mention made of a messianic figure.[13]

Regarding the first argument, God is not the only figure in the biblical and post-biblical tradition who administers fiery judgment. As J. A. T. Robinson observed, just as with John's Coming One, Elijah was closely associated with fire and the final judgment: "it is the *character* of the coming one which is the real indication that John may have seen him as Elijah *redivivus*. For he is before anything else to be a man of fire. And the man of fire *par excellence* was Elijah."[14] Moreover, as we have observed in our discussion of the literature of Second Temple Judaism, intermediary figures such as angels could serve as the deliverers of fiery judgment.[15] More significant, however, are the two most frequent objections to the thesis that

---

12. Loisy, *Birth of the Christian Religion*, 65.

13. Whether or not Elijah was expected to precede the coming of the Messiah is the subject of considerable debate. See Allison, "Elijah Must Come First"; Faierstein, "Why Do the Scribes Say Elijah Must Come First?"; Fitzmyer, "More about Elijah"; Öhler, "Expectation of Elijah."

14. Robinson, "Elijah, John and Jesus," 30. There are, of course, problems with the thesis that John expected Elijah, especially if one accepts the tradition that John styled himself as Elijah and Jesus associated John with the Tishbite.

15. See pages 28, 38, 40-43, 67, 92-93, 105 above.

John expected the Coming One to be God. First, it is commonly argued that the image of God wearing sandals would be too flagrantly anthropomorphic for any Jew of the Baptist's era. There may be some truth to this claim; however, Hughes responds to the criticism that this image of God wearing sandals is too anthropomorphic by citing as evidence to the contrary Pss 60:8 and 108:9, both of which state, "Moab is my washbasin; on Edom I hurl my shoe." Hughes argues, "[j]ust as God's shoes are cited metaphorically in these psalms in order to describe the appropriation of Edom by Yahweh, so also John's reference to the Coming One wearing sandals must not be pressed beyond the realm of metaphor where it belongs."[16] As there is little else in the Hebrew Bible that Hughes can draw upon to make his argument, however, he is far from making a decisive case that this sort of anthropomorphic language was common. He does demonstrate, nevertheless, that it was not unheard of and at least opens up the possibility that John could have spoken in such a manner.

The second criticism of the position that the Coming One is God is weightier. John is reported to have said that the Coming One is "mightier" than he. It is highly improbable that the Baptist would have used this sort comparative language to describe his relationship to God. As Kraeling points out, "[t]he fact of the comparison shows that the person in question is not God, for to compare oneself with God, even in the most abject humility, would have been presumptuous for any Jew in John's day."[17] And according to Brownlee, "God would not naturally be referred to in this way, but rather as the Almighty."[18] Hughes's attempt to disarm this argument with the assertion, "[i]t is entirely possible that John would have made a humble comparison, or rather contrast, between himself and God in order to reinforce the substantial difference between his own water baptism and God's baptism with holy spirit and with fire," fails to compel, for he cannot adduce any parallels to support this claim.[19] Thus, while it is slightly possible that John could have used metaphor to speak of God's sandals, the probability that he, even with the utmost humility, considered it appropriate to place himself within the realm of comparison with God seems unlikely.

Lastly, we must consider the tradition in Matt 11:2–6//Luke 7:18–23, according to which John sent disciples to inquire of Jesus whether he was the Coming One. According to this tradition, Jesus accepted the title of the Coming One, though not without equivocation. We are inclined to believe

---

16. Hughes, "John the Baptist," 196.
17. Kraeling, *John the Baptist*, 54.
18. Brownlee, "John the Baptist," 41.
19. Hughes, "John the Baptist," 196–97.

this represents a historical memory, for it passes the criterion of embarrassment: as Wink argues, "[t]he early church would scarcely have ascribed uncertainty to John and then answered his uncertainty with baffling ambiguity."[20] Thus, given the cumulative weight of the above arguments, we are left to conclude that John the Baptist could not have been referring to God when he spoke of the Coming One. Given the complexities involved regarding the other candidates—Elijah, whom the tradition associates more with John the Baptist than with the Coming One, the Son of Man, and the Messiah, titles concerning which New Testament scholars continue to debate—we cannot make a positive identification and must content ourselves with the observation that John expected an intermediary figure of some sort and at least entertained the possibility that Jesus would fill that role. This identification also fits the criterion of plausibility quite easily, for as we have observed in chapter 3, intermediary figures could be depicted as the bearers of fiery judgment in the literature of Second Temple Judaism. Thus, the identification of the Coming One with one other than God fits quite comfortably in the context of Second Temple Judaism.

Baptism in Holy Spirit and/or Fire?

Related to the question of the function of the eschatological baptism prophesied by John is whether he proclaimed a baptism in Holy Spirit, a baptism in Holy Spirit and fire, or simply a baptism in fire. Mark, John, and Acts agree over against Q in having the Baptist proclaim that the Coming One would baptize with the Holy Spirit alone. Do they provide us with the more historically accurate prophecy that can be traced back to the Baptist himself? Some have argued this to be so, suggesting that "and fire" καὶ πυρί, was added in light of Pentecost, in which case Mark 1:8 would represent the original phrasing of our saying. John P. Meier makes just such an argument on the basis of the criterion of multiple attestation, suggesting that in addition to Mark and John, Acts represents an independent tradition (despite Luke's dependence upon Mark). He takes these three purportedly independent witnesses to indicate that the earliest prophecy mentioned only a "baptism with the Holy Spirit," whereas the reference to the fire is a *vaticinium ex eventu* written in light of Pentecost: "Divided tongues, as of fire, appeared among them, and a tongue rested on each of them. All of them were filled with the Holy Spirit . . ." (Acts 2:3–4). Meier proposes, "it may well be that the Baptist's statements in Acts are neither mere Lucan redaction nor a sudden strange preference for Mark over Q. Rather, they

---

20. Wink, "Jesus' Reply to John," 125.

may represent a distinct stream of tradition in the sources of Acts closer to John [the Baptist] than to Q."[21]

However, there is reason to doubt that the clause καὶ πυρί, "and with fire," was an addition written in light of Pentecost, even if we are arguing against the criterion of multiple attestation. It is clear that Luke, the author of Acts, knows the tradition regarding the baptism of fire, for he has used it in his Gospel at 3:16. That Luke, who knows John's prophecy of the baptism with fire, does not see that prophecy fulfilled at Pentecost is manifest in the saying ascribed to the risen Jesus at 1:5: "for John baptized with water, but you will be baptized with the Holy Spirit not many days from now." If, as Meier suggests, "the relevant 'and with fire,' [corresponds] perfectly to the tongues of fire at Pentecost," it is difficult to understand why Luke does not make reference to the baptism of fire in Acts 1:5. Indeed, if the reference to the baptism with fire is a *vaticinium ex eventu* written in light of Pentecost, it is more than a little strange that Luke, who has knowledge of the longer form of the saying from Q, which includes the "and with fire" clause, does not use it in Acts 1:5, the context for which it was purportedly created, but resorts to the Markan form (or some supposedly independent tradition). This is not to suggest that Q had any knowledge of the Pentecost story, but that Luke interpreted the baptism in fire of Q 3:16 in such a way so as to preclude it from being understood as a reference to the tongues of fire of Acts 2:3–4. Only this can explain its absence from Acts 1:5.

The most significant argument against the proposal that John did not proclaim a baptism in fire, however, is that there is absolutely no need to posit the Pentecost experience as the source for the saying about the baptism with fire, for the Baptist's preaching is already rife with fiery imagery. As with the other references to fire in John's proclamation, the fire in Q 3:16 is an image of judgment. Much more probable, then, is that the baptism with fire was understood as an unfulfilled prophecy. As such, it is much easier to accept that at a relatively early stage it fell out of the tradition than that it was created out of the Pentecost experience.[22] Moreover, in light of the events surrounding Pentecost, it is far easier to understand why the reference to the baptism with the Holy Spirit, the fulfillment of which Luke clearly does see at Pentecost as Acts 1:5 indicates, was given prominence over the baptism with fire. Indeed, to turn Meier's own argument against him, the baptism in Holy Spirit makes far more sense as a *vaticinium ex eventu* in light of Pente-

---

21. Meier, *Marginal Jew*, 2:36.

22. So also Schweizer ("πνεῦμα," 398), who writes, "Omission of the difficult καὶ πυρί, is readily understandable, but not its addition, for there was never any baptism with fire. Lk. himself did not see any such in Ac. 2:3, since he quotes the saying only in the Marcan form in 1:5. Hence it must go back to the Baptist."

cost than does the "and with fire" clause. Thus, despite its poor attestation, the reference to the baptism with fire was more likely than not a component of John's preaching.

If the reference to fire is historical, can the same be said for the reference to the Holy Spirit? Nothing precludes this option *a priori*, for the Holy Spirit was not a solely Christian concept. As Scobie notes, "to suggest that John may nevertheless have used the words does not imply that he was anticipating the doctrine of the Trinity."[23] Rather, as many have noted, the outpouring of the Spirit may be understood as an eschatological gift in light of Joel 2:28–29: "Then afterward I will pour out my spirit on all flesh; your sons and your daughters shall prophesy, your old men shall dream dreams, and your young men shall see visions. Even on the male and female slave, in those days, I will pour out my spirit."

Dunn has argued that the fact that the sect at Qumran could speak of the Spirit as a cleansing and purifying power (1QS III.7–10; IV.30–21; 1QH XVI.12) lends substantial support to the prospect that John may have, in fact, spoken of baptism in the Holy Spirit in conjunction with baptism in fire.[24] We have noted that fire in the Hebrew Bible is frequently a symbol of both judgment and purification. In connection with this, Dunn has observed the following:

> the river and the flood offered themselves as obvious metaphors for being overwhelmed by calamity (P. 69:23, 15; Isa. 43.2a). But what was more distinctively apocalyptic was the combination of the two—judgment, purification in a river of fire (e.g., Dan 7:10; 1 *Enoch* 14:19; 67:13; *Sib. Or.* 3.54; 4 *Ezra* 13:10–11; 1QH 3:29–33). Spirit (*ruah*) also served to denote judgment as well as blessing (e.g., Isa. 4:4; Jer. 4:11–12; 1QSb 5:24–25), and it too had been used with the metaphor of the river of judgment, as in the apocalyptic type imagery of Isa. 30:27–28, while John's near neighbors at Qumran gave some prominence to the Spirit as a purifying, cleansing force (particularly 1QS 4:21).[25]

Moreover, "the fact that 'liquid' verbs are one of the standard ways of describing the gift of the Spirit in the last days would make it very easy for John to speak of the messianic gift of the Spirit in a metaphor drawn from the rite which was his own hall-mark."[26]

---

23. Scobie, *John the Baptist*, 70.
24. Dunn, "Spirit-and-Fire Baptism."
25. Dunn, "Birth of a Metaphor," 136.
26. Dunn, *Baptism in the Holy Spirit*, 12–13.

Following Dunn, Fitzmyer cites this text from Qumran as a demonstration of how Holy Spirit, water, and refining could all be juxtaposed in a way that forms "a plausible matrix for John's own utterance"[27]:

> Then God will purge by his truth all the deeds of man, refining for himself some of mankind in order to remove every evil spirit from the midst of their flesh, to cleanse them with a holy Spirit from all wicked practices and sprinkle them with a spirit of truth like purifying water (1 QS 4:20–21).

While the word "refining" does not require such an interpretation, it is just possible that implicit in it is the notion of refining *by fire*. More importantly, however, it is apparent that there is, then, no need to exclude the reference to the Holy Spirit from the Baptist's preaching solely on the basis that it held widespread significance in the early church.[28]

Nonetheless, some have argued that the original prophecy only referred to a baptism in fire. The recognition that we are here dealing with a Q saying has led some to suggest that Matthew and Luke both independently conflated Mark's version with the version in Q which read, "He will baptize you with fire," without any mention of "the Holy Spirit" πνεύματι ἁγίῳ whatsoever.[29] Such an argument is made on the grounds that this hypothetical reconstruction coheres admirably with what we already know of John's preaching, which nowhere else mentions the Holy Spirit but makes repeated reference to the fire of judgment.[30] The baptism with the Holy Spirit is, moreover, easily understood as a later development inspired by the pneumatic experience at Pentecost. Furthermore, in seeming support of such a reading, those mentioned in Acts 19:1–7, who had received the baptism of John claim, "we have not even heard that there is a Holy Spirit." According to this position, Mark replaced the reference to fire with his reference to the Holy Spirit in an attempt to domesticate or "Christianize" the

---

27. Fitzmyer, *Gospel of Luke*, 1:474.

28. We shall consider in our examination of Mark 9:49 and Luke 12:49–50 whether or not those sayings have any bearing on our reconstruction of the original version of John's saying.

29. See Manson, *Sayings of Jesus*, 41; Lüdemann, *Jesus*, 129.

30. We must admit, however, that we know relatively little concerning the preaching of John. As Dale Allison notes, "Unless he was exceedingly boring or was akin to the Jesus, son of Ananias, in Josephus, *War* 6.301–2, who uttered the same refrain over and over again, the Baptist must have said much more than the few utterances our sources have preserved," in Allison, "Continuity," 19.

Baptist.³¹ Matthew and Luke, it is asserted, then subsequently brought these two elements together in their gospels.³²

At first, this position seems plausible; however, it would be an unlikely coincidence if both Matthew and Luke had conflated Mark and Q in precisely the same manner. They agree to the word: αὐτὸς ὑμᾶς βαπτίσει ἐν πνεύματι ἁγίῳ καὶ πυρί. This is in fact the longest stretch of verbatim agreement between Matthew and Luke in this entire verse. To suggest that this point of similarity did not exist in Q runs counter to the evidence. If ever the prophecy existed with only the mention of the baptism in fire and not in the Holy Spirit, it must have been a pre-Q tradition.

Fledderman proposes that there was in fact, prior to Q, a tradition which made no reference to the Holy Spirit but only to the fire. He asserts:

> We can mark three stages in the development of the saying. Originally the saying did not mention the Spirit; it contrasted baptism in water with baptism in fire. In a second stage, reflected in Q, the reference to the Holy Spirit was added to make the saying more explicitly Christian. In the third stage, reflected in Mark, the spirit baptism has crowded out the fire baptism which now appeared superfluous. Mark reflects the latest stage in the development of the saying.³³

As speculative as this reconstruction is, it is a possibility that cannot be ignored. Against it, however, is the fact that the reference to the Holy Spirit is attested to in every one of our sources. If the presence of the words "Holy Spirit" is due to a Christian addition, it is one that occurred at a very early date. Furthermore, the Holy Spirit need not be understood in explicitly Christian terms, and without further evidence against it, it cannot be ruled out. We shall return to this question in chapter 5 when we can consider it in light of similar traditions attributed to Jesus (Mark 9:49; Luke 12:49–50).

---

31. Bultmann (*History of the Synoptic Tradition*, 246), asserts, "it is Christianizing editing when Mark fails to mention the fire of judgement in the Baptist's messianic preaching, though Q has preserved it."

32. For arguments that Q spoke only of the baptism with fire and that Matthew and Luke both independently conflated this tradition with the baptism in the Holy Spirit tradition found in Mark 1:8, resulting in the pairing of "the Holy Spirit and fire," see Harnack, *Sayings of Jesus*, 3; Manson, *Sayings of Jesus*, 41; Bovon, *Luke*, 126; Luz, *Matthew*, 138.

33. Fleddermann, *Mark and Q*, 37.

The Function of Baptism in Fire

Typically, the function of the fire in Matt 3:11/Luke 3:16 is contrasted with that of the Holy Spirit. As Lang has it, "[t]he logion describes the gathering of the eschatological community in grace and judgment (see the purging of the floor in Mt. 3:12). The coming Messiah will give penitents the Spirit promised for the last time (see 1 QS 4:20–22) and judge the recalcitrant with fire."[34] The problem with this approach, as has frequently been observed, is that it posits that two distinct baptisms are being administered to two separate groups. The grammar of the Baptist's proclamation as it is expressed in the Greek of Matt 3:11 and Luke 3:16 does not allow this interpretation. As Dunn observes, "the πνεύματι ἁγίῳ and the πυρί are united into a single baptism both by the ὑμᾶς and by the solitary ἐν. The recipients of John's baptism ἐν ὕδατι will also receive the Coming One's baptism ἐν πνεύματι ἁγίῳ καὶ πυρί."[35] If John had spoken of two distinct baptisms, it would have been far more natural for the Greek translation of this saying to use the conjunction ἤ rather than καί in order to differentiate between the two distinct baptisms. As it is, however, the direct address to one ὑμᾶς "you" and the conjunction καί indicate that—at least in the Greek translation—all who are addressed will be baptized in *both* Holy Spirit *and* fire, not Holy Spirit *or* fire. Thus the distinction between two baptisms, one of grace and the other of judgment, cannot be supported by the grammar of the pronouncement. Dunn is surely correct that if John spoke of a baptism in Holy Spirit and fire, the Spirit and the fire are not antithetical in their function, the former cleansing and the latter destroying. Rather, it appears that they are combined as *one baptism* of both "Holy Spirit *and* fire" and that this single baptism exhibits a dual function. In his words, "the Baptist expected that all people would have, as it were, to pass through the stream of fiery *ruah*, which would destroy the impenitent and purge the penitent."[36]

While we have found it possible that John spoke of the future judge baptizing with both Holy Spirit and fire, it is likewise possible that he spoke only of a baptism *in fire*.[37] Even if John spoke only of a baptism with fire, part of Dunn's argument remains relevant. The ὑμᾶς "you" to whom he promises a future baptism in fire are the same group whom he has baptized in water, the baptism in fire stands in close parallel to John's baptism in water, and if John baptized in *a river of* water, then perhaps he envisioned

---

34. Lang, "πῦρ," 943.
35. Dunn, "Spirit-and-Fire Baptism," 83–84.
36. Dunn, "Birth of a Metaphor," 136.
37. When we consider the evidence from the Jesus tradition in the next chapter, we shall see that this position has considerable merit.

the Coming One baptizing in *a river of* fire. This interpretation was perhaps first suggested by Origen.³⁸ Origen's insight is instructive here, for John's reference to the baptism with fire can be identified with a motif found in the apocalyptic literature of Second Temple Judaism: an eschatological river or flood of fire.

We have already encountered the eschatological river of fire in the literature surveyed in chapters 2 and 3. We find fiery streams or the combination of fire and water in Daniel 7; *1 En.* 17:5; 67:4-7; 1QHa XI.27b-37 1QHa XIV.16b-19a 1QHa XVI.11-20; 1QS IV.20b-22a; *4 Ezra* 7:6-11, and while in some of these passages the fire clearly plays a destructive role, in others the fire plays either a purifying role or it is a flame that is endured by the righteous. In our survey of the above texts, we saw that the fate of the persons subjected to the fire is determined by their moral standing: the wicked are destroyed while the righteous are refined or preserved. With this in mind, it bears repeating that in Q 3:16, the ὑμᾶς "you" who stand on the banks of the Jordan to receive John's baptism are the same ὑμᾶς who will receive the baptism with fire at the hands of the Coming One. This accords well with the tradition that *all* must pass through the same river of fire, but that the fire affects those immersed in it differently, as if there is wisdom in the fire which can discern who is penitent and who is not. As we have already discussed, the notion that fire can be both punitive and purgative has its antecedents in the Hebrew Bible and the intertestamental literature and was taken over by early Christian writers as well. Richard Bauckham explains that this idea goes back to "the notion of a judicial ordeal which distinguishes the innocent from the wicked," such as that prescribed in Num 5:11–31.³⁹ He also notes that "[a]n ordeal by plunging in a river was actually an ancient judicial practice: it occurs in the code of Hammurabi, for example."⁴⁰

In addition to the texts cited above, in many of the later Jewish and Christian apocalyptic visions the river of fire performs a dual function. It is an ordeal that must be endured, and those who are righteous will be saved while the wicked will be destroyed. In *Apoc. Pet.* 6, a text written sometime around 135 CE, a river of fire judges both the righteous and the unrighteous—the elect "will not see death by the devouring fire. But the evil creatures, the sinners and the hypocrites will stand in the depths of the darkness that passes not away, and their punishment is the fire."⁴¹ Similarly, *Sib. Or.* 2.252-54, which postdates the texts from the *Sibylline Oracles*

---

38. See pages 3–4 above.
39. Bauckham, *Fate of the Dead*, 203.
40. Ibid.
41. Trans., C. Detlef G. Müller in *NTApoc*, 628.

discussed in chapter 3, emphasizes that, although their fates differ, *all* are immersed: "And then all will pass through the blazing river and the unquenchable flame. All the righteous will be saved, but the impious will then be destroyed for all ages."[42] In *T. Isaac*, a Christian composition produced sometime in the second century CE, the seer reports seeing a river of fire at the last judgment: "And that river had wisdom in its fire: It would not harm the righteous, but only the sinners by burning them. It would burn every one of them because of the stench and repugnance of the odor surrounding the sinners" (5.21–5).[43] Clearly the river of fire was a popular motif and played a central role in eschatological judgment in the Jewish and Christian apocalyptic literature around the turn of the era. This concept can be found in texts that both antedate and postdate the Gospels, and very significantly, it does justice to the inclusive "you" with which John addresses all who have come to the Jordan, whether they have come as penitents for baptism, as hostile opponents, or as indecisive spectators, for according to this eschatological concept, *all* will be immersed in the river of fire. And if *all* are to endure the same baptism, we may infer that the fire functions differently upon the penitent than it does upon the recalcitrant.

As Dahl observes, this may point to the purificatory quality of fire:

> The original idea will here be that of an immersion in fire, analogous to the immersion in water. Fire is a means of purification of a higher order than that of water. . . . . The baptism in fire which John announced, therefore not only predicts the annihilation of sinners in fire, but also the purification and renewal of those who are to be members of the heavenly, eschatological assembly of the people of God.[44]

In this regard John is, although perhaps without explicit allusion, echoing the idea behind Num 31:23 in which both water and fire are seen to be cleansing—only fire is more so: "everything that can withstand fire, shall be passed through fire, and it shall be clean. Nevertheless it shall also be purified with the water for purification; and whatever cannot withstand fire, shall be passed through the water." The contrast John seemingly draws between his own baptism and that of the Coming One is not one of opposition but one of intensity.

According to Reid:

---

42. Trans., Collins in Charlesworth, *OTP*, 1:351.
43. Trans., Stinespring in Charlesworth, *OTP*, 1:909.
44. Dahl, "Origin of Baptism," 44–45.

> John saw likeness, and not opposition, in the relation between his work and that of the Messiah. To him "water" and "fire" were agents of cleansing, and the difference in their efficacy was a measure of the difference in the work which both were to do. The Baptist and the Messiah were seeking the moral purification of the people, but the cleansing of the one was the cleansing of water, and the cleansing of the other was as the cleansing of fire. Water washes away the outer defilement, and leaves the substance unchanged; fire penetrates and transforms what it cleanses.[45]

This interpretation satisfactorily recognizes that those who submitted to John's water baptism would also be subjected to the coming fire baptism. Perhaps we can suggest that John saw his own baptism as initiating the purification process of the penitent and at the same time marking them off as members of the eschatological community. Nonetheless, they, too, along with the rest of humankind, would be immersed in the eschatological river of fire which would complete the purification process already initiated with John's water baptism and which would be experienced as punitive judgment by those who did not receive John's ritual ablution.[46] Perhaps John even foresaw the possibility that some whom he baptized may have received his baptism without true repentance and that for them the baptism of fire would be not remedial but ruinous. Upon this we can only speculate, but such a proposal would seem to cohere with what we can gather concerning John's own expectations.

Conclusion

In closing this discussion of John the Baptist's proclamation of the Coming One's baptism in fire, I wish to make one suggestion that to my knowledge has not previously been made. Just prior to comparing the wicked, who will be consumed with fire, to barren trees and worthless chaff, John utters these words: "Do not presume to say to yourselves, 'We have Abraham as our ancestor'; for I tell you, God is able from these stones to raise up children to Abraham" (Matt 3:9//Luke 3:8). It is not unreasonable to infer that John attributed stone-like qualities to the true children of Abraham who accepted his baptism of repentance in the same way that he attributed the

---

45. Reid, "Baptism of Water," 306.

46. The contrasting view, that the fiery baptism played only a destructive role and that those who were baptized by John would be spared from it, is expressed by Reiser, *Jesus and Judgment*, 185. This perspective, however, fails to account for the single "you."

qualities of barren trees and chaff to the unrepentant.⁴⁷ In contrast to chaff and dead wood, which are easily consumed by flames, stone withstands the fiery judgment. This combination of images in Q 3:8–12—fire (πῦρ), stone (λίθων), roots (ῥίζαν), trees (δένδρων), and chaff (ἄχθρον)⁴⁸—recalls similar associations we find elsewhere. Malachi, for instance, when speaking of the Day of the Lord, juxtaposes the image of fire (MT: אֵשׁ; LXX: πῦρ) with silver (כֶּסֶף; ἀργύριον) and gold (זָהָב; χρυσίον) (3:2–3) and later with stubble (קַשׁ; καλάμη), roots (שֹׁרֶשׁ; ῥίζα), and branches (עָנָף; κλῆμα) (4:1).⁴⁹ While Malachi mentions precious metals and not stones, when we turn to 1 Cor 3:12, we find stones being tested by fire alongside these precious metals as well as flammable substances: "gold (χρυσόν), silver (ἄργυρον), precious stones (λίθους), wood (ξύλα), hay (χόρτον), straw (καλάμην)" (vv. 14–15).⁵⁰ Explicit in both Malachi 3–4 and 1 Cor 3:10–15 is the fact that the combustible stubble, roots, branches, wood, hay, and straw will be consumed by the fire while the silver, gold, and precious stones will be refined or tested. It is just possible then—though far from certain—that John was employing the images of stone, trees, and chaff in a similar manner, indicating that in the baptism of fire the wicked will fare as do the chaff and the dry and dead trees that fail to produce fruit while the penitent children of Abraham will endure as stone withstands fire.

---

47. This need not conflict with the view that John was also alluding to the stones that were "raised up" (הֲקִימִים) to mark the Israelites' crossing of the Jordan in Josh 4:1–9. See Seitz, "What Do These Stones Mean?" 254.

48. Typically the word ἄχυρον is translated as chaff, but Reiser (*Jesus and Judgment*, 177), suggests that since the chaff would be blown away on the wind, John is more likely referring to stubble here, for it is the stubble, which resonates with the Hebrew קַשׁ; in Mal 4:1, that would land nearby and be thrown into the fires.

49. The prophecies of Malachi are clearly in the background of the New Testament descriptions of John the Baptist, and this supports our argument that there may be further allusions to that prophet in John's proclamation. For the influence of Malachi in the preaching of John, see especially Trumbower, "The Role of Malachi." A word should be said here on our use of the LXX in this and any following discussions of New Testament references to the Old Testament. I accept the general consensus that the authors of the New Testament relied primarily on Greek translations of the Hebrew Bible when quoting or alluding to the Old Testament. When considering potential allusions to the Old Testament, therefore, I shall prefer to make use of the LXX over the MT. I recognize, however, that the authors of the New Testament may have had different versions of the Old Greek to hand. On the the use of the Old Greek in the New Testament and all the complexities that attend these matters, see especially McLay, *Use of the Septuagint*, 17–36. On the authority of the LXX in early Christianity, see Hengel, *Septuagint*. I provide the Hebrew text here to demonstrate that the MT corresponds to the LXX at this point.

50. We shall return to this matter in our discussion of 1 Cor 3:10–15 below.

CHAPTER 5

The "Fire Words" of Jesus of Nazareth

DESPITE A STRONG TENDENCY in contemporary scholarship to remain silent on the subject of judgment in general and of eschatological fire in particular in the Jesus tradition, a noteworthy portion of the preaching of Jesus includes mention of eschatological judgment by fire, particularly with mention of hell or Gehenna. Indeed the word "fire" occurs twenty-five times in the Synoptic tradition, and in all but two cases (Mark 9:21; Luke 22:25) it is in reference to eschatological judgment.[1] Most of these passages, to be sure, have to do with the fire of hell or Gehenna, particularly those in Mark and Matthew.[2] Alongside the sayings of hell-fire, however, Friedrich Lang has identified other sayings that have to do with "the final judgment," and he suggests that the context of eschatological judgment is "the best starting point for an interpretation of the difficult sayings Mk. 9:49 and Lk. 12:49."[3] While the sayings concerning "the hell of fire" are certainly a subject worthy of consideration, this chapter focuses only on the fire sayings where the context is eschatological judgment (in the forensic sense) and the function is ambiguous, that is, where the fire may play a dual function, both purgative and punitive and thus may be interpreted in terms of a fiery ordeal.[4] Five logia, some of which Dunn refers to as the "fire words" of Jesus, fit this description: Mark 9:49, which is unique to Mark; Luke 12:49 (see also *Gos. Thom.* 10; 16); Luke 17:26–30; the enigmatic verse in Luke 23:31; and *Gos. Thom.* 82.

---

1. Allison, *End of the Ages*, 125.
2. See Lang, "πῦρ," 945.
3. Ibid., 943.
4. For a recent treatment of the subject of hell/Gehenna see Allison, "Problem of Gehenna."

Having discussed John's proclamation concerning the Coming One's baptism in fire in our last chapter, it is well worth noting that one of the least disputed events in the life of the historical Jesus is his baptism by John the Baptist. While Mark narrates this event with no apparent anxiety, it caused such obvious discomfort to the later evangelists, that it is often used as a test case to illustrate "the criterion of embarrassment."[5] It also passes the criterion of multiple attestation (Q 3:21–22; Mark 1:9–11; John 1:29–34; *Gos. Heb.* 2)[6] as well as the criterion of dissimilarity, for it is difficult to imagine early Christian authors creating a tradition that has Jesus submitting to John the Baptist and receiving his "baptism of repentance for the forgiveness of sins" (Mark 1:4). The very high probability of the historicity of Jesus' baptism by John provides us with much more than a record of an event that occurred in the Jordan River sometime in the mid-to-late 20s CE. Much more importantly, it indicates a strong link between John the Baptist and Jesus of Nazareth. We should not be surprised, therefore, to find elements of John's teaching regarding the eschatological functions of fire in the teachings of Jesus. Like John, Jesus also proclaimed a fiery judgment on the wicked. Again, however, while Jesus' teaching on the subject of Gehenna would make for a fascinating study, we will maintain our focus on those sayings where fire plays at the very least an ambiguous role, where both the wicked and the righteous are confronted by the fire though the effect of this fire may differ between these two groups. We begin, therefore, with one of the most opaque sayings of Jesus—Mark 9:49.

Mark 9:49—"Salted with Fire"

"For everyone will be salted with fire" (Mark 9:49). These words, attributed to Jesus by the second evangelist, constitute what is no doubt one of the most perplexing pronouncements in the Synoptic tradition. To be sure, two prominent commentators have recently deemed it "one of the New Testament passages that defy interpretation"[7] and "perhaps the most enigmatic logion of Jesus in the NT."[8] These judgments do not, however, reflect a solely modern sentiment, for it is apparent that the apprehension that attends this peculiar saying is nearly as ancient as the verse itself. Mark 9:49 is, for instance, one

---

5. See, for instance, Meier, *Marginal Jew*, 1:168–69.

6. For arguments in favor of the independence of these sources, see Webb, "Jesus' Baptism by John," 95–108. Crossan (*Historical Jesus*, 234) lists *Ign. Eph.* 18:2 as a fifth potentially independent source.

7. Boring, *Mark*, 284.

8. Marcus, *Mark*, 2:698.

of the few Markan texts that neither Matthew nor Luke incorporated into his gospel.[9] Furthermore, the Greek manuscript tradition contains a number of textual variants of this logion, providing a strong indication that even those ancient scribes, whose mother tongue was Greek, were uncertain what to make of it. And strikingly, the history of the interpretation of our verse in the patristic period is, so far as I can tell, a history of silence.[10]

This exegetical unease has lead many modern scholars to conclude that the Greek text is corrupt, thus prompting the proliferation of several Semitic retroversions, which, having been misheard or mistranslated at some stage in the saying's oral or textual history, may have given rise to the Greek text now preserved in Mark's Gospel. The most noteworthy of these reconstructions is arguably that of T. J. Baarda, who suggests that the Greek verb ἁλισθήσεται ("will be salted") may derive from confusion between the Aramaic verbs תְּבַל ("to spice, season")[11] and טְבַל ("to immerse, to bathe for purification"),[12] which may have occurred at the oral stage.[13] Baarda also allows that it may have been the result of a scribal mistranslation of the verb טְבַל, which could mean "to season" in addition to its more common meaning "to immerse, bathe for purification."[14] In either case, Baarda proposes that the original Aramaic form of our saying may have been די כל אנש בנורא יטבל, which should have been translated into Greek as πᾶς γὰρ πυρὶ βαπτισθήσεται ("For everyone will be baptized in fire").[15]

---

9. See Kummel, *Introduction to the New Testament*, 45. Kummel observes, "Only three short reports (Mk. 4:26–29, parable of the self-growing seed; 7:31–37, healing of a deaf mute; 8:22–26, the blind man of Bethsaida) and three very short texts (3:20f., the relatives regard Jesus as beside himself; 9:49, salt with fire; 14:51f., the fleeing youth) of Mark are found neither in Matthew nor Luke." It is interesting to note, however, that Tatian does preserve this curious verse in his *Diatessaron* (25.23).

10. Ambrose of Milan, *Exp. Ev. Sec. Luc.* 5, 8 (CCSL 14, ed. M. Adriaen, 137), who cites the scribal gloss in Mark 9:49b—"*omnis ... victima sale salietur*" (every sacrifice will be salted with salt)—is, to my knowledge, the only patristic exegete who comes close to breaking this silence. See *T. Levi* 9.14 and *Gos. Phil.* 35 for other possible allusions to Mark 9.49b of an early date.

11. Jastrow, *Dictionary*, s.v. תְּבַל.

12. Ibid., s.v. טְבַל.

13. Baarda, "Mark IX.49."

14. Ibid., 320.

15. Ibid. I have elected to translate the dative πυρί as "*in* fire" rather than "*with* fire." Although it is the latter that is typically found in translations of Mark 9.49, the preposition "in" is just as likely as "with"—if not more so—for the Hebrew and Aramaic verb טְבַל, like the Greek verb βαπτίζω, most often suggests immersion *in* something, usually a liquid; see Delling, βάπτισμα βατισηῆναι, esp. 95–102; Oepke, βάπτω, κτλ. Moreover, the Greek dative πυρί presupposes the Semitic preposition בְּ, which may be translated as "in" just as readily as "with." This last judgment may find support in mss א C 1342,

Despite the fact that Baarda's retroversion can be shown to fit quite plausibly within the context of the life and preaching of the historical Jesus, his attempt at unraveling this mysterious saying has gone largely unappreciated. A more thorough consideration of the merits and implications of his reconstruction is, therefore, in order. As I shall seek to demonstrate, whereas the Greek text of this fiery logion generates more smoke than light, Baarda's proposal not only illumines this previously dark saying, but in heretofore unexplored ways, it may also throw new light on the similar pronouncement attributed to John the Baptist—"He [the Coming One] will baptize you with the Holy Spirit and fire" (Q 3:16)—some of the difficulties regarding which we have already begun to scrutinize.

Text-Critical Issues

Since the very need for such a reconstruction hangs on the assumption that the Greek text is corrupt, it is necessary first to consider the manifold difficulties that accompany the present form of our saying before entering into an evaluation of Baarda's proposal. As is well known, the Greek manuscript tradition attests to three principal forms of Mark 9:49:

1. πᾶς γὰρ πυρὶ ἁλισθήσεται ("For everyone will be salted with fire," B L Δ 0274 $f^{1.13}$ 28*. 565. 700 *pc* sy^s bo^{pt} (ℵ: ἐν πυρί));

2. πᾶσα γὰρ θυσία ἁλὶ ἁλισθήσεται ("For every sacrifice will be salted with salt," D it$^{b,c,d,ff^2,i}$), which alludes to Lev 2:13 LXX: καὶ πᾶν δῶρον θυσίας ὑμῶν ἁλὶ ἁλισθήσεται ("And every one of your sacrificial gifts will be salted with salt"); and

3. πᾶς γὰρ πυρὶ ἁλισθήσεται καὶ πᾶσα θυσία ἁλὶ ἁλισθήσεται ("For everyone will be salted with fire, and every sacrifice will be salted with salt", A K Π (2427) 𝔐 lat sy$^{p.h}$ bo^{pt} (C: ἐν πυρί)).

Other less well-attested variants include the following:

4. πᾶς ἄρτος γὰρ πυρὶ ἁλισθήσεται ("For every loaf of bread will be salted with fire," 11. 88. 124. 220. 230);

5. πᾶς γὰρ πυρὶ ἀναλωθήσεται πᾶσα δὲ θυσία ἁλὶ ἁλισθήσεται ("For everyone will be consumed with fire and every sacrifice will be salted with salt," Θ);

---

which supply the preposition ἐν before πυρί.

6. πᾶς γὰρ πυρὶ ἁλισθήσεται πᾶσα δὲ θυσία ἁλὶ ἀναλωθήσεται ("For everyone will be salted with fire and every sacrifice will be consumed with salt," Ψ);

7. πᾶς γὰρ πυρὶ ἁλισγηθήσεται ("For everyone will be polluted with fire," W); and

8. πᾶς γὰρ πυρὶ δοκιμασθήσεται ("For everyone will be tested by fire," 1195).

Paul-Louis Couchoud, moreover, has argued that the original form of our saying may be preserved in the Latin Codex Bobiensis:

9. omnia autem substantia consumitur ("And all [their] substance will be consumed," it^k), which Couchoud translates into Greek as πᾶσα δὲ οὐσία ἀναλωθήσεται. In their Markan context these words would refer to the fate of those thrown into Gehenna.[16]

Last, at least two hypothetical Greek variants for which there are no extant textual witnesses have been proposed:

10. πᾶσα πυρία ἁλισθήσεται πᾶσα δὲ θυσία ἁλὶ ἁλισθήσεται ("Every burning [that is, every offering made by fire] shall be salted, and every sacrifice will be salted with salt"), as proposed by Joseph Scaliger;[17] and

11. πᾶς γὰρ πυρὶ ἁγνισθήσεται ("For everyone will be purified with fire"), as suggested by Alexander Pallis.[18]

Most textual critics agree that of these eleven extant and hypothetical variants, the earliest Greek form of our saying is reading (1), "For everyone will be salted with fire," for it is attested by some of the best and earliest uncials (Vaticanus and Sinaiticus), and it alone could have given rise to all of the variants enumerated above.[19] Regarding the textual history of Mark 9:49, Bruce Metzger ventures the following explanation:

---

16. Couchoud, "Was the Gospel of Mark Written in Latin?" 47–48; Couchoud, "Notes," 124. See Lohmeyer, *Das Evangelium des Markus*, 197.

17. Scaliger, "Notae," 380.

18. Pallis, "A Few Notes," 18.

19. See, however, Zimmermann ("Mit Feuer," 28–31), who offers the dissenting opinion that the longer reading (3) is the original and that the shorter reading (1) resulted from an instance of *homoeteleuton*, in which the copyist's eye slid from the ἁλισθήσεται (will be salted) at the end of the first clause to the same verb at the end of the second, thus omitting the second clause altogether. See also Elliott, "Eclectic Textual Commentary," 199. Elliot argues for the longer reading on the basis of Mark's frequent use of repetition, which he takes to indicate that "[t]he so-called conflate reading is characteristic of Mark."

> At a very early period a scribe, having found in Lv 2:13 a clue to the meaning of Jesus' enigmatic statement, wrote the Old Testament passage in the margin of his copy of Mark. In subsequent copyings the marginal gloss was either substituted for the words of the text, thus creating reading (2), or was added to the text, thus creating reading (3).[20]

Metzger's conjecture provides a plausible explanation for the genesis of the principal variants, and one can perceive how the minor variants might have originated from the most primitive reading (1) or from the later conflate reading (3). For instance, while reading (4) is an intentional modification of (1), several of the other variants can be explained either as scribal errors in which copyists mistook the original verb ἁλισθήσεται ("will be salted") for a similar looking verb, or as scribal attempts to correct what they perceived as a previously committed scribal error. In either case, the verb in question was replaced with ἀναλωθήσεται ("will be consumed") in variants (5) and (6), and with ἀλισγηθήσεται ("will be polluted") in variant (7). Variant (8), however, cannot be explained in this manner, for the verb δοκιμασθήσεται ("will be tested") could never have been confused with the verb ἁλισθήσεται. Rather, this variant probably represents one of the earliest attainable interpretations of reading (1).[21] Variant (9) is probably a Latin translation of a variant closely related to reading (2) in which the Greek noun θυσία ("sacrifice") was mistaken for οὐσία ("substance"). As James Morison observes, Scaliger's proposal (10) requires the invention of a sacrificial word, πυρία ("burning," i.e. "offering made by fire"), a fact which tells strongly against its authenticity.[22] Finally, while Pallis's suggested reading (11) is attractive and its meaning clear, the absence of a single supporting textual variant anywhere in the manuscript tradition is detrimental to the claim that it represents the original Greek text. It seems most probable, following Metzger, that reading (1) preserves the earliest attainable Greek form of our saying.

---

20. Metzger, *Textual Commentary*, 102.

21. That the reference to salting with fire could be interpreted as a fiery test is confirmed by the comment of Theophylact in *Ennaratio in Ev. Marci* (PG 123.70): Πᾶς γὰρ, φησὶ, πυρὶ ἁλισθήσεται, τουτέστι, δοκιμασθήσεται ("He says, 'For everyone will be salted with fire,' that is, will be tested"). The saying is similarly interpreted by the anonymous ancient commentator cited in Cramer, *Catenae*, 368–69. The commentator likens Mark 9:49 to 1 Cor 3:13, according to which "the fire will test (τὸ πῦρ δοκιμάσει) what sort of work each one has done." See *Did.* 16.5: "Then all humankind will come to the fiery test (τὴν πύρωσιν τῆς δοκιμασίας)," (trans. Holmes, *The Apostolic Fathers*, 208–9.) See also *T. Ab.* 13.11: "and he tests (δοκιμάζει) the works of men through fire (πυρός)" (trans. Sanders in Charlesworth, *OTP*, 1:890).

22. Morison, *Practical Commentary*, 284.

History of Interpretation

Despite the fairly widespread consensus regarding the original form of this saying in the Greek, there is almost complete disarray when it comes to its interpretation. Indeed, the history of the interpretation of Mark 9:49 is as convoluted as the history of its textual transmission. Unlike Matthew, Luke, and the early church fathers, who were perhaps allergic to this saying's ambiguity and were thus inclined to evade the enigmatic expression, exegetes from the medieval period to the present offer up a cacophony of dissonant voices, which only succeed in drowning out one another. As evidence of this exegetical turmoil, Heinrich Meyer cites no fewer than fourteen differing readings of our saying in his treatment of this text.[23]

It is important to note that prior to the widespread influence of textual criticism, the accepted text of Mark 9:49 was reading (3), which included both the original clause ("For everyone will be salted with fire") and the scribal gloss ("and every sacrifice will be salted with salt"). Moreover, this longer form of the saying was often read in the context of the preceding verses (Mark 9:43–48), which prescribe the amputation of offending limbs for the sake of preserving oneself from the fires of Gehenna. Assuming the longer reading to be correct and the Markan context to be original, exegetes of an earlier era typically took one of two approaches: they read the two clauses in either synonymous or antithetical parallelism.

When read in synonymous parallelism, the "*every one* in the former part of the verse is the same with *every* sacrifice in the latter part."[24] Consequently, the salting is understood positively and is thought of as being metaphorically applied to the faithful elect, who present themselves as living sacrifices willing to forfeit eye, hand, and foot for the kingdom of God. They are those who are spiritually salted by the fire of the gospel or the Holy Spirit, which preserves from sin and ultimately from the future fires of eschatological punishment. For these exegetes the purifying, preservative, and sacrificial connotations of salt are central to the exegesis of Mark 9:49, which is interpreted as commending such salting as an essential element of discipleship.[25] In contrast to those who read the two clauses in synonymous parallelism, those who look to our verse and see antithetical parallelism suggest that in the first clause "Πᾶς, *all*, is not to be understood of every man,

---

23. Meyer, *Critical and Exegetical Hand-Book*, 120–23.

24. Poole, *Commentary*, 169.

25. See *Expositio Evangelii Secundum Marcum* 9, 48 (CCSL 82, ed. Michael Cahill, p. 44); Bedae Venerabilis, *In Marci Evangelium Expositio* 9, 49 (CCSL 120, ed. D. Hurst, p. 555–56); Bengel, *Gnomon*, 542–43; Gibson, *Commentaries of Isho'dad of Merv*, 136–37; Calvin, *Harmony of the Gospels*, 176.

but of every one of them 'whose worm dieth not'" and has as its antecedent those who in the preceding verses are threatened with the unwelcome fate of being cast into Gehenna.[26] According to some, moreover, the purpose of the salting is not for the preservation of believers *from* corruption, but for the preservation of sinners *in* their state of eschatological punishment, with the effect of prolonging their unbearable torment unto eternity.[27]

These classical interpretations bring to light the extent to which nearly all exegesis of our verse up to the past century has been influenced by both the scribal gloss in Mark 9:49b and the Markan context. However, just as the textual critics have informed us that the scribal gloss is probably not original to our saying, the form critics have suggested that the Markan context (vv. 33–50) is most probably secondary.[28] Modern exegesis, therefore, typically begins with the recognition that Mark 9:49 owes its present position to the verbal similarities it shares with the preceding and following verses. Six clearly discernible catchwords hold together Mark 9:33–50, suggesting that the passage is not organically unified but is composed of several originally independent logia: ὄνομα ("name") links vv. 37, 38, 39 and 41; παιδίον ("child") links vv. 37 and 42; σκανδαλίζω ("to cause to stumble") links vv. 42, 43, 45, and 47; πῦρ ("fire") links vv. 48 and 49; and ἁλίζω ("to salt") and its cognate ἅλας ("salt") link vv. 49 and 50.[29]

Those who have insisted upon holding these verses together in their exegesis of this text have produced dubious interpretations. J. D. M. Derrett, for instance, suggests that since salt and fire were used in the ancient world to cauterize and cleanse flesh after the amputation of limbs, the reference to salt and fire in vv. 49–50 stands in a natural relationship to vv. 42–48, which speak of cutting off one's hand or foot to prevent oneself from sinning.[30] Against such exegesis, it seems that too obvious a disjuncture exists between vv. 43–48; 49 and 50 to maintain their unity. First, in vv. 43–48 fire refers to the fires of Gehenna and clearly has a negative meaning. Next, in v. 49 salt and fire are functionally equated with one another, thereby attributing

---

26. Lightfoot, *Commentary on the New Testament*, 425. See Grotius, *Operum Theologorum*, 316–17.

27. Gill, *Commentary*, 367–68; Wesley, *Commentary*, 430.

28. See Wellhausen, *Das Evangelium Marci*, 415; Taylor, *Mark,* 413; Nineham, *Gospel of St. Mark*, 255; Schmidt, *Der Rahmen*, 233; Hooker, *Gospel according to St. Mark*, 230–31.

29. Taylor (*Formation of the Gospel Tradition*, 92) notes, "This kind of artificial arrangement appears to be especially characteristic of the sayings in Mark." So also Neirynck, "Tradition of the Sayings of Jesus," 65; Best, "Mark's Preservation of the Tradition," 159.

30. Derrett, "Salted with Fire."

to salt the negative connotations already associated with fire. Last, in v. 50 salt is described in an unambiguously positive sense: "Salt is good." This incongruity strongly indicates that Mark 9:49 was an originally independent saying which was found to be a convenient, albeit poorly integrated, transition from the fire sayings in vv. 43–48 to the salt sayings in v. 50.[31] On the other hand, Urban von Wahlde believes v. 49 was originally "a paradoxical answer to the question of how to salt the flavorless salt" mentioned in v. 50 and that the Markan redactor transposed the sayings in vv. 49 and 50 so as to establish the catchword "fire" between vv. 48 and 50.[32] This is, however, an unlikely proposition, for it would suggest that the redactor was more concerned with creating catchwords than with maintaining meaning. Any historically oriented consideration of Mark 9:49 must, therefore, provide a plausible meaning for this saying that is independent of its secondary Markan context.[33] It is also probable that the postpositive γάρ ("for") in v. 49 was introduced by the compiler of these disparate logia in order to link it to the preceding verses.[34]

This recognition makes interpretation of our saying all the more difficult, for "if we disengage v 49 from its present context . . . almost any guess is as good as another."[35] It is for this very reason that conflicting interpretations abound among modern exegetes. Some have continued to see here a reference to Lev 2:13 and suggest that the saying stresses the sacrificial

---

31. Sharman (*Teaching of Jesus*, 74) asserts that v. 49 is a Markan invention created for the sole purpose of linking vv. 42–48 and v. 50. For more recent proponents of this position see Fleddermann, "Discipleship Discourse," 71; Funk, Hoover and the Jesus Seminar, *The Five Gospels*, 87; Lüdemann, *Jesus*, 65. This, however, is highly improbable, for if Mark had wished to create a saying merely for transitional purposes, it seems that he would have devised something more easily intelligible and contextually relevant.

32. Von Wahlde, "Mark 9:33–50," 61.

33. See Cranfield, *Gospel according to Saint Mark*, 315. Although historically speaking Mark 9:33–50 was probably not a unified whole, the attempt to discern Mark's redactional purposes in organizing this cluster of sayings in his Gospel is still a valid pursuit; see Henderson ("Salted with Fire," 47), who argues that Mark 9.33–50's pivotal location and uncharacteristic length (it is the third longest discourse in Mark) must indicate that Mark's redactional intentions are at work in this passage. He goes on to suggest, however, that we should take "seriously the possibility that Mark's rhetorics failed here. Such a possibility is suggested by the radical dismemberment of the Markan Jesus' speech in both Matthew's and Luke's redactions." If he is correct, the "failure" of Mark's rhetorics to persuade Matthew and Luke may be a further indication that these originally distinct sayings were too unwieldy to be integrated effectively into a cohesive discourse.

34. See also Mark 4:21–25 (vv. 22, 25); Mark 8:35–38, which similarly use γάρ to link clusters of pre-Markan tradition.

35. Gundry, *Mark*, 527.

nature of Christian discipleship.[36] Others find here a cryptic allusion to the eschatological tribulation which all must face.[37] A handful of scholars continue to believe it to be a reference to the fires of Gehenna, but some take the salt imagery to indicate the "purificatory character of the final fire of judgment."[38] Still others see in Mark 9:49 a reference to the end-of-the-world cosmic conflagration that dissolves the elements with fire (see 2 Pet 3:10–12).[39] Last, at least one exegete has discerned in this saying a metaphor for persecution with "no eschatological significance" whatsoever.[40] This brief foray into the history of the interpretation of Mark 9:49 should suffice to underscore Meyer's conclusion, which confirms what the textual variants have already shown us, namely that "[t]his great diversity of interpretation is a proof of the obscurity of the utterance."[41] Indeed, it would be an understatement to say that the Greek text of this saying is inherently ambiguous.

The Problem of Polyvalent Images

It is difficult to avoid the inference that at the root of the confusion evinced by so many textual variants and conflicting interpretations of Mark 9:49 is the juxtaposition of two highly polyvalent images in our saying: salt and fire. In the literature ranging from the Hebrew Bible to the earliest Christian writings and beyond, the images of both fire and salt carry a surplus of meanings which at times seem antithetical to one another and thus contribute to the cryptic nature of this text. Indeed, as with the image of fire, which we have already discussed at great length, the image of salt evokes an array of symbolic meanings.[42] To begin with, Lev 2:13b commands, "with all your offerings you shall offer salt" (see Ezek 43:24; 1 Esd 6:24, 30; Jos. Ant. 3.9.1; 12.3.3; 11Q18 21; 11Q19 XX.13; XXXIV.11–14; 11Q20 7; Gos.

---

36. See Hauck, "ἅλας"; Gnilka, *Das Evangelium nach Markus*, 66; France, *Gospel of Mark*, 383–84.

37. See Cranfield, *Gospel according to Saint Mark*, 316; Schweizer, *Good News according to Mark*, 199; Allison, *End of the Ages*, 127; Hooker, *Gospel according to Saint Mark*, 233.

38. Gould, *Mark*, 181; Jeremias, "γέεννα."

39. See Montefiore, *Synoptic Gospels*, 223.

40. Cadoux, *Historic Mission*, 242. Cadoux does, however, concede that the verse's present location in Mark suggests that the evangelist interpreted the saying in an eschatological sense.

41. Meyer, *Critical and Exegetical Hand-Book*, 123.

42. For the many uses of salt in the ancient world see Pliny, *Natural History* 31.34–35. See Coleman, "Note"; Coleman, " 'Salt' and 'Salted'"; Nauck, "Salt as a Metaphor"; Latham, *Religious Symbolism of Salt*.

Phil. 35; *T. Levi* 9:14), from which salt derives its sacrificial connotations (see *Iliad* 1.449; *Aeneid* 2.133). According to Philo's interpretation of this commandment, salt was employed to preserve the sacrifice from decay just as the soul preserves the body from moral corruption (*Spec. Laws* 1.53).[43] A meal of bread and salt played a role in the sealing of covenants in the Covenant of Salt and by extension symbolized hospitality (Num 18:19; 2 Chr 13:5). Additionally, salt was believed to have purifying qualities, and thus all things from newborn babies (Ezek 16:4) to "the spiritual works of the marvelous firmament are purified with salt" (4Q405 19 ABCD).[44] In the New Testament, Col 4:6 alludes to the idea that speech could be "seasoned with salt," an idiom that draws upon the observation that just as salt seasons food and makes its flavor more interesting, speech should likewise be seasoned with wit and wisdom to prevent it from becoming insipid and boring.[45] Mark 9:50 (see Matt 5:13//Luke 14:34) gives a manifestly positive description of salt—"Salt is good"—and Jesus exhorts his disciples to have salt in themselves. Precisely what positive quality the salt symbolizes, however, is unclear.[46] Negatively, it was recognized in the ancient world that too much salt, like too much water or fire, could be destructive, particularly given that soil which has been sown with salt becomes sterile (Ps 107:33–34; Jer 17:6; Judg 9:45).[47] These negative qualities, moreover, are often associated with the destruction of Sodom and Gomorrah (Deut 39:23; Zeph 2:9), where Lot's wife was transformed into a pillar of salt (Gen 19:24–26).[48]

Given the multi-faceted nature of both of these images, it is not surprising that James Latham concludes his exhaustive survey of the religious symbolism of salt in the ancient world with the observation that Mark 9:49 "is too rich in symbolism to settle on any one meaning. This is especially evident because the two symbols, salt and fire, contain, each one, contrary meanings."[49] It is understandable that the rich salt symbolism paired with the equally evocative image of fire in Mark 9:49 has led to many diverse and conflicting interpretations. And while it is evident that Jesus was

---

43. The power of salt to preserve from corruption was also apparently important in earliest Christianity. See, for instance, Ign., *Magn.* 10: "Have yourselves salted in [Christ], and then there will be no scent of corruption about any of you—for it is by your odour that you will be proved" (trans. Holmes, *Apostolic Fathers*, 208–9).

44. Trans. Vermes, *Complete Dead Sea Scrolls*, 337.

45. See Lohse, *Colossians and Philemon*, 169; Dunn, *Epistles to Colossians and to Philemon*, 266–67.

46. See Davies and Allison, *Matthew*; 1:473.

47. See also *Gos. Phil.* 36, where salt is associated with the barrenness of Sophia.

48. See page 39–52 above.

49. Latham, *Religious Symbolism of Salt*, 239.

fond of employing provocative images in his parabolic speech, the text we are investigating seems particularly recondite, and one is forced to ask, "[m]ay we not say that in the womb of this verse are two parables striving to be born, one of fire and the other of salt?"[50] In light of such extreme ambiguity, one is left to surmise either that it was the intention of Jesus (or whoever first formulated this saying) to flummox his hearers with such an esoteric utterance, or that our saying has been imperfectly handed down through its oral and scribal transmission.[51]

Semitic Retroversions

Given the great many interpretive difficulties inherent in Mark 9:49, it is no surprise that Baarda is not the only modern scholar to have assumed that our Greek text represents a mistranslation of an Aramaic or Hebrew original. Indeed, several have attempted to reconstruct Semitic retroversions that circumvent the difficulties we have observed in the received text. These proposed reconstructions, some of which are more commendable than others, are almost as numerous as the Greek textual variants surveyed above. According to Hirsch Perez Chajes, the Aramaic original of Mark 9:49 may have been כי כל אש באש ימלח, which he translates as "denn jedes Feuer wird mit Feuer gesalzen" ("For every fire will be salted with fire").[52] Such a reconstruction is unfortunately "still more obscure than the Greek" and therefore no more plausible.[53] Charles Torrey, who deems the Greek text of our saying "pure nonsense,"[54] proposes that in Aramaic the verse "Every one (כֹּל) with fire (בָּאֵשׁ) will be salted" could also be translated as, "Anything spoiling is salted."[55] It is difficult, however, to imagine why such a colorless and pedestrian pronouncement would have been committed to memory by Jesus' disciples or preserved by the early Church.[56] Only slightly better is Günther Schwarz's proposal. He presupposes that behind the Greek lies an Aramaic text, which read כּוֹלָא יִתְמְלֵיחַ מַנּוּרָא ("Everyone will be salted with fire"). This he believes to derive from a misreading of a more original כּוֹלָא

---

50. Coleman, "'Salt' and 'Salted,'" 361.

51. For the view that Jesus did intend his teachings, especially his parables, to confound his hearers, see Davies, *Jesus the Healer*, 128–36. See Mark 4:11–12.

52. Chajes, *Markus-Studien*, 53.

53. Baarda, "Mark IX.49," 318.

54. Torrey, *The Four Gospels*, 302.

55. Torrey, *Our Translated Gospels*, 11.

56. See Taylor, *Gospel according to St. Mark*. Taylor deems Torrey's translation "prosaic" (413).

יִתְמְלֵיט מַנּוּרָא, which he renders into Greek as πᾶς ἐκ πυρὸς σωθήσεται and translates as "Jeder wird aus dem Feuer gerettet werden" ("Everyone will be saved from the fire").⁵⁷ In support of his reading, Schwarz appeals to similar notions found in 1 Cor 3:15 and Jude 23.⁵⁸ Notably lacking, however, are any corroborating declarations within the Jesus tradition that could substantiate his proposal. Perhaps more persuasive is the contribution of Weston Fields, who argues that when Mark 9:49 is translated into Hebrew (כָּל אִישׁ בָּאֵשׁ יִמְלָח), an alternative and much preferred interpretation presents itself. Fields observes that while in both Hebrew and Aramaic the verb מָלַח most frequently means "to salt," it can also mean "to destroy."⁵⁹ Fields seizes upon this second meaning and contends, "[i]t would fit this context perfectly to translate 9:49, 'everyone [who is sent to hell] will be *completely destroyed* (destroyed by fire).'"⁶⁰ While it is true that Fields's reconstruction does fit its Markan context better than many other proposed renderings of our saying, the qualifier "who is sent to hell" with which he feels compelled to modify his translation indicates that his reconstruction does not fit its context as perfectly as he would like. Moreover, we have already observed the substantial arguments suggesting that the Markan context is secondary. If this is the case, there is no reason to believe that the indefinite pronoun πᾶς ("everyone"), which Fields renders as כָּל אִישׁ ("every man"), refers to those who have been thrown into Gehenna, for there is no such indication within Mark 9:49 itself. When this is taken into consideration and Fields's reconstruction is divorced from its Markan context, his reading amounts to nothing less than a nihilistic proclamation of universal destruction, which is a very improbable candidate for the saying's original meaning.

All of the above proposed solutions fail to resolve adequately the mystery of Mark 9:49.⁶¹ While the proposals of Chajes and Torrey are no

---

57. Schwarz, "πας πυρι αλισθησεται," 45.

58. See, however, my argument below regarding the interpretation of 1 Cor 3:15.

59. Fields, "Everyone Will Be Salted with Fire," 302. See Grotius, *Operum Theologorum*, 316; Parker, "The Posteriority of Mark," 71–72.

60. Fields, "Everyone Will Be Salted with Fire," 302; bracketed and parenthetical text original.

61. Although it is perhaps the least likely retroversion of Mark 9:49, that of Thomas F. McDaniel, "Clarifying Mark 3:7 and 9:49," Online: http://tmcdaniel.palmerseminary.edu/Mark_3&9.pdf [accessed 15 November 2008], is, for the sake of thoroughness, worthy of mention. McDaniel proposes that when our saying is reconstructed in Hebrew as כי הכל ימלח בבער, it can be translated not only as "for everyone will be salted with fire," but also as "for everyone will be dragged through the muck." He takes this to be a description of the fate of those who are stoned as a punishment for causing the little ones in v. 42 to stumble and whose bodies are consequently "dragged" through the garbage in גֵּי־הִנֹּם (the Valley of Hinnom), which McDaniel mistakenly interprets not

less problematic than the Greek text, the meaning of Fields's retroversion remains captive to its secondary Markan context. And while Schwarz's proposal may be intelligible and its meaning may be independent of its secondary context, it can find no sure footing within the Jesus tradition. Strikingly, however, when we return to Baarda's reconstruction, די כל אנש בנורא יטבל ("For everyone will be baptized in fire"), we confront none of these shortcomings.

Historical Plausibility

There is, in fact, much to commend in Baarda's proposal. Its meaning is clear and in no way dependent upon its Markan context, and perhaps most importantly, it finds a very plausible context within the Jesus tradition. John the Baptist preached a baptism in fire (Q 3:16), and Jesus was baptized by John and was probably, at least for some time, one of his disciples.[62] As one who submitted to the water baptism of John, Jesus would have undoubtedly heard and known well John's message. To be sure, his willingness to receive John's baptism indicates, at the very least, a tacit endorsement of his proclamation. Jesus, moreover, makes his approval of John's message explicit elsewhere when he deems John "more than a prophet" (Q 7:26) and the greatest "among those born of women" (Q 7:28) and indicates that John's baptism is "from heaven" (Matt. 21:25//Mark 11:30//Luke 20:4).[63] It would, therefore, be surprising if in his own preaching ministry Jesus had not adopted elements of John's message for which he clearly had profound respect.

Several scholars have made a case for a high degree of continuity between the preaching of John the Baptist and that of the historical Jesus.[64] For our purposes, the most significant of these points of continuity is that which exists between Q 3:16, John's prophecy of the Coming One who would baptize with the Holy Spirit and fire, and Luke 12:49–50: "I came to bring fire to the earth, and how I wish it were already kindled! I have a baptism with

---

as a metaphor for hell but as a literal reference to the terrestrial region known by that name (6–12).

62. See Sanders, *Jesus and Judaism*, 11. Sanders lists Jesus' baptism at the hands of John as an "indisputable fact." On the question of whether Jesus was a disciple of John, see Becker, *Johannes der Täufer*, 12–15; Hollenbach, "The Conversion of Jesus," 196–219; Meier, *Marginal Jew*, 2:116–30; Dapaah, *Relationship between John the Baptist and Jesus of Nazareth*, 93–96.

63. On the historical plausibility of Jesus' praise of John the Baptist, see Meyer, *Aims of Jesus*, 124–28; Meier, *Marginal Jew*, 2:130–71; Taylor, *The Immerser*, 299–316.

64. See Webb, "John the Baptist and His Relationship to Jesus," 227–29; Taylor, *The Immerser*, 261–316; Allison, "Continuity," 16–22.

which to be baptized, and what stress I am under until it is completed!"[65] In this last saying Jesus combines the images of fire and baptism in what appears to be an intentional allusion to John's proclamation about the Coming One's baptism with fire.

If Jesus was baptized by John, as few would doubt, and if the saying preserved in Luke 12:49–50 goes back to Jesus, the plausibility of which we shall discuss shortly, we find a very plausible framework within which to locate Baarda's reconstruction of Mark 9:49, for the pronouncement "Everyone will be baptized in fire" comports with the proclamation of the Baptist in Q 3:16, and Luke 12:49–50 provides independent attestation that Jesus deliberately echoed precisely this saying of John. Thus, although certainty regarding the historical authenticity of Baarda's retroversion is beyond our grasp, the revised saying harmonizes remarkably well with John the Baptist's proclamation of the Coming One's baptism with fire (Q 3:16) and with the preaching of Jesus attested to elsewhere in the tradition (Luke 12:49–50). The retroversion thus presents a plausible—and more coherent—alternative to the problematic logion found in the Greek text of Mark 9:49. However, before unpacking the full significance of this reconstruction, we would do well to consider a closely related text, which similarly speaks of baptism and fire.

Luke 12:49–50 "Fire upon the Earth"

Perhaps more clearly than any other saying of Jesus in its present form, Luke 12:49–50, with its pairing of the words fire and baptism, indicates the influence that John's preaching about the Coming One who would baptize with fire may have had on Jesus. John proclaimed that the Coming One would baptize with fire, and his disciple, Jesus, who elsewhere implicitly identifies himself with John's Coming One (Matt 11:2–6//Luke 7:18–23), now proclaims that he has come to cast fire on earth and that he has a baptism with which to be baptized. We shall discuss the origin of these two verses, considering whether they were originally independent of one another or if they constituted a unity from their earliest stage. We shall also consider the imagery of baptism as it is used in this saying and how that image and the imagery of fire being cast upon the earth are to be interpreted in light of one another.

---

65. On the historicity of Luke 12:49–50 and its original unity, see below.

Sources and Historical Plausibility

By way of introduction we note that a parallel to Luke 12:49 appears in *Gos. Thom.* 10 and that we have a saying similar to Luke 12:50 in Mark 10:38. The fact that *Gos. Thom.* 10 speaks of fire but not baptism and that Mark 10:38f. includes the saying on baptism but has no reference to fire in its immediate context has led some to suggest that Luke is responsible for combining these two originally independent logia to create his own hybrid saying, in which case neither saying independent of the other would cause us to suspect an explicit connection with the preaching of the Baptist. Such an argument typically is based upon the notion that 12:49 existed in Q without 12:50 and that 12:50 is essentially a Lukan rewriting of Mark 10:38b. Further, the suggestion that 12:49 has its textual origin in Q is frequently derived from the observation that Matt 10:34 (parallel to Luke 12:51), like Luke 12:49, contains the formula ἦλθον βαλεῖν . . . ἐπί τὴν γῆν, which is lacking in Luke's version of this saying (12:51). This is taken to suggest not only that Luke 12:49 belonged to Q, but also that Matthew applied the formulaic structure found in that verse to the saying now found in Matt 10:34.[66] In addition, as Kloppenborg submits, "[n]one of the vocabulary of v. 49 is Lukan and the verse is thematically coherent with other parts of Q."[67] Arens arrives at a similar conclusion if only by a different route. He proposes that Matt 10:34 is closer to the Q version of that saying, the Lukan version of which has been heavily redacted. He further suggests that Q 12:49 (πῦρ ἦλθον βαλεῖν ἐπὶ τὴν γῆν) and the Matthean form of Q 12:51 (ἦλθον βαλεῖν εἰρήνην ἐπὶ τὴν γῆν ἀλλὰ μάχαιραν) originally stood together in Q "tied by the hook-expression(s) ἦλθον βαλεῖν and ἐπὶ τὴν γῆν."[68] It is then asserted that Matthew omitted Q 12:49 and that Luke 12:50, while inspired by Mark 10:38, is in its present form the product of Luke's own hand and could not have stood in Q, for it interrupts the parallelism between Q 12:49 and 51.

In support of this last claim, that Luke 12:50 is a Lukan formulation, Stephen Patterson notes the following distinctively Lukan linguistic usages: "ἔχειν ('to have') with the infinitive in 50a, and πῶς ('how') used as an exclamation, συνέχειν ('to be in anguish'), τέλειν ('to finish'), and ἕως ὅτου ('until') in 50b."[69] Of the above linguistic features, it is Luke's frequent use of verb συνέχω in particular, which occurs nine times in Luke-Acts (Luke 4:38; 8:37, 45; 12:50; 19:43; 22:63; Acts 7:57; 18:5; 28:8) but only three times in the rest

---

66. These arguments are succinctly summarized in Kloppenborg, *Formation of Q*, 142.

67. Ibid.

68. Arens, *ΗΛΘΟΝ-Sayings*, 69.

69. Patterson, "Fire and Dissension," 124. See März, "Feuer auf die Erde," 483.

of the New Testament (Matt 4:24; 2 Cor 5:14; and Phil 1:23), that suggests to Köster that "only Luke himself, who almost alone in the NT uses συνέχω, can have formulated the expression."[70]

Lastly, Lüdemann has registered the claim that despite the parallelism one may initially perceive between Luke 12:49 and 50, they do not cohere well together:

> v. 49 is evidently about the goal of Jesus' activity, whereas v. 50 speaks of a temporary personal experience of Jesus, namely his death, about which he is anxious (Wellhausen). Verse 50 must therefore be a *vaticinium ex eventu* like Mark 10.38. It was either attached to Jesus' saying in v. 49 after the death of Jesus or Luke has formulated it himself in retrospect. In the latter case v. 49 may also come from Luke.[71]

As do Bultmann and others, Lüdemann finds it incomprehensible that Jesus could have foreseen his own imminent demise. For Lüdemann, even if v. 49 is authentic, of which he is dubious, the prophecy of Jesus' own death in v. 50 cannot be.

In sum, the argument that Luke 12:49 and 50 were originally discreet sayings and ought not be held together in our exegesis runs like this: Matt 10:34 demonstrates apparent knowledge of the formula used in 12:49 and thus suggests that 12:49, which shows no signs of Lukan redaction, was part of Q. Luke 12:50, on the other hand, particularly given the occurrence of the verb συνέχω, manifests several distinctively Lukan linguistic features and, therefore, is said to be Luke's rendering of Mark 10:38. The parallel to Luke 12:49, which appears in *Gos. Thom.* 10 without any parallel to Luke 12:50, further suggests that Luke 12:49 was originally independent of 12:50. Moreover, 12:49 and 50 appear to have differing emphases; the former stresses the goal of Jesus' ministry, the latter his individual fate. And finally, 12:50, with its allusion to Jesus' death must be a *vaticinium ex eventu*, for Jesus could not have known what the future held for him.

Prior to entering into debate with the position outlined above, we wish to acknowledge one point of agreement: it is highly improbable that Luke is the creator of 12:49, for this saying has more in common with verifiably Q material than it does with material that is distinctively Lucan. For instance, Jesus' announcement that he came to cast fire on the earth "appears

---

70. Köster, "συνέχω."

71. Lüdemann, *Jesus*, 250. Bultmann asserts, "these sentences in parallel do not really match each other properly, since v. 49 is clearly referring to the aims of Jesus' ministry, while v. 50 speaks of 'a passing personal experience' (Wellhausen)" (*History of the Synoptic Tradition*, 154).

contrary to the intent of 9:54f., where Jesus reproaches his followers for thinking about calling down fire from heaven upon the unresponsive Samaritan village."[72] Further, as Tuckett has affirmed, if the verse belonged in Q, "it would tie in well with John's prediction in 3:16: the 'fire' predicted by John is now seen to be administered by Jesus, and Jesus, as John's Coming One, looks forward here to fulfilment of John's prophecy."[73] Similarly, Hiers maintains, "[i]t also corresponds, in substance, to the several 'Q' sayings in Luke that warn of a fate worse than that of Sodom for the cities that fail to repent (Luke 12:12–15)."[74] Although we can never be certain, it is possible that 12:49 originally stood in Q. At the very least, we can be quite confident that Luke did not create this tradition *ex nihilo*; if it is not from Q, it comes from his special L source.

That point of agreement stated, at least three objections against the position that Luke 12:49 and 50 originally circulated independently and were only later crafted into a unified and parallel saying by Luke require our attention. First, the evidence drawn from *Thomas* is highly debatable, for that gospel has a tendency to fragment and reorder the tradition it has received.[75] Even if the author of *Thomas* independently knew a saying parallel to Luke 12:49–50, he may have broken it up for his own purposes. Moreover, while the tradition history of *Thomas* and questions regarding its dependence upon or independence from the Synoptic tradition remain contested, Tuckett has demonstrated that there is sufficient reason to believe that at some points *Thomas* reflects certain redactional tendencies of the Synoptic Gospels, particularly those of Luke.[76] Thus, *Thomas*'s isolated version of our saying more likely reflects his own treatment of the Lukan version rather than an independent witness to its original form. Indeed, *Thomas*'s version exhibits that text's realized eschatology with the second aorist verb ἔβαλον "I have cast" in contrast to Luke's phrasing, πῦρ ἔλθον βαλεῖν ἐπὶ τὴν γῆν, καὶ τὶ θέλω εἰ ἤδη ἀνήφθη, the second clause of which in particular indicates that this action is as of yet unfulfilled.

Secondly, as is easily observed, in its present form Luke's dual saying exhibits a significant degree of literary parallelism: "I have come . . . , and how I wish . . . . I have a baptism . . . , and how I am constrained . . . ." This

---

72. Hiers, *Jesus and the Future*, 35.

73. Tuckett, *Q and the History of Early Christianity*, 157–58.

74. Hiers, *Jesus and the Future*, 35.

75. See Tuckett, "Thomas and the Synoptics." Tuckett notes, "[t]he evidence of the POxy fragments shows that, at least at one point, a Coptic translator (or a later editor) changed the order of the text for no better reason than that of a Stichwort link" (139).

76. Ibid., 145–57.

may indicate an original unity between these sayings.⁷⁷ According to Beasley-Murray, "[t]he closeness of connection between vv. 49 and 50 indicates that the two sayings have been handed down together in the tradition and were originally spoken together."⁷⁸ Moreover, Burney avers that Luke 12:49-50 is easily translated back into Aramaic and represents an instance of the poetic form known as *synthetic parallelism*.⁷⁹ In making these observations, we are not innocent of the above linguistic arguments made by Köster and others. If they are correct in attributing the phrase "how I am constrained until it is completed" in particular to Lukan redaction, our saying would be stripped of its parallelism.

We must take seriously, however, the frequently made observation that this is not the type of prophecy that is likely to be the product of early Christian communities. Despite the fact that many have judged the presence of the verb συνέχω to be an indication of Lukan redaction, it is difficult to attribute such a tradition to the creative hand of Luke or any other early Christian. Indeed, it is the very idea behind Luke's verb συνέχω ("how great is my distress, what vexation I must endure")⁸⁰ that speaks against the likelihood that it can be attributed to Lukan creativity, for "the hint that Jesus shrinks from the suffering shows that this saying too can hardly be a *vaticinium ex eventu*."⁸¹ Moreover, as Jeremias observes, there is reason to believe that this particular instance of the verb συνέχω is traditional, for "[t]he only place in the NT, in which the verb us used absolutely, is found in the non-Markan material of Luke (12:50). It could hardly be Lukan."⁸² Thus, while the verb συνέχω may be characteristically Lukan, this particular usage is distinctively not. It probably comes from Luke's special L material or possibly Q.⁸³

---

77. See Rothschild, *Baptist Traditions and Q*, 122.

78. Beasley-Murray, *Baptism in the New Testament*, 74.

79. Burney, *Poetry of Our Lord*, 89–90.

80. BDAG s.v. συνέχω. Italics removed.

81. Kümmel, *Promise and Fulfillment*, 70. See Otto, *Kingdom of God*. Although it was not yet articulated as such, Otto essentially appeals to the criterion of embarrassment: "one can feel how he shakes with the same inner shuddering as that which later broke forth once more in Gethsemane. No later person would have invented a vaticinium ex eventu in such a form. In particular, the words: 'how am I straitened till it be accomplished,' were beyond the invention of any later person equally with those spoken by Jesus in Gethsemane itself" (360).

82. Jeremias, *Die Sprache des Lukasevangeliums*, 223.

83. While it is difficult to imagine this tradition as a Lucan creation, Luke's willingness to preserve similar traditions is apparent in Luke 22:42, which he has taken over from Mark 14:36.

Additionally, when contrasted with the detailed passion predictions one finds in Mark 8:31; 9:12; and 9:31, the prophecy in Luke 12:50 and Mark 10:38 "is quite vague, and indicates nothing but a premonition."[84] Mark 10:33 provides the clearest example of what we might expect from a *vaticinium ex eventu*:

> See we are going down to Jerusalem, and the Son of Man will be handed over to the chief priests and the scribes, and they will condemn him to death; then they will hand him over to the Gentiles; they will mock him, and spit upon him, and flog him, and kill him; and after three days he will rise again.

It is precisely the lack of this sort of specificity in Luke 12:50 that commends its historical plausibility. Indeed, as Keck puts it, "[s]o imprecise an *interpretation* of his death would scarcely have been invented by the church, since the 'Passion predictions' show that the trend was to have Jesus speak explicitly about his death."[85] On the other hand, to espouse the notion that Jesus could not have foreseen in any way an ominous conclusion to his strife-ridden ministry is to portray him as exceedingly naïve, especially after he had witnessed the fate of John, his predecessor.[86]

Third, we have repeatedly observed that there is precedent for the imagery of fire and water being brought together in the Hebrew Bible and Jewish apocalyptic literature as symbols of judgment.[87] Such a pairing of fire and flood imagery is likewise attested elsewhere in the Jesus tradition. In Luke 17:26–30, which shall be treated below, Jesus threatens that "the days of the Son of Man" will be like the days of Noah and the days of Lot, which are characterized by water and fire respectively. To be sure, if Jesus— a disciple of the Baptist—was influenced by his teacher, as his baptism by and praise of John indicate he was, we should not be surprised to recognize echoes of John's proclamation in the words of Jesus. To the contrary, we should be perplexed if it were otherwise. If we may add to this Baarda's retroversion of Mark 9:49, the argument gains considerable strength.

Finally, we must address Ludemann's above suggestion that vv. 49 and 50 have differing emphases: the former being concerned with the goal of Jesus' ministry and the latter with his individual fate. Such a distinction is only superficial. There is no reason to insist that Jesus could not have seen his mission of ushering in a period of eschatological tribulation for all, while

---

84. Otto, *Kingdom of God*, 360.

85. Keck, *A Future for the Historical Jesus*, 228–29.

86. On the prospect that Jesus anticipated an untimely end, see especially McKnight, *Jesus and His Death*, 121–56.

87. See Allison, *End of the Ages*, 126–27.

at the same time recognizing that he, too, would be required to face the fiery trials associated with that tribulation. The emphasis does not shift; rather, as we proceed from v. 49 to v. 50, the significance of the pronouncement comes into sharper focus—all, including even Jesus himself, are to face this fiery ordeal. Even if one were to maintain a shift in emphasis, which I do not, one could argue that v. 49 reflects an earlier stage in Jesus' thinking, according to which all would face a fiery ordeal, while v. 50 reflects a later refinement in his thought, according to which Jesus accepts the baptism in fire for himself alone.

In the foregoing discussion we have observed the following: (1) the sayings found in Luke 12:49–50 express a high degree of parallelism, which accords with Aramaic poetic formulations and suggests an original unity; (2) although some would suggest that Luke 12:50 is redactional and later than Luke 12:49, it seems highly improbable that early Christians, and Luke in particular, would create a tradition such as Luke 12:50 in which Jesus is depicted as being in great distress about his fate; moreover, the prophecy recorded here is so wrought with ambiguity that we are inclined to believe this to be a very early one, which Luke took over either from Q or L in its entirety;[88] and (3) given this saying's resemblance to the preaching of John the Baptist, whose influence upon Jesus can scarcely be denied, and the possible parallel in Baarda's reconstruction of Mark 9:49, such a saying can easily be accounted for in the preaching of the historical Jesus. If Jesus did speak these words, however, their meaning is no less obvious to us than if they originated with Luke, perhaps even less so. In what follows, therefore, it is with the interpretation of this saying within the preaching of Jesus that we must concern ourselves.

If Luke 12:49 and 50 do indeed belong together, our exegesis must reflect that unity. As noted above, there is precedent in the Hebrew Bible for the unity of fire and flood, and so fire and baptism likewise ought to fit comfortably together in our exegesis. In his treatment of Luke 12:49, A. J. Mattill identifies three distinct interpretive categories into which Luke 12:49 has been classified throughout the history of interpretation. These are (1) spiritualizing interpretations; (2) sociological interpretations; and (3) literal interpretations.[89] As we shall see, some of these approaches integrate the fire and baptism symbolism more satisfactorily than do certain others.

---

88. I am inclined to favor the suggestion that the saying was taken over from Q given Luke 12:49's correlation with that material.

89. Mattill, *Luke and the Last Things*, 210–20.

Spiritualizing Interpretations

Beginning with the church fathers, one can note a tendency for the fire of Luke 12:49 to be interpreted as a reference to spreading the gospel or as the fiery outpouring and indwelling of the Holy Spirit. When commenting upon Luke 12:49, Cyril of Alexandria wrote,

> That henceforth not in Judaea only should the saving message of the Gospel be proclaimed—comparing which to fire he said, "I am come to cast fire on the earth"—but that now it should be published even to the whole world. ... Behold therefore, yea, see, that throughout all nations was that sacred and divine fire spread abroad by means of the holy preachers.[90]

Likewise, Bengel is representative of such an interpretation when he writes that the fire which Jesus wishes for is "the fire of spiritual ardour [which is] kindled on Pentecost."[91]

We note, however, that such a reading typically interprets the fire as referring to something other than that to which the baptism in 12:50 refers. The fire is the gift of the Spirit or the gospel whereas the baptism is a painful and horrific death to be endured. Not only are the verses interpreted in isolation from one another, but the very approach to their individual interpretations is varied: while the fire is given a *spiritual* interpretation, the term baptism is employed to designate a *historical* event, namely the passion which "must precede the fire, and the kindling of it."[92] This will not do. It fails to respect the strong parallelism between the two verses, and as Creed suggests, the verb "βαλεῖν is not appropriate to spiritualizing interpretations of the 'fire.'"[93] Let us, then, consider the sociological interpretation.

Sociological Interpretations

In its present context in Luke, our saying about fire and baptism precedes a word about the division and strife among families which Jesus says he came to instigate:

> Do you think that I have come to bring peace to the earth? No, I tell you, but rather division! From now on five in one household will be divided, three against two and two against three;

---

90. Cyril of Alexandria, *Comm. on Luke*, II, 94.
91. Bengel, *Gnomon*, 114.
92. Ibid.
93. Creed, *Gospel according to St. Luke*, 178.

> they will be divided: father against son and son against father, mother against daughter and daughter against mother, mother-in-law against her daughter-in-law and daughter-in-law against her mother-in-law. (Luke 12:51–53)

The sociological interpretation, then, reads Luke 12:49 and 12:51–53 in parallelism and, skipping over v. 50, interprets v. 49 in light of what follows. According to Creed, "[t]he fire must be a symbol for the division of which the subsequent verses speak."[94] Thus the fire anticipated in v. 49 metaphorically refers to the familial discord that arises as family members side themselves either with or against the Messiah. The sociological reading also finds voice among the church fathers. Tertullian writes that when Jesus speaks of coming to cast fire on earth in Luke 12:49, "he meaneth the fire of turmoil and upheaval" (*Against Marcion* IV.29).

According to Mattill, however, there is no use of the image of fire to be found in the Hebrew Bible that would suggest such a reading of our text. The only text that comes remotely close to supporting the sociological interpretation would be Prov 26:20–21: "As charcoal is to hot embers and wood to fire, so is a quarrelsome person for kindling strife." As Mattill maintains, however, "there is hardly any analogy between a gossipy troublemaker fanning the flames of arguments and strife, on the one hand, and the Messiah creating eschatological division within families, on the other."[95] Moreover, as Allison affirms, "everywhere else in the Synoptics, with the exception of Mark 9:21 (the possessed boy who throws himself into the fire) and Luke 22:25 (Peter warming himself in the courtyard), *pyr*, 'fire', has to do with the last assize."[96] Why should we interpret Luke 12:49 any differently? What is more, in the sociological interpretation the supposed parallelism between vv. 49 and 51–53 is emphasized at the expense of the parallelism that exists between vv. 49 and 50. The only way that the baptism saying can be worked into this interpretation is by the suggestion that Jesus believed that the strife which he came to initiate would inevitably lead to his own death.[97]

Having briefly surveyed the spiritual and sociological approaches, we must take issue with the position assumed in the above exegesis that the baptism Jesus predicts he must undergo should be interpreted as an unambiguous reference to his imminent crucifixion and death. This interpretation is arrived at through reading the baptism mentioned in Luke 12:50 in light of Mark 10:38f., in which Jesus speaks of the cup (τὸ ποτήριον) he must

---

94. Ibid.
95. Mattill, *Luke and the Last Things*, 211–12.
96. Allison, *End of the Ages*, 125.
97. See Otto, *Kingdom of God*, 360.

drink, in synonymous parallelism with the baptism which he must undergo. This is probably a correct reading of Mark 10:38; however, the confusion arises when this saying is in turn read in light of Jesus' anguished words in Gethsemane where he, in apparent reference to his imminent crucifixion, pleads, "Abba, Father, for you all things are possible; remove *this cup* (τὸ ποτήριον) from me; yet, not what I want, but what you want" (Mark 14:36 par.). The interpretive chain is, then, from baptism to cup to crucifixion, and the prophecy about baptism is thus equated with a prediction of crucifixion. Moreover, the interpretation of the second clause of Luke 12:50, "how I am constrained until *it is finished* (τελεσθῇ)," as pointing to Jesus' crucifixion may be influenced by Jesus' cry from the cross as it is found in John's Gospel: "*It is finished* (τετέλεσται)" (19:30). According to Cunningham, "[t]his thought will be reinforced in the subsequent chapter [of Luke], where the accomplishment (τελειοῦμαι) of Jesus' journey refers to his death as a prophet (13:32–33)."[98] These factors naturally lead some exegetes to believe that in 12:50 the baptism with which to be baptized is an unambiguous reference to Jesus' crucifixion. Such a reading of our saying may be legitimate within its canonical and Lukan contexts; it may even be Luke's intended meaning. We should, however, be wary of projecting evangelists' thoughts onto the historical Jesus.

Indeed, nothing in the immediate context of Luke 12:49–50 itself suggests that Jesus is speaking explicitly of crucifixion. More likely this is a general reference to imminent judgment and tribulation, which is not given a clear form as a definite reference to crucifixion. It is, therefore, unlikely that the spiritual reading of this text, according to which the fire refers to the gospel's spread, "spiritual ardor," or the Holy Spirit and the expected baptism points explicitly towards Jesus crucifixion, was the intended meaning of this saying. Such an interpretation fails to respect, on the one hand, the parallelism between vv. 49 and 50 and, on the other hand, the ambiguous nature of the reference to the baptism with which Jesus anticipates being baptized. We now turn to the literal interpretation of our verse to determine if it shines a brighter light upon our saying.

Literal Interpretation

Reginald Fuller has suggested that in Luke 12:49 Jesus is speaking of "the fire of eschatological judgment, the negative aspect of the coming of the Kingdom (see Mark 9:49)."[99] This coheres with the findings of our survey of the

---

98. Cunningham, *Through Many Tribulations*, 111.
99. Fuller, *The Mission and Achievemen of Jesus*, 62.

Hebrew Bible in which we observed that fire was a widely expected feature of the coming judgment on the Day of the Lord. In support of this reading, our verse has several close cousins in the eighth chapter of the Apocalypse of John in which angels of judgment *cast fire upon the earth*. What is most striking in this parallel is the recurrence of the very same Greek vocabulary that we find in Luke 12:49: βαλεῖν (to cast, hurl, throw), πῦρ (fire), and the prepositional phrase εἰς τὴν γῆν (upon the earth), although in the last example cited here the object of the preposition is τὴν θαλάσσαν (the sea), not τὴν γῆν (the earth).[100]

> 8:5: Then the angel took the censer and filled it with fire (τοῦ πυρός) from the altar and threw it on the earth (ἔβαλεν εἰς τὴν γῆν).
>
> 8:7: The first angel blew his trumpet, and there came hail and fire (πῦρ), mixed with blood, and they were hurled to the earth (ἐβλήθη εἰς τὴν γῆν).
>
> 8:8: The second angel blew his trumpet, and something like a great mountain, burning with fire (πυρί), was thrown into the sea (ἐβλήθη εἰς τὴν θαλάσσαν).

I do not wish to suggest any theory of dependence between Luke 12:49 and Rev 8:5-8. Rather, what is more likely is that what we have here is a shared form of apocalyptic speech. Revelation 8 clearly depicts an eschatological scenario, and one would be hard pressed to find exegetes who would wish to argue that what is envisioned here is the outpouring of the Holy Spirit or a metaphor for familial discord. Yet, when these very words are found on the lips of Jesus in Luke 12:49, interpreters have hastened to find an alternative to the literal reading of the text.

The literal interpretation of 12:49 coheres very agreeably with Luke 12:50 when that verse is properly understood. As we have suggested above, v. 50 is too vague to refer explicitly to Jesus' impending crucifixion. Moreover, as Delling has shown, although the word βάπτισμα is not found in the LXX, the notion expressed by the image of baptism does appear as one of being overwhelmed by distress and life-threatening danger.[101] This interpretation of βάπτισμα as being overcome by anguish is enforced by the pairing of the images of the cup and baptism in Mark 10:38. While in its Markan context the cup may foreshadow Jesus' words in Gethsemane, which are closely linked with Jesus' imminent crucifixion, in the saying's original context, it probably referred generally to the cup of wrath (see Ps 75:9; Isa 51:17;

---

100. See Mattill, *Luke and the Last Things*, 214.
101. Delling, "βάπτισμα βαπτισθῆναι," 101-2.

Jer 25:28; 49:12; Ezek 23:31–34; Hab 2:16).[102] As Beasley-Murray notes, "[a]t once therefore it should be observed that in Mk. 10.38 the two clauses are parallel in meaning: a *cup* is to be drunk, a *baptism* is to be endured. The idea of drinking a cup of suffering is frequent in the Old Testament, but significantly it is most commonly used of the cup of wrath which God apportions to sinful peoples."[103] Such imagery could be employed to depict universal eschatological judgment far more readily than it would be called upon to prophesy one's individual death.

Moreover, there is an impressive correlation between this saying and John the Baptist's preaching of the Coming One and his fiery baptism. As Dunn affirms,

> Most striking of all is the echo of the distinctive metaphor coined by the Baptist: "He will baptize you with the Holy Spirit and fire" (Matt. 3.11/Luke 3.16). It was the Baptist, we may recall, who brought the metaphor of baptism into play as an image for the great tribulation to come, in which he expected his hearers to be immersed. That two of the three key images in the Baptist's prediction (baptism, fire) should reappear here with similar effect and in a not dissimilar combination (both predictive of intense tribulation) can hardly be dismissed as merely coincidental. More likely, Jesus was remembered as taking up and echoing (deliberately) the Baptist's metaphor.[104]

And if, as we have argued, the Baptist envisioned the Coming One's baptism in fire as a literal immersion in the eschatological river of fire as a fiery ordeal that would accompany the last judgment, and if Jesus is here deliberately echoing the Baptist's words, as suggested above, would he not have imbued those words with the same meaning that the forerunner did? If when speaking of the future baptism in fire John envisioned an immersion in a river of fire, would Jesus not have had in mind the same idea when he spoke of fire and baptism in the same breath? In the words of Jeremias, "Jesus is the bringer-in of the New Age. But he knows—and this troubles him deeply—that the way to New Creation lies through disaster and destruction, through purging and judgement, through the deluge of fire and water."[105] Like John, Jesus, in pairing the imagery of fire and baptism here,

---

102. Ibid., 93–95.
103. Beasley-Murray, *Baptism in the New Testament*, 73.
104. Dunn, *Jesus Remembered*, 804.
105. Jeremias, *The Parables of Jesus*, 164.

probably envisions the eschatological river of fire through which all must pass at the last judgment.[106]

And yet, while there is continuity with the Baptist here, there is also an apparent variation on this theme of judgment with fire. Jesus was influenced by John's message, to be sure, and in large part he adopted the proclamation of this predecessor, but he did not merely imitate him. John spoke of the Coming One administering a fiery baptism; Jesus, who has implicitly claimed to be the Coming One (Matt 11:2–6//Luke 7:18–23), says he has a fire which he not only came to cast upon the earth but with which he must also be baptized. Lang puts it as follows:

> the One who baptises with the Spirit and fire must first tread the path of suffering himself.... Thus v. 49 says that Jesus will bring a judgment of fire on the earth in which He Himself will be implicated. The meaning of πῦρ here is controlled by the basic sense of the eschatological judgment of fire, but the judgment is present in and with Jesus.[107]

Whereas the Baptist, as far as we can tell, expected the Coming One to baptize others in fire, Jesus sees himself as being immersed in the eschatological river of fire, of having the cup of wrath poured out upon him. There is perhaps a shift in Jesus' use of the Baptist's language, for "the fire and the baptism of Luke 12:49–50 appear to be headed at Jesus rather than coming from Jesus."[108] However, while here Jesus speaks only of his own baptism which he must endure, we should not too hastily jump to the conclusion that this saying suggests that his own expiatory death would remove the cup of suffering from all others and excuse them from the baptism with fire. As Beasley-Murray perceives, in Mark 10:38f., in which Jesus also speaks of a baptism with which to be baptized,

> the disciples are called to drink the cup of woe and be baptized with the baptism of Jesus—they are to suffer along with Him. This is not solitary in the teaching of the New Testament, for the Lord called on men to take a cross with Him to Jerusalem (Mk

---

106. Unlike Conzelmann's argument (*Theology of St. Luke*, 109) that "Luke interprets the saying as referring to the eschatological conflagration, for he does not draw a parallel between vv. 49 and 50, but contrasts them: the End has not yet come, but instead there is the baptism of death," our exegesis holds together the parallelism apparent in this saying.

107. Lang, "πῦρ," 944.

108. McKnight, *Jesus and His Death*, 147. See Caird, *Gospel of St. Luke*, 167.

8.34) and Paul sought to enter into the sufferings of the Christ for the sake of the Body (Col 1.24).[109]

Indeed, Jesus' prediction that the sons of Zebedee will walk the path of suffering with him may be taken as support for the historical plausibility of Mark 10:48f., for it is difficult to imagine Christian tradition inventing an unfulfilled prophecy and putting it in the mouth of Jesus. Moreover, as we have already seen, Mark 9:49 may bolster the view that Jesus expected everyone to endure the judging and purifying baptism of fire which Jesus himself must endure.

In sum, we have argued against those who would assert that the sayings found in Luke 12:49 and 50 were originally autonomous sayings that circulated independently of one another. To the contrary, their parallelism with one another and the continuity they express between Jesus and the Baptist indicate their original unity. As a cohesive logion, therefore, the two verses should be interpreted in light of one another. Whereas spiritualizing and sociological interpretations of this saying fail to hold its two parts together in close parallelism, a literal interpretation of their meaning, which interprets the fire and baptism imagery in accord with the sense in which these two images are used in the biblical tradition to refer to judgment and with John the Baptist's use of that very same imagery does respect the unity and parallelism of this saying. Jesus, like John before him, here anticipates a final judgment by immersion in the eschatological river of fire, but perhaps to the Baptist's surprise Jesus believed that he himself must submit to the very baptism in fire that John expected him to administer. The Coming One who was to baptize with fire now joins with his baptisands to endure the tribulation that precedes the coming of the kingdom.

The Coming One's Baptism Revisited

A few details from our discussions of Mark 9:49 and Luke 12:49–50 are worth noting in connection with Q 3:16, for if one accepts Baarda's retroversion of Mark 9:49, fruitful avenues for exploring several of the vexing questions surrounding Q 3:16 reveal themselves.[110] First, the most readily apparent difference between John's proclamation in Q 3:16 and Baarda's retroversion of Mark 9:49 ("For everyone will be *baptized* in fire") is the presence of the Holy Spirit in the former and its absence in the latter. It is

---

109. Beasley-Murray, *Baptism in the New Testament*, 75. Perhaps the obscure saying in *Gos. Thom.* 82 ("Whoever is near me is near the fire, but whoever is far from me is far from the kingdom"), which we shall address below, similarly attests to this expectation.

110. See our discussion of these questions above.

frequently asserted that John the Baptist originally spoke of a baptism with fire without mention of the Holy Spirit; however, while it is certainly imaginable that the prophecy concerning the Holy Spirit was added to John's proclamation in the wake of the early Christians' pneumatic experience at Pentecost, James D. G. Dunn notes that there exists "no text which speaks of baptism in fire [alone]; it is a purely hypothetical construction."[111] We have only the tradition in Mark 1:8; John 1:33; Acts 1:5; 11:18 that attests to a baptism of the Holy Spirit and the tradition in Matthew and Luke (Q 3:16) that attests to a baptism with both the Holy Spirit and fire. It is, however, possible that Baarda's reconstruction furnishes us with the missing testimony to a baptism in fire alone and offers a window into the earliest tradition. If we may presume that in his own preaching Jesus faithfully represents the message of his mentor, John, Jesus' proclamation of an imminent baptism in fire, which makes no mention of the Holy Spirit, may correspond to the tradition he received from the Baptist. This possibility finds further support in Luke 12:49–50, which similarly links baptism and fire while excluding any mention of the Holy Spirit.[112] Judging from the witness of his most famous disciple, then, it may be possible to maintain that John proclaimed the Coming One's baptism with fire alone.

Second, if Jesus' words, "Everyone will be baptized in fire," faithfully convey the message of John, it would confirm the arguments of those who believe that when the Baptist said, "He [the Coming One] will baptize *you* (ὑμᾶς) in (the Holy Spirit and) fire," he addressed both the penitent baptisands whom he immersed in water and the unrepentant scoffers standing on the banks of the Jordan whom he decried as a "brood of vipers" (Q 3:7). That is, *everyone* would be subjected to the same baptism in fire. As Dunn observes, "[i]ts effect would then presumably depend on the condition of its recipients: the repentant would experience a purgative, refining, but ultimately merciful judgment; the impenitent, the stiff-necked and hard of heart, would be broken and destroyed."[113]

---

111. Dunn, "Spirit-and-Fire Baptism," 84.

112. Commenting elsewhere on Luke 12.49–50 and its relationship to Q 3.16, James Dunn (*Jesus Remembered*, 804) notes, "[t]hat two of the three key images in the Baptist's prediction (baptism, fire) should reappear here with similar effect and in a not dissimilar combination (both predictive of intense tribulation) can hardly be dismissed as merely coincidental." Dunn is surely correct, but the absence of the third element (the Holy Spirit) cannot be so easily dismissed, for if baptism in the Holy Spirit had figured prominently in the Baptist's proclamation, one would expect it to appear in those sayings of Jesus that are most clearly indebted to John. On the contrary, the only instance in which Jesus links baptism and the Holy Spirit is in Matt 28:19, which is widely believed to be a late tradition.

113. Dunn, "Spirit-and-Fire Baptism," 86; Davies and Allison, *Matthew*, 1:316.

This expectation coheres well with several apocalyptic texts that imagine the eschatological judgment of all humankind taking place through immersion in a river of fire (see *Apoc. Pet.* 6; *T. Isaac* 5:21–25). Strikingly similar to Baarda's reconstruction of Mark 9:49 ("Everyone will be baptized in fire") is *Sib. Or.* 2.252–54: "all will pass through the blazing river and the unquenchable flame. All the righteous will be saved, but the impious will then be destroyed for all ages."[114] Notably, in all of these texts both the elect and the wicked are immersed in the selfsame river of fire; it is only the effect of the fire upon them that differs. Luke 12.49–50 may again be of some relevance here, for according to this tradition, Jesus anticipated fire being cast upon the earth and that even he would not be exempted from this fiery lustration. What is more, if Mark 10:39b is historical ("with the baptism with which I am baptized, you will be baptized"), Jesus also expected the same eschatological ablution for his closest followers. All of this would be exceedingly odd if John, from whom Jesus received this tradition, had believed the baptism in fire to be a punishment reserved only for the wicked as so many have alleged.[115]

Last, regarding the identity of the Coming One, it may be significant that some scholars have understood the passive verb ἁλισθήσεται ("will be salted") in Mark 9:49 as an instance of the "divine passive," implying God as the active subject.[116] If this is the correct interpretation of the Greek text of Mark 9:49, it would suggest that the passive verb in Baarda's reconstruction should be interpreted similarly. Thus, if Jesus spoke the words "Everyone will be baptized in fire," it would at first glance appear that he was employing the "divine passive" to indicate that he understood the Coming One whom

---

114. These texts are admittedly all from the second century CE or later; the motif, however, antedates the texts in which it is found. Bauckham (*Fate of the Dead*, 204) observes, "this idea of the eschatological river of fire which distinguishes the righteous from the wicked is a genuinely old Iranian one, which is found already in the *Gathas*. The Apocalypse of Peter seems to be the earliest Jewish or Christian text in which it occurs, but it presumably was already to be found in Jewish apocalyptic tradition."

115. Dunn perceives Jesus' expectation that he too must endure the eschatological baptism (Luke 12.49–50) as an indication that Jesus has reinterpreted the message of the Baptist: "the baptism is one which is to be accomplished *on himself* rather than by himself—he sees himself as the baptisand rather than as the baptizer" (Dunn, "Birth of a Metaphor," 137). Given the dearth of sources at hand for reconstructing the Baptist's original expectation, it seems unwarranted to assume that John could not have anticipated the Coming One as being both the baptizer and a recipient of the fiery baptism. The parallelism between Luke 12:49 and 50 indicates that Jesus did not perceive these as mutually exclusive roles, unless we assume that a radical change of mind has taken place between the two verses.

116. See Jeremias, *New Testament Theology*, 11, n. 2; Dunn, "Birth of a Metaphor," 137; Pesch, *Das Markus-Evangelium*, 117.

John the Baptist expected to administer the baptism in fire to be none other than God.[117]

In favor of this interpretation, one may appeal to the fact that the "divine passive" form occurs frequently in the words attributed to Jesus when he speaks of the activities of God.[118] This is, however, only one factor within a much more complex discussion, and several objections have been raised against the proposition that John identified the Coming One as God.[119] Many have judged the reference to the Coming One's sandals (Matt 3:11//Mark 1:7//Luke 3:16) to be too anthropomorphic an image for John to have applied to God.[120] Perhaps even more compelling is the argument that John's use of comparative language in calling the Coming One ἰσχυρότερός ("mightier") weighs heavily against the likelihood that God is the referent, "for to compare oneself with God, even in the most abject humility, would have been presumptuous for any Jew in John's day."[121] Last, the question put by John's disciples to Jesus, "Are you the one who is to come (ὁ ἐρχόμενος), or are we to wait for another?" (Q 7:19), suggests that John did not anticipate the coming of God but of a human figure, for implicit in the question is the possibility that Jesus could have answered in the affirmative.[122]

In light of these substantial arguments against the probability that John envisioned God as fulfilling the role of the Coming One, it is perhaps necessary to question the absolute nature of the "divine passive." There are naturally instances in which the passive verb without expressed agency may be used without the purpose of identifying the subject as God.[123] More to the point, the Synoptic episode in which Jesus pronounces over the para-

117. For a parallel argument concerning the similarly phrased and clearly germane prophecy of the risen Jesus in Acts 1:5; 11:16 ("John baptized with water, but you will be baptized with the Holy Spirit"), see Dunn, *Christology in the Making*, 143. Dunn asserts, "whether they [the first Christians] thought of Jesus as the baptizer in Spirit is put in doubt by the 'divine passive' form."

118. See Jeremias, *New Testament Theology*, 11. Jeremias notes, "[t]he 'divine passive' occurs round about a hundred times in the sayings of Jesus." He does go on to caution, however, that "there are a number of borderline cases in which it is not certain whether the passive is intended as a circumlocution for an action on the part of God or whether it is used without this consideration."

119. See our discussion on pages 133–35 above.

120. Scobie, *John the Baptist*, 66–67. Above we have noted the counterargument by Hughes, "John the Baptist," 196.

121. Kraeling, *John the Baptist*, 54. See Brownlee, "John the Baptist," 41.

122. Manson, *Sayings of Jesus*, 67. On the historicity of this encounter, see Wink, "Jesus' Reply to John"; Meier, *Marginal Jew*, 2:136; Dunn, *Jesus and the Spirit*, 55–60. Wink ("Jesus' Reply to John," 125) writes, "[t]he early church would scarcely have ascribed uncertainty to John and then answered his uncertainty with baffling ambiguity."

123. Callow, "Some Initial Thoughts."

lytic, "Son, your sins are forgiven (ἀφίενταί)" (Matt 9:2//Mark 2:5//Luke 5:20), provides an instance of Jesus using the passive form to describe his own actions, which "may be a case in which the passive is chosen as a circumlocution for the first person singular."[124] Hence, a possible use of the passive verb, particularly in reference to eschatological activities of a human agent, may be to draw attention away from oneself as the subject in humble recognition that it is God who is the ultimate subject behind such activities.

There is, in fact, further evidence from within the Synoptic tradition itself indicating that Jesus did in certain cases employ the passive form when speaking of his own deeds, particularly deeds of healing that were typically considered the eschatological activities of God or the Messiah (see Mark 1:12; 4:11; 8:12; Matt 9:29; Luke 7:47–48; 13:32). Of obvious relevance to our discussion is Jesus' response to the question posed above by John's disciples. In his reply, Jesus makes what appears to be an implicit claim to be John's expected Coming One by answering with a catalogue of his own activities, many of which are listed in the passive form: "the blind receive their sight, the lame walk, the lepers are cleansed (καθαρίζονται), the deaf hear, the dead are raised (ἐγείρονται), the poor have good news brought to them (εὐαγγελίζονται)" (Q 7:22).[125] It is particularly striking that it is precisely in the context of identifying himself with ὁ ἐρχόμενος ("the Coming One")—the one whom John expected to execute the divine activity of judgment by means of baptizing in the river of fire—that Jesus uses the supposed "divine passive" form to describe several of his own activities.[126]

If, then, Jesus identified himself as the Coming One of John's proclamation with whom God's eschatological activities were associated, and believed God to be the ultimate source of those eschatological deeds, it is plausible that he employed the so-called "divine passive" to describe the actions he himself had performed and would perform in the role of God's

---

124. See the "Excursus on 'The So-called *Passivum Divinum*'" in Reiser, *Jesus and Judgment*, 270–71. That this is not, strictly speaking, an instance of the "divine passive," as is often asserted, is suggested by the fact that the Synoptic evangelists do not interpret it as such. They present the scribes as accusing Jesus of blasphemy (Matt 9:3//Mark 2:7//Luke 5:21) and asking, "Who can forgive sins but God alone?" (Mark 2:7//Luke 5:21 only), indicating that at least Mark and Luke understood Jesus' words to be an implicit claim to the authority to forgive sins.

125. That the passive verbs alternate with intransitive verbs may also be significant. Sidebottom ("So-Called Divine Passive," 202) observes, "[t]here are a great number of intransitives in the gospels, and often they are taken by the evangelists themselves as equivalent to passives."

126. See also Collins, "The Works of the Messiah." Collins judges the similar catena of healings recorded in 4Q521 to be particularly indicative of the prophetic Messiah's participation in the divine activity.

eschatological agent. This position is strengthened by the observation that "in the eschatological expectation of contemporary Judaism there were notions according to which God and the eschatological bearer of salvation work together very closely, where their respective work in fact merges."[127] Thus, it is possible that Jesus saw his own eschatological activities as indistinguishable from the eschatological work of God and for this reason used the passive form when speaking of his own deeds. Hence, given 1) that weighty arguments have been adduced against the probability of understanding the Coming One as God, 2) that Jesus used the passive form to allude to his own eschatological activities on other occasions, and 3) that it was believed that exalted or human agents could partake in the divine activity, it follows that the possible "divine passive" in Baarda's reconstruction of Mark 9.49 need not and should not lead us to the conclusion that when John the Baptist spoke of the Coming One who would baptize with fire, he had in mind God. This is not to say precisely whom John anticipated, only whom he did not.

In the preceding pages we have observed that Jesus took up the message of John the Baptist and proclaimed a baptism in fire. In so doing he probably imagined an immersion in a river of fire similar to John's immersion in a river of water. Given his insistence that *"everyone"* (Mark 9:49), including himself (Luke 12:49–50), would endure the same baptism, we can infer that the baptism would affect the righteous differently than it would impact the wicked. We cannot conclude definitively that he believed the fire would play a purgative role upon the elect, but the parallel with the cleansing baptism in water may point in that direction. At the very least, we can imagine that he believed the righteous would be preserved through such an ordeal whereas the wicked would be destroyed. We now turn to a text that similarly combines the imagery of fire and water—Luke 17:26–32.

Luke 17:26–32—Noah and Lot

We have been considering a complex of sayings (Q 3:16; Mark 9:49; and Luke 12:49–50) all related to the dual motifs of baptism and fire (if one accepts Baarda's retroversion of Mark 9:49). These sayings, we have suggested, are all interrelated in that they are mutually interpretative and they support one another's historicity. We now turn to another tradition that likewise bears some relationship to these motifs. While an explicit reference

---

127. Gnilka, *Jesus of Nazareth*, 74–75; see Volz, *Die Eschatologie*, 224–26; Webb, *John the Baptizer and Prophet*, 219–60.

to baptism is lacking, in Luke 17:26–32 judgment by water and judgment by fire are similarly juxtaposed:

> καὶ καθὼς ἐγένετο ἐν ταῖς ἡμέραις Νῶε, οὕτως ἔσται καὶ ἐν ταῖς ἡμέραις τοῦ υἱοῦ τοῦ ἀνθρώπου· ἤσθιον, ἔπινον, ἐγάμουν, ἐγαμίζοντο, ἄχρι ἧς ἡμέρας εἰσῆλθεν Νῶε εἰς τὴν κιβωτὸν καὶ ἦλθεν ὁ κατακλυσμὸς καὶ ἀπώλεσεν πάντας. ὁμοίως καθὼς ἐγένετο ἐν ταῖς ἡμέραις Λώτ· ἤσθιον, ἔπινον, ἠγόραζον, ἐπώλουν, ἐφύτευον, ᾠκοδόμουν. ᾗ δὲ ἡμέρᾳ ἐξῆλθεν Λὼτ ἀπὸ Σοδόμων, ἔβρεξεν πῦρ καὶ θεῖον ἀπ' οὐρανοῦ καὶ ἀπώλεσεν πάντας. κατὰ τὰ αὐτὰ ἔσται ᾗ ἡμέρᾳ ὁ υἱὸς τοῦ ἀνθρώπου ἀποκαλύπτεται. ἐν ἐκείνῃ τῇ ἡμέρᾳ ὃς ἔσται ἐπὶ τοῦ δώματος καὶ τὰ σκεύη αὐτοῦ ἐν τῇ οἰκίᾳ, μὴ καταβάτω ἆραι αὐτά, καὶ ὁ ἐν ἀγρῷ ὁμοίως μὴ ἐπιστρεψάτω εἰς τὰ ὀπίσω. μνημονεύετε τῆς γυναικὸς Λώτ.

> Just as it was in the days of Noah, so too it will be in the days of the Son of Man. They were eating and drinking, and marrying and being given in marriage, until the day Noah entered the ark, and the flood came and destroyed all of them. Likewise, just as it was in the days of Lot: they were eating and drinking, buying and selling, planting and building, but on the day that Lot left Sodom, it rained fire and sulphur from heaven and destroyed all of them—it will be like that on the day that the Son of Man is revealed. On that day, anyone on the housetop who has belongings in the house must not come down to take them away; and likewise anyone in the field must not turn back. Remember Lot's wife.

Here we have the "days of Noah" and "the days of Lot" both serving as prototypes for "the days of the Son of Man." A few preliminary comments are in order. First, the plural ἡμέραις τοῦ υἱοῦ τοῦ ἀνθρώπου "*days of the Son of Man*" is a unique phrase, given that we are more accustomed to the singular "day of the Son of Man"; it was apparently introduced by Luke to parallel the phrase ταῖς ἡμέραις Νῶε "the days of Noah" and ταῖς ἡμέραις Λώτ "the days of Lot." Second, this brings us squarely up against the perennial "Son of Man problem." Giving this problem the attention it deserves far exceeds the scope of this study, so I can only relate that I am inclined to accept the position that Jesus himself may have employed this phrase not solely in an idiomatic, non-titular sense—whether as a circumlocution for the first person pronoun, as a generic reference to humankind, or as an allusion to class of individuals—but also in an apocalyptic and titular sense with reference to Daniel 7.[128]

---

128. The "Son of Man Problem" has a notoriously long and complex history. See especially the discussions in Burkett, *Son of Man Debate*; Casey, *Solution to the "Son of*

Sources

The first substantial issue we must address is the question of Luke's sources. The first section regarding "the days of Noah" has a close parallel in Matt 24:37–39, thus suggesting an origin in Q:

> For as the days of Noah were, so will be the coming of the Son of Man. For as in those days before the flood they were eating and drinking, marrying and giving in marriage, until the day Noah entered the ark, and they knew nothing until the flood came and swept them all away, so too will be the coming of the Son of Man. (Matt 24:37–39)

Certain elements in Matthew's version, however, betray his own redactional interests. Matthew emphasizes "the coming of the Son of Man" (vv. 37, 39) and the ignorance of those who were swept away: "*they did not know* until the flood came" (v. 39) in contrast to Luke's simple description that "the flood came and destroyed them all" (v. 27). Matthew, following Mark, has placed this material immediately after Jesus' pronouncement that "about that day and hour no one knows, neither the angels of heaven, nor the son, but only the Father" (v. 36), and has followed it with "Keep awake therefore, for you do not know on what day your Lord is coming" (v. 42).[129] It seems more probable, then, that Matthew has altered the material to fit his overall theme and that Luke retains the earlier form.

More significant for our present discussion is the fact that a Matthean counterpart to Luke's section regarding "the days of Lot" is lacking, raising the question whether or not this logion derived from Q.[130] One of the most important observations related to this question is the fact that in Jewish and early Christian tradition Noah and Lot are frequently associated with one another.[131] This observation has led to at least two contradictory historical

---

Man" Problem; Müller, *The Expression "Son of Man"*; Owen and Hurtado, *Who is this Son of Man?* The contributions to Hurtado and Owen's work provide a strong defense of the apocalyptic titular use of "Son of Man" by the historical Jesus.

129. Gregg, *Historical Jesus*, 254. See Davies and Allison (*Matthew*, 3:374), who write regarding v. 36: "Its declaration of eschatological ignorance grounds the entire section [24:36—25:30]."

130. In favor of its existence in Q are, among others, the following: Schnackenburg, "Der Eschatologie Abschnitt Lk 17, 20–37"; Polag, *Fragmenta Q*, 78–79; Kloppenborg, *Formation of Q*, 157; Davies and Allison, *Matthew*, 3:381–82. Opposed to it are: Holtzmann, *Die synoptischen Evangelien*, 202; Zmijewski, *Die Eschatologiereden des Lukas-Evangeliums*, 454; Fitzmyer, *Gospel according to Luke*, 2:1165; Guenther, "A Fair Face is Half the Portion."

131. See Lührmann, *Die Redaktion der Logienquelle*, 75–83; Schlosser, "Les jours de Noe et de Lot." See Wis 10:4–6; *T. Naph.* 3.4–5; 3 Macc. 2.4–5; Philo, *Mos.* 2.10–12;

conclusions. On the one hand, some have suggested that the association of Noah and Lot was one that Luke exploited to expand on the Q tradition about Noah by adding the complementary Lot material, which he patterned after the Noah section. One strong argument in favor of this approach is the observation that Matthew has no aversion to the theme of fiery judgment, so if it had been in his source, we would expect him to have incorporated it. On the other hand, some have taken it to imply that since these traditions are so frequently linked, it would have made perfect sense for Jesus to have combined them as well. Jesus, a Jewish prophet of judgment who was saturated in the Jewish tradition, was drawing on a well-known juxtaposition of two prototypes of eschatological judgment. This, of course, bypasses the question of whether the Lot tradition was in Q, but if it can be traced back to Jesus, the probability that it was also in Q may potentially be increased, for it does not ascribe the pairing of Lot and Noah to Luke. Still others suggest the Lukan form of Q contained it while the Matthean form did not.[132]

A number of factors incline me to favor the view that we are here dealing with a Q tradition shared by Matthew and Luke. While the absence of the Lot material from Matthew is evidence that must be given serious consideration, several other factors suggest its presence in Q. First, there is no sign of Lukan redaction in vv. 28–29, where the days of Lot are introduced. The language is not typically Lukan,[133] and while the comparisons with the days of Noah and Lot focus on the unexpected nature of the coming of the Son of Man, "Luke's redactional interest lies in a parenetic application: attachment to possessions is pointless and indeed dangerous since the judgment will be sudden and unheralded."[134] Moreover, the following verses, which are redactional, shift the focus away from Lot and onto Lot's wife.[135] Luke is unlikely to have created the relatively cumbersome Lot parallel merely as a transition to his redaction of Mark 13:15–16.[136]

Further, both vv. 26 and 28 are instances of what Edwards has identified as "the eschatological correlative," which employs the following pattern:[137]

---

2 Pet 2:5–7.

132. Tödt, *Son of Man*, 51; Bultmann, *Histyory of the Synoptic Tradtion*, 87–88.

133. Jeremias, *Die Sprache des Lukasevangeliums*, 269.

134. Kloppenborg, *Formation of Q*, 157.

135. Tuckett, *Q and the History of Early Christianity*, 158.

136. Geiger, *Die lukanischen Endzeitreden*, 93. See Kloppenborg, *Formation of Q*, 157.

137. Edwards, *The Sign of Jonah*, 49.

Protasis: καθὼς (ὥσπερ, ὡς)—verb in the past or present tense

Apodosis: οὕτως (κατὰ τὰ αὐτά)—ἔσται—ὁ υἱὸς τοῦ ἀνθρώπου

According to Edwards, the eschatological correlative occurs four times in Q (Edwards counts Luke 17:28, 30 as one of these) and only once outside of Q (Matt 13:40f.). It is thus more in keeping with Q's style than with Luke's.[138] Moreover, Luke typically excludes rather than introduces parallels such as that seen in Luke 17:26–30.[139] Third, the pairing of fire and water is a motif we have already encountered in Q 3:16 and Luke 12:49–50 (which possibly derives from Q). Fourth, despite the initial impression that Matthew is unlikely to have excluded the Lot material, several possible explanations can be imagined.[140] Matthew may have wished to abbreviate his inherited material since the Lot illustration adds little not already present in the Noah illustration. He may have felt that it stood in conflict with Matt 11:23–24, which suggests some measure of leniency towards Sodom.[141] The omission may have resulted from an instance of *homoioteleuton* (compare the ends of Luke 17.29 and Luke 17.27).[142] Or, and this argument is perhaps most compelling, "the evangelist may have deemed only the flood story—of universal scope—truly parallel to the *parousia*: the disaster at Sodom was local."[143] In connection with this last observation—that the parallel with the *parousia* is inexact—it is worth noting that "[t]he main thrust of both sayings in Q is similar: the SM will come without warning suddenly in the midst of everyday activity."[144] Thus, on the whole, the arguments in favor of inclusion in Q slightly outweigh those against it.

---

138. Edwards argues that the eschatological correlative is unique to Q and material dependent on Q, which he takes to indicate that the Q community created these sayings. Others, however, have demonstrated similar forms in the Septuagint and the Dead Sea Scrolls.

139. Cadbury, *Style and Literary Method of Luke*, 85–88. See Kloppenborg, *Formation of Q*, 157.

140. The following are taken from Davies and Allison, *Matthew*, 3:381.

141. However, it must be conceded that Luke apparently felt no such tension between this passage and the saying in 10:12 that "on that day it will be more tolerable for Sodom than for that town."

142. Similarly, Beasley-Murray, *Baptism in the New Testament*, 340. He notes that the Matthean form of our passage curiously contains two occurrences of the clause "so will be the coming of the Son of Man" (vv. 37 and 39) and suggests that the second occurrence may have derived from the comparison with "the days of Lot."

143. Davies and Allison, *Matthew*, 3:381.

144. Tuckett, *Q and the History of Early Christianity*, 159.

Interpretation

Having considered the question of sources and tentatively concluded that both traditions derived from Q, another question must be considered: what is the main point of the juxtaposition of the days of Noah and the days of Lot? Some have suggested that it is the wickedness of "this generation" while others have argued it is the imminence of judgment on that wicked generation that is in the fore. Surely, these are common features of both. The generation of Noah was notoriously wicked as were the men of Sodom, and it is possible that "eating and drinking" alludes to morally questionable behavior and that, at least in the context of the Noah saying, "marrying and being given in marriage" may refer to the sons of God who lay with the women (Gen 6:1–4). More probably, however, it alludes to the unpreparedness of the generation of Noah and the people of Sodom.

Moreover, in the context of the saying, it is Noah and Lot who are the center of attention.[145] As Glasson observes:

> it is significant that in each case attention is drawn to *the survivors*. The time of crisis is described, not as the day when the flood came, but "the day that Noah entered into the ark," not the day when it rained fire from heaven, but "the day that Lot went out from Sodom." Had it been the purpose of Jesus merely to illustrate sudden judgment, there would have been no need to mention Lot at all.[146]

The emphasis is laid on Noah, who escaped the flood, and on Lot, who was rescued from the rain of fire. Strikingly, this introduces a distinctive note in the material we have been considering. In the preceding material (Q 3:16; Mark 9:49; Luke 12:49–50), we have noted that both the righteous and the wicked are expected to endure the baptism in fire. In the Noah and Lot analogies, the righteous are expected to escape from any sort of fiery judgment.

Historical Plausibility

Lastly, we turn to the question of whether or not this saying can be attributed to the historical Jesus. It is the view of Davies and Allison that "Lk 17.26–30 (Q) preserves words of Jesus. He warned of eschatological judgment, spoke of its suddenness (see Lk 12.39 (Q)), and likened his own generation to sinful generations of the past. . . . Also, the tradition elsewhere

---

145. Lövestam, *Jesus and 'this Generation'*, 62.
146. Glasson, *The Second Advent*, 82–83.

has him associating Sodom with the great assize (Lk 10.12 (Q))."[147] We may add to this the observation that this passage juxtaposes fire and water in a manner similar to Q 3:16 and Luke 12:49-50 (as well as Baarda's retroversion of Mark 9:49), which we have argued have a plausible origin with the historical Jesus. Luke 17:26-30 thus coheres to a degree with previously established traditions. On the other hand, as we have just noted, it introduces a new element—the righteous are delivered from, rather than tested by, the eschatological fire of judgment. The saying itself may only be singly attested; however, in addition to the above arguments offered by Davies and Allison, the motif of fire combined with water at the last judgment enjoys multiple attestation and thus coheres nicely with the traditions for which we have already rendered a favorable judgment. We may thus offer a cautiously positive judgment on the historicity of this saying. We now proceed onto less firm ground in our consideration of the singly-attested Luke 23:31.

Luke 23:31—The Green Wood and the Dry

In Luke 23:31, at the conclusion of his woes on the daughters of Jerusalem, Jesus cries out, ὅτι εἰ ἐν τῷ ὑγρῷ ξύλῳ ταῦτα ποιοῦσιν, ἐν τῷ ξηρῷ τί γένηται; ("For if they do this when the wood is green, what will happen when it is dry?").[148] Some have protested that this proverb is "not to be pressed to mean 'judgment of fire.'"[149] However, while it is true that fire is not explicitly mentioned, the motif of judgment by fire stands clearly in the background. We have already encountered the green tree/wood and dry tree/wood motif in Ezek 20:45-47 and 1QHa XI.27b-37, in both of which the judgment of fire is explicit and prominent.[150] Here, however, the green wood and dry wood imagery is employed in a more subtle proverbial form. Frequently cited is the rabbinic parallel from Jose ben Joezer (second cent. BCE), who, on his way to being crucified, reportedly stated, "If such things (as crucifixion) happen to those who do His will, how much more (and worse will hap-

---

147. Davies and Allison, *Matthew*, 3:382.

148. While it is possible to read this verse as a continuation of the lament of the daughters of Jerusalem, in its present context this saying is most likely spoken by Jesus himself on his way to be crucified and thus refers to his own fate.

149. Ellis, *Gospel of Luke*, 266.

150. Brant Pitre ("Blessing the Barren," 71) identifies an intertextual link not with Ezek 20:45-47, but with Hos 9:16, which mentions Ephraim's dry root, leading Pitre to read this verse as an allusion to a time of barrenness during the period of eschatological tribulation. While Pitre's exegesis is suggestive, his elevation of Hos 9:16 over Ezek 20:45-47 (of which he makes no mention) leads him to favor a very distant intertextual link over one that shares much more vocabulary in common with our verse.

pen) to those who offend Him."[151] Similarly, Jesus likens himself to a green tree, moist, slow to kindle, and undeserving of the fire, and implies that those who will be subjected to the future judgment will be like dry trees, easily combustible and quickly consumed.[152] Whereas the fate awaiting him is unjustified in light of his righteousness, the future judgment of fire will be fittingly administered to the wicked. Note the similar idea present in 1 Pet 4:17–18, which is notably in the context of the "fiery ordeal": "For the time has come for judgment to begin with the household of God; if it begins with us, what will be the end for those who do not obey the gospel of God? And 'If it is hard for the righteous to be saved, what will become of the ungodly and the sinners?'"

Interpretation

Nearly all interpreters agree that we are dealing with a proverb of judgment, the gist of which is this: "The fate of Jesus, like that of the prophets, is sure to befall his enemies."[153] However, the identities of the judge and, in the case of the dry tree, the judged remain disputed. It has become customary among commentators to follow Plummer's outline of the various positions:[154] (1) If the Romans are willing to crucify Jesus whom they found innocent of insurrection, how will they treat those who participate in armed resistance? One proponent of this position is George Caird: "Israel's intransigence has already kindled the flames of Roman impatience, and if the fire is now hot enough to destroy one whom Roman justice has pronounced innocent, what must the guilty expect?"[155] This position has the obvious benefit of fitting the context, for it was, of course, the Romans who crucified Jesus. (2) If the Jews treat Jesus, the one who came to bring them salvation, in such a manner, how will God treat them?[156] This position is problematic for the reason that the subject shifts from "the Jews" to "God" halfway through the saying.[157] (3) If humans behave in this way before their cup of wickedness is full, what will they do when it overflows? This interpretation is even less satisfactory, for as Darrel Bock notes, it "sees the green tree as a negative

---

151. Cited in Manson, *Sayings of Jesus*, 343. See Strack and Billerbeck, *Kommentar zum Neuen Testament*, 263–4.

152. Fitzmyer, *Gospel according to Luke*, 2:1498.

153. Schweizer, *Good News according to Luke*, 358.

154. Plummer, *Luke*, 529–30.

155. Caird, *Gospel of St. Luke*, 249–50.

156. See Green, *Gospel of Luke*, 816.

157. Bock, *Luke*, 2:1847.

reference, which is unlikely."[158] (4) Fitzmyer adds to Plummer's list the common view here given expression by Marshall: "If God has not spared the innocent Jesus, how much more severe will be the fate of guilty Jerusalem?"[159] (5) Nolland adds to this list the possibility that the proverb is intentionally vague with no particular referent in view.[160] This, however, is highly unsatisfying, for it is difficult to imagine why Luke would record a saying of judgment with no apparent person or people in mind.

I am left favoring the first and fourth interpretations. On the one hand, the first reading, as noted above, is appealing, for it makes sense in the context of Jesus' crucifixion, for it was the Romans who condemned Jesus and administered his punishment. More, from Luke's perspective sometime in the 80s, the Romans were likewise responsible for the destruction of Jerusalem, and notably the burning of the temple. Fitzmyer makes the intriguing suggestion that "[t]he contrast is further between the wood on which Jesus is crucified (not consumed by flames) and the wood of Jerusalem (consumed by flames) in its destruction," by which he obviously has in mind destruction by the Romans.[161] On the other hand, the fourth reading, which is by far the most popular interpretation, suggests itself for several reasons. First, it is possible that the personal pronoun "they" is used as a circumlocution for God. This is supported by the fact that in the two other known texts that employ the green/dry motif, Ezek 20:45-47 and 1QHa XI.27b-37, the judgment of fire is administered by supernatural beings. In the former, it is God who kindles the fire; in the latter, it is "the torrents of Belial." Second, in all of the other passages we have been examining that have to do with the judgment of fire, the fire is of divine origin.

It is possible that, in attempting to decide between these two interpretations, we may be falling prey to a false dichotomy, and that these two interpretations can be held together. We know that Josephus interpreted the destruction of Jerusalem as God's will, even though it was carried out by the Romans, in the same way that Isaiah viewed Assyria as the rod of the Lord's anger (10:5). So it is possible that Luke held the same view; that God's will was being carried out by the Romans in their destruction of Jerusalem and he thus interpreted the personal pronoun "they" inclusively, referring both to God and to the Romans, the agents of the divine will. What is more significant for us is the fact that in this saying, Luke's Jesus envisions the judgment of fire for all. It is not reserved for the dry trees only, but the green as

---

158. Ibid.
159. Marshall, *Gospel of Luke*, 865.
160. Nolland, *Luke*, 2:1138.
161. Fitzmyer, *Gospel according to Luke*, 2:1498-99.

well, including even, or perhaps especially, himself. This all-inclusive aspect is the major theme of both Ezek 20:45–47 and 1QHa XI.27b–37.[162]

Lucan Context

In fact, Ezek 20:45–47 appears to play a significant role in Luke's casting of this material, and from our present vantage point, a look back over Luke's Gospel with this Old Testament passage in mind proves instructive. In light of the apparent allusion to the green and dry trees of Ezek 20:47 in Luke 23:31, it is interesting to note other possible allusions to this passage in Luke's Gospel, particularly in his travel narrative. Ezekiel 20:45–47, in which Ezekiel is addressed as son of man, a common title of Jesus in the Synoptic tradition,[163] begins with the proclamation, "The word of the LORD came to me: Mortal, set your face toward the south (MT: תֵּימָנָה; LXX: Θαιμαν), preach against the south (MT: דָּרוֹם LXX: Δαρωμ), and prophesy against the forest land in the Negeb (MT: נֶגֶב; LXX: Ναγεβ)" (MT 20:46; LXX: 21:2). From Ezekiel's perspective in exile in Babylon the most obvious referent for "the south" would be Judah, and particularly its capital, Jerusalem; even "the Negeb" need not be taken literally, but may be used in the directional sense to indicate the south, again pointing towards Jerusalem[164]—this is especially the case given Ezekiel's frequent dire warnings of destruction against Jerusalem. With this in mind, it is very likely that God's command that Ezekiel set his face towards the south/Negeb is picked up by Luke in 9:51, where he says that Jesus had set his face towards Jerusalem.[165] Notably, this occurs in a context of fiery judgment, for James and John have just requested permission to call down fire from heaven to punish the Samaritans. Jesus, however, denies their request, for in Luke's narrative, such judgment is reserved for Jerusalem.

162. See pages 57–58 and 97–103 above.

163. See Mattill, *Luke and the Last Things*, 221. Mattill concedes that in Ezekiel the title is meant in the sense of "mortal man" whereas in Luke it becomes an apocalyptic title.

164. As Zimmerli (*Ezekiel*, 423) notes, "[h]ere, as in 47:19; 48:28, תימן must be understood as a geographical direction. The same is true of דרום, which is used only as a geographical direction in all its Old Testament occurrences . . . נגב on the contrary, in the Old Testament first denotes 'Land that is dried out,' i.e., the region south of the Judean mountain range with its poor rainfall, the hollow of Beersheba and the desert region which extends south of it. From this the expression has become in the Old Testament a straightforward geographical direction."

165. This potential intertextual link is much stronger than the one adduced by Mattill (*Luke and the Last Things*, 221), who calls attention to the reference to the south wind in Luke 12:55, not least because in that verse the wind comes from the south.

Further, in v. 47 the word of the Lord commands Ezekiel to proclaim, "I will kindle a fire in you." The verb used for "kindle" in the Septuagint of this verse is ἀνάπτω (LXX: 21:3), which is the very word that Jesus uses in Luke 12:49, when he says, "I came to bring fire to the earth and how I wish it were already kindled (ἀνήφθη)!" It is possible, therefore, that we have here a second intertext between Luke and Ezek 20:45-47.[166] Reading Luke 12:49 in this context would indicate that Jesus intended to kindle this fire in Jerusalem. Moreover, Luke would almost certainly have understood the baptism with which Jesus must be baptized in 12:50 as a reference to his crucifixion (see Mark 10.38), and as kindling the fire and being baptized stand parallel to one another, both expressions point to his crucifixion. This brings us back to Luke 23:31, where Jesus again alludes to Ezek 20:45-47 with reference to the green wood and the dry wood as he makes his way to the cross. These intertextual links can be seen quite clearly when laid out side by side:[167]

Ezekiel	Luke
20:46: Mortal [son of man], set your face toward the south [i.e. Jerusalem]	9:51: he [Jesus, the Son of Man] set his face to go to Jerusalem
LXX 21:2: υἱὲ ἀνθρώπου στήρισον τὸ πρόσωπόν σου ἐπὶ Θαιμαν	GNT: αὐτὸς τὸ πρόσωπον ἐστήρισεν τοῦ πορεύεσθαι εἰς Ἰερουσαλήμ
20:47: I will kindle a fire in you	12:49: I came to bring fire to the earth, and how I wish it were already kindled!
LXX 21:3: ἐγὼ ἀνάπτω ἐν σοὶ πῦρ	GNT: πῦρ ἦλθον βαλεῖν ἐπὶ τὴν γῆν, καὶ τί θέλω εἰ ἤδη ἀνήφθη

---

166. So also Mattill, *Luke and the Last Things*, 221.

167. I offer here only the Greek of the LXX for comparison with the Greek text of Luke's Gospel. While there are no major discrepancies between the Greek and the Hebrew, Luke the Evangelist appears to have relied on the LXX or an Old Greek text very similar to the LXX here, for if he were making his own translation of the Hebrew of this text, it is difficult to explain why he did not translate דֶּרֶךְ "way, road, journey" in Ezek 20:46, given the centrality of ὁ ὁδός "the way" in his own Gospel and in Acts.

Ezekiel	Luke
20:47: it shall devour every green tree in you and every dry tree	23:31: For if they do this when the wood is green, what will happen when it is dry?
LXX 21:3: καταφάγεται ἐν σοὶ πᾶν ξύλον χλωρόν καὶ πᾶν ξύλον ξηρόν	GNT: ὅτι εἰ ἐν τῷ ὑγρῷ ξύλῳ ταῦτα ποιοῦσιν, ἐν τῷ ξερῷ τί γένηται;

These intertextual links strongly affirm the reading that in Luke 23:31 Jesus is anticipating judgment on Jerusalem, whether at the hands of God or the Romans with divine decree, as a direct result of his undeserved crucifixion, which served to kindle the fire there.[168] In the words of Joachim Jeremias, "The fire of judgment will pass on from the green wood to the dry, that is, Jesus' suffering will be the prelude to the collective suffering of the tribulation."[169]

Historical Plausibility

Judgment regarding the historicity of this saying is extremely difficult, especially now that we have seen how intentionally Luke has integrated this material into his travel narrative. On the one hand, the saying is only attested here and has no obvious parallel elsewhere in the Jesus tradition. Moreover, the Rabbinic parallel may indicate that Luke has taken a known saying associated with an individual wrongly condemned to crucifixion and attributed it to Jesus. On the other hand, while it is a singly-attested saying, it may echo Jesus' warning that trees that do not bear fruit will be cast into the fire (Matt 7:19; Luke 13:7–9; see Q 3:9). It also coheres with other aspects of the tradition we have been examining, particularly the notion that all will be subjected to the eschatological judgment of fire (Q 3:16; Mark 9:49; Luke 12:49–50). Thus, while the logion is certainly imaginable on the lips of the historical Jesus, and its historicity cannot be ruled out, there is insufficient evidence to grant a positive judgment of historical plausibility.

---

168. A rough parallel to this may be found in Matthew's parable of the wedding feast (22:1–10), according to which the king's son is unjustly killed by those who refuse to attend the feast, as the result of which "The king was angry, and he sent his troops and destroyed those murderers and burned their city" (v. 7).

169. Jeremias, *New Testament Theology*, 284.

Thomas 82—"Near the Fire"

While *Gos. Thom.* 82 falls outside the canon of the New Testament, it is so remarkably similar in content and style to the sayings we encounter in the canonical Gospels that it has struck many as being one of the "Synoptic cousins,"[170] and it ranks among the few logia in the *Gospel of Thomas* without a Synoptic parallel that is frequently given serious consideration as a saying of the historical Jesus.[171] Thus, despite our primary focus on the New Testament texts dealing with the eschatological functions of fire, we would be remiss if we did not take into consideration this "Synoptic cousin."

Sources

Until the past century, the enigmatic logion, "Whoever is near me is near the fire; whoever is far from me is far from the kingdom," was known as one of the *agrapha*—that category of sayings attributed to Jesus by the church fathers, but not recorded in any known gospel—and was attested by Origen, Didymus the Blind, and Pseudo-Ephraem. However, when the Nag Hammadi Library was unearthed in 1945, the saying was discovered in the Coptic *Gospel of Thomas* (82), and more recently it has been discovered in the *Gospel of the Savior* (71).[172] The saying takes the following forms (arranged in order of probable chronology):

1. ⲠⲈⲦϨⲚ ⲈⲢⲞⲈⲒ ⲈϤϨⲚ ⲈⲦⲤⲀⲦⲈ ⲀⲨⲰ ⲠⲈⲦⲞⲨⲎⲨ ⲘⲘⲞⲈⲒ ϤⲞⲨⲎⲨ ⲚⲦⲘⲚ ⲦⲈⲢⲞ. "Whoever is near me is near the fire; whoever is far from me is far from the kingdom" (*Gos. Thom.* 82).

2. [ⲈⲦϨⲚ] ⲈϨⲞⲨⲚ ⲈⲢⲞ[Ⲓ ⲈϤ]ϨⲚ ⲈϨⲞⲨⲚ Ⲉ[ⲠⲔ]ⲰϨⲦ ⲠⲈⲦⲞⲨⲎⲨ ⲈⲂⲞⲖ ⲘⲘⲞⲒ ⲈϤⲞⲨⲎⲨ ⲈⲂⲞⲖ ⲘⲘⲰⲚϨ Ⲡ. "[Whoever is close] to [me] is close to [the fire]. Whoever is far from me is far from life" (*Gos. Sav.* 71b).

3. *Que iuxta me est, iuxta ignem est; qui longe est a me, longe est a regno.* "Whoever is near me is near the fire; whoever is far from me is far from the kingdom" (Origen, *Hom. in Jer. Lat.* 23).

---

170. On the term, see Patterson, *Gospel of Thomas*, 18. Patterson, however, does not include Logion 82 in this category, for in his view, "in terms of content there is really nothing like it this [sic] in the early Christian corpus" (89). As we shall see, however, other scholars have taken a different view.

171. Jeremias, *Unknown Sayings*, 66–73.

172. See Hedrick and Mirecki, *Gospel of the Savior*, 41 (107.43–48 according to Hedrick and Mirecki's versification).

4. ὁ ἐγγύς μου ἐγγὺς τοῦ πυρός· ὁ δὲ μακρὰν ἀπ'ἐμοῦ μακρὰν ἀπὸ τῆς βασιλείας. "Whoever is near me is near the fire; whoever is far from me is far from the kingdom" (Didymus the Blind, *Comm. in Ps.* 88.8).

5. ὃς ἐγγίζει πρὸς ἐμέ, πρὸς (τὸ) πῦρ ἐγγίζει, καὶ ὃς (or ὃς δὲ) μακρὰν ἐστιν ἀπ'ἐμοῦ ἐγγίζει, μακράν ἐστιν ἀπό (τῆς) ζωῆς. "He who joins me, joins with fire, and he who is far from me, is far from life" (Pseudo-Ephraem, *Exposition of the Gospel*).[173]

The saying thus appears to have been quite popular, and was widely attributed to Jesus. Despite these multiple attestations, however, problems persist.

First, sources range from as early as the late first century (*Gospel of Thomas*) to the early fifth century (Pseudo-Ephraem). Second, there is the question of dependence. Origen was familiar with the *Gospel of Thomas*, and it is quite possible that he knows this saying from the Greek version of that text. Jeremias observes that although Origen elsewhere condemns the *Gospel of Thomas* (*In Luc. Hom.* I), and here introduces the saying with some ambivalence,[174] on another occasion he introduces the first half of it with "scriptum est" (*In lib. Jesu Nave Hom.* IV.3), suggesting its authoritative standing. Jeremias takes this to indicate that despite Origen's disregard for the *Gospel of Thomas*, he accepted the dominical nature of this saying.[175] It is often asserted that Didymus took his version of the saying from Origen,[176] but as Jeremias points out, this cannot be demonstrated. It is just as possible that Didymus independently found this saying in the Greek *Gospel of Thomas* given Thomas's circulation in Egypt.[177] Regardless, Thomas, Origen, and Didymus all present us with one form of the tradition, which is identical in content despite the different languages in which it is recorded (Coptic, Greek, and Latin). Alongside this tradition, Jeremias espied an independent tradition in Pseudo-Ephraem's *Explication of the Gospel*, which he dated to sometime prior to 430 CE. The reason Jeremias believed this to be independent is because it offers a slightly variant form, which concludes, "whoever

---

173. Unfortunately this version survives only in Armenian, which is cited in Jeremias as follows: Or mer̄jenay ar̄ is, ar̄ hur mer̄jenay. Ew or her̄i ê yinên, her̄i ê i kenaç (*Unknown Sayings*, 68, n. 4). Nearly a century ago Shäfers did us the service of rendering the Armenian into the above Greek translation. See Schäfers, *Eine altsyrische antimarkionitische Erklärung*, 185.

174. He introduces the saying with this: "I have read somewhere that the Saviour said—and I question whether someone has assumed the person to the Saviour, or called the words to memory, or whether it be true that is said—but at any rate the Saviour says there . . . ."

175. Jeremias, *Unknown Sayings*, 66–67.

176. Bauer, "Das Jesuswort"; Grant and Freedman, *Secret Sayings*, 174.

177. Jeremias, *Unknown Sayings*, 67–68.

THE "FIRE WORDS" OF JESUS OF NAZARETH   191

is far from me is far from *life*." Significantly, Jeremias's suggestion may be upheld by the recently discovered *Gospel of the Savior*, a document that has been given little consideration due to its relatively recent discovery.[178] If scholars are correct in dating the *Gospel of the Savior* to the second century, it provides us with much earlier attestation of the alternative ending ". . . far from life," which was previously only known from Pseudo-Ephraem.[179] As noted by Hedrick and Mirecki, although the version of our saying in the *Gospel of the Savior* is preserved in Coptic, it appears to be independent of *Gos. Thom.* 82: "The sayings are the same except for certain syntactic variants (omission of ⲉϩⲟⲩⲛ and ⲉⲃⲟⲗ in the Thomas version of the saying), and one variant in substance. The Gospel of the Savior reads at the end of the second stich 'life,' instead of 'kingdom' as it appears in the Gospel of Thomas."[180] While one might guess that the linguistic similarities between the *Gospel of the Savior* and Pseudo-Ephraem suggest dependence, Hedrick and Mirecki aver "there is no indication of a literary relationship between the Gospel of the Savior and the Armenian text."[181] At any rate, we have here a tradition which concluded ". . . far from life," and which appears to be independent of the version known to Thomas, Origen, and Didymus. Bauer, who knew the variant from Pseudo-Ephraem, argued that both forms may have gone back to the historical Jesus[182] while Jeremias argued "life" is secondary.[183] Regardless, with the discovery of the Gospel of the Savior, we may conclude that this form was known as early as the second century, if not earlier, which is a considerable advancement since Jeremias was only able to date it to the early fifth century.

Significantly, many have observed a probable allusion to the saying in Ign. *Smyrn.* 4:2: "Why, moreover, have I surrendered myself to death, to fire, to sword, to beasts? But in any case, 'near the sword' means 'near to God' [ὁ ἐγγὺς μαχαίρας, ἐγγὺς θεοῦ]; 'with the beasts' means 'with God.'"[184]

---

178. The Gospel of the Savior, also known as the Unknown Berlin Gospel, was acquired in 1961 by the Egyptian Museum of Berlin, but was neglected until 1991 when it was rediscovered by Paul Mirecki. It was first made accessible to the world of biblical scholarship in Hedrick and Mirecki, *Gospel of the Savior*. However, it has been reedited and corrected in Emmel, "The Recently Published Gospel of the Savior." While Hedrick and Mirecki offer an excellent introduction to and commentary on the text, as well as the Coptic text, I here follow Emmel's corrected order and translation.

179. Note how closely this also parallels b. Qid 66b; B Zeb. 13a, which is quoted below.

180. Hedrick and Mirecki, *Gospel of the Savior*, 22.

181. Ibid., 23.

182. Bauer, "Das Jesuswort," 447.

183. Jeremias, *Unknown Sayings*," 69, n. 4.

184. Trans. Holmes, *Apostolic Fathers*, 252–53.

Grant and Freedman are indecisive regarding dependence: "Perhaps Ignatius alludes to the saying; on the other hand, this saying may be based on the words of Ignatius."[185] Schoedel is skeptical that Ignatius offers an allusion to our saying, but he makes too much of the absence of the converse "far from the sword means far from God," an absence which can easily be explained by Ignatius's confidence in his faith.[186] He had every intention of dying a martyr's death, which may have precluded him from entertaining the thought that he would be far from God, the kingdom, or life. In light of the mention of fire in the first part of this saying, it appears that Ignatius presents an adaptation of our verse which has been applied to his own situation. Grant and Freedman are open to the notion that Ignatius is the source of *Gos. Thom.* 82; however, it is difficult to explain why Thomas would have excluded the other elements (death, sword, and beasts), for the gospel shows no aversion to such themes,[187] whereas, given Ignatius's position, it is easy to imagine why he would have added them.

However, matters are complicated somewhat by the existence of a proverb attributed to Aesop (*Aesop. Prov.* 7): ὁ ἐγγὺς Διός, ἐγγὺς κεραθνοῦ (he who is near Zeus is near the lightning). Elsewhere, we read "far from Zeus and the lightning" (Diogenianus *Cent.* 7.77b; Apostolius *Cent.* 14:65). It is extremely difficult to discern the relationship between the pagan proverb and the saying attributed to Jesus. Bauer argued that the pagan proverb was known by Jesus and reapplied by him.[188] Jeremias, on the other hand, noted that the pagan parallels are all of a late date and suggested that they derive from the Jesus saying. Schoedel rightly warns that Jeremias is too bold in claiming "that a saying so thoroughly pagan owed its form to its Christian parallel."[189] Yet, even if the pagan proverb antedates the Jesus saying, that is no reason to insist that it is simply "another example of the attribution of popular secular wisdom to Jesus,"[190] for the following Rabbinic parallel adduced by Burchard, indicates how Jewish tradition may have applied such a proverb to a Jewish teacher: "Akiba, he that separates himself from you separates himself from life" (b. Qid 66b par. B Zeb. 13a).[191] The statement closely parallels the second half of the Jesus saying and the pagan proverb,

---

185. Grant and Freedman, *Secret Sayings*, 175.

186. Schoedel, *Ignatius of Antioch*, 232.

187. See *Gos. Thom.* 16 in which Jesus says he has come to cast fire, sword, and war upon the earth. See also *Gos. Thom.* 7: ". . . cursed is the human whom the lion will eat."

188. Bauer, "Das Jesuswort"; Bauer, "Echte Jesuswort," 123–24.

189. Schoedel, *Ignatius of Antioch*, 232.

190. Patterson, *Gospel of Thomas*, 89. See Fieger, *Das Thomasevangelium*, 224.

191. Cited in Hofius, "Unknown Sayings," 356. Note the concluding "from life" which parallels the form of our saying in the Gospel of the Savior and Pseudo-Ephraem.

providing evidence of a Jewish appropriation of such a tradition. There is thus no reason to insist *a priori* that Jesus could not have similarly adopted the proverb. Schoedel likewise suggests Ignatius was drawing upon the pagan proverb rather than a saying of Jesus. However, in light of the partial Rabbinic parallel, which demonstrates the currency of similar proverbs in Jewish tradition, I am inclined to follow Jeremias in seeing Ignatius's sentence as an allusion to the saying attributed to Jesus, which Ignatius has reformulated to apply to his impending fate, and reflecting his confidence that he is indeed near to Jesus, God, and the kingdom.

Interpretation

Before turning to the question of historicity, it must be noted that "the reference to 'fire' eludes specific definition,"[192] a fact which has led to some disagreement among exegetes concerning the meaning of this logion. Broadly speaking, two main lines of interpretation have been followed.[193] The first is to situate the saying within the context of apocalyptic eschatology, in which case "[t]he fire is that which Jesus came to cast on the earth."[194] The fire thus takes on negative connotations. As Hofius put it, "[i]n content this agraphon, like the logion in Luke 12:49 (see Mark 9:49) brings to expression that the eschatological fire of tribulation and eschatological salvation (βασιλεία) appeared with the coming of Jesus and that the issue of fellowship with God or distance from God hinges on one's attitude to Jesus."[195] Those willing to follow Jesus are warned of the dangers of doing so,[196] yet they are at the same time promised the kingdom. Indeed, it is only through the fire of eschatological tribulation, which Jesus ushers in, that any will enter the kingdom.[197]

The second approach emphasizes the parallelism between Jesus, the fire, and the kingdom, and contextualizes the saying within the mystical tradition according to which theophanies are associated with fire and light. According to this reading, "fire is seen as a positive symbol. In fact, it is

---

192. Valantasis, *Gospel of Thomas*, 162.

193. Gathercole (*Gospel of Thomas*, 505–6) identifies six distinct interpretations.

194. Grant and Freedman, *Secret Sayings*, 174.

195. Hofius, "Unknown Sayings," 356.

196. Zöckler, *Jesu Lehren*, 59: "The addressee should keep in mind that they expected pain and distress when they entered into him."

197. See Lelyveld (*Les Logia*, 68 and 93), who links the saying with the fire of judgment and eschatological purification.

identified with the kingdom."[198] One of the foremost recent proponents of this position is April DeConick. She explains:

> When encountering God in a heavenly ascent experience, the mystic expected that the hidden *kavod* would be revealed through its light. In the Jewish mystical tradition, encountering the light of God was a transforming experience. Thus, Jewish tradition taught that the righteous will be transformed into beings of light or fire resembling or superior to the angels.[199]

In this view, the saying can be paraphrased as follows: "the person who is near to Jesus has ascended to the place of the light or fire, where Jesus now is, the heavenly Kingdom; the person who is not near to Jesus, has not yet ascended there."[200] DeConick admits that *Gos. Thom.* 82 "does not mention the transforming power of the fire directly," but goes on to suggest, "it is most probable that the early Thomasites were familiar with the fire transformation motif from their Jewish and Hermetic heritage. Once in the presence of Jesus and the fire of God, they could expect to be transformed into an angelic-like figure, resembling the light of God himself."[201] DeConick's last comment makes clear that the mystical reading is far more plausible in the context of Thomas's community than in the context of the historical Jesus. It is, however, questionable whether this is the best interpretation of the saying even within the context of the *Gospel of Thomas*. For DeConick prefers to interpret logion 82 in light of the even more enigmatic logion 13, where fire comes out of the stones and burns up Judas—she assumes this is a positive thing—rather than in light of sayings 10 and 16, where fire conveys eschatological connotations.[202] Thus, even in the context of the *Gospel of Thomas* the apocalyptic interpretation, which sees the fire as a tribulation or ordeal through which one must pass in order to enter the kingdom, seems just as likely as, if not more likely than, the theophanic interpretation, which sees the fire positively as standing in synonymous parallelism with the kingdom in the second half of the saying. Further, in light of our saying's

---

198. Marjanen, "Is Thomas a Gnostic Gospel," 125. See Valantasis, *Gospel of Thomas*, 162. Fieger, *Das Thomasevangelium*, 224: "Only he who has recognized the spark of light in himself is near the light (=Jesus = the fire) and finds his own salvation. . . . The one who is far from Jesus, however, does not always resist the temptations and passions of the material world."

199. DeConick, *Seek to See Him*, 105. See DeConick, *Recovering the Original Gospel of Thomas*, 137; DeConick, *Original Gospel of Thomas in Translation*, 247.

200. DeConick, *Seek to See Him*, 109.

201. Ibid.

202. Ibid., 114.

similarities with sayings such as Q 3:16; Mark 9:49; and Luke 12:49—in addition to *Gos. Thom.* 10 and 16—the apocalyptic interpretation appears far superior to the theophanic.

Historical Plausibility

We have thus far refrained from commenting on the question of whether this saying goes back to the historical Jesus. Perhaps the most thorough attempt at establishing its historicity is that made by Edwin K. Broadhead, who builds on the arguments of Bauer, Jeremias, and Hofius. Bauer put much weight on the observation that in our two versions of this saying "the kingdom" and "life" are interchangeable, and noted how this is also the case in some Synoptic sayings (Mark 9:43, 47). He thus argued that both versions may go back to the historical Jesus, who drew it from the Greek proverbs we have noted above.[203] Jeremias contended that the multiple attestations in *Thomas*, Ignatius, and Pseudo-Ephraem were independent of one another; that the saying has formal similarities with Synoptic traditions (see below); and that when translated back into Aramaic, the saying forms two lines with four beats each, a pattern which Jeremias and others have found elsewhere in the Jesus tradition:[204]

> *Man diqerib ʿimmi, qerib ʿim nura;*
> *man direchiq minni, rechiq mimmalkuta.*[205]

The Aramaic verse also contains rhyme and the repetition of the letter *mem*, thus raising the possibility of an Aramaic original form.[206] Jeremias concludes that "the most important indication of authenticity is the purpose of the saying, which is to convey a stern warning, to frighten people off (Matt. 8.19f. par.; 16.24f par.)."[207]

Let us now consider the case for historicity made by Edwin K. Broadhead. Broadhead first points out that *Gos. Thom.* 82 employs an antithetical parallelism similar to that which we find in several Synoptic traditions (Q 6:46-49; 10:16; 12:8-9; 12:10; 13:30; 14:11; 16:18; 18:14b; 19:26; Matt 10:37-38, 39).[208] He draws specific attention to the following:

---

203. Bauer, "Das Jesuswort," 447.

204. See Burney, *Poetry of Our Lord*, 112–30.

205. Jeremias, *Unknown Sayings*," 71; Jeremias's retroversion is adopted by Ménard, *L'Évangile selon Thomas*, 182–85.

206. Jeremias, *Unknown Sayings*, 71.

207. Ibid., 72.

208. Broadhead, "An Authentic Saying of Jesus," 134.

> Q 12:8–9: Everyone who acknowledges me before others, the Son of Man will also acknowledge . . . but whoever denies me before others, will be denied . . .
>
> Q 14:11 (18:14b): Everyone who exalts self will be humbled, and whoever humbles self will be exalted.
>
> Q 17:33 (Matthean form): Whoever finds their life will lose it, and the one who loses their life for my sake will find it.

As noted by Broadhead:

> these examples suggest that the Sayings Tradition (Q) contained numerous sayings in which a "whoever . . ." parallelism is used to compare or contrast two different groups. Of these instances, a few build the contrast distinctly around how one relates to Jesus (Q 6.47–49; 10.16; 11.23; 12.8–9; 12.10; 14.26–27; 17.33). Saying 82 of GThom employs this same form.[209]

In addition to formal similarities, the content of *Gos. Thom.* 82 parallels that of the Synoptic tradition, for as we have been noting throughout this chapter, several sayings in the Synoptic tradition call on the motif of fire as a symbol of eschatological judgment. Beginning with the preaching of John (Q 3:16–17), and continuing through the teaching of Jesus, fiery judgment is a common motif. Broadhead calls special attention to John's proclamation of a baptism in fire (Q 3:16), Jesus' saying that he came to cast fire upon the earth (Luke 12:49), and the enigmatic statement that "everyone will be salted with fire" (Mark 9:49). Regarding these two Jesus sayings he notes:

> These two enigmatic uses of the fire imagery show key traits of authenticity: (1) they focus almost exclusively and without polish upon the dangers of discipleship; (2) they are isolated and enigmatic sayings credited solely to Jesus; (3) in their present form they are largely ignored or abandoned by the rest of the tradition; (4) there is a tendency within the tradition to qualify these sayings through other connections and to transform them toward general metaphors. Saying 82a of GThom shares similar characteristics.[210]

Broadhead's case may be buttressed by the argument we have made for the historical plausibility of Luke 12:49–50 and Mark 9:49 and their relationship to Q 3:16.[211]

---

209. Ibid., 135.
210. Ibid., 137.
211. As we have proposed regarding Luke 12:49–50 and Mark 9:49, Broadhead

Broadhead goes on to demonstrate that the second half of our saying, "whoever is far from me is far from the kingdom," also resonates with the Synoptic sayings according to which one's nearness to the kingdom is emphasized (Mark 12:34; Matt 5:19) and one's entrance into the kingdom is contingent upon one's standing in relation to Jesus himself (Q 12:8-9; 11:23). Just as in the Synoptic tradition, then, *Gos. Thom.* 82 emphasizes the imminence of the kingdom and the centrality of Jesus. He then concludes by demonstrating that "[s]aying 82 contributes little to the goals and designs of GThom. Saying 82 does not deny the world, and it does not proclaim the need for illumination." In short, the saying does not fit the literary environment of the *Gospel of Thomas* as neatly as we might expect if it were a product of the Thomasine community. Broadhead has provided the most rigorous application of the traditional criteria to our saying. When we add the *Gospel of the Savior* 71 as a potential second-century witness as well as the early allusion of Ignatius, the argument is even stronger. While certainty always eludes us in such matters, I am inclined to favor the historicity of this saying given its formal similarities with other sayings attributed to Jesus, its multiple attestations, and particularly in light of its coherence with other sayings in the tradition that we have been examining (esp. Q 3:16; Mark 9:49; Luke 12:49-50).[212]

Conclusion

In the preceding pages we have considered the individual logia attributed to John the Baptist and Jesus that deal with the fire of eschatological judgment (though excluding those sayings specifically associated with hell or Gehenna). These sayings belong to the category of traditions that Dunn referred to as "fire words" and that Lang incorporated under the category "judgment by fire." While we have sought to demonstrate that there is reason to believe at least some of these traditions can individually be traced back to the historical Jesus (and to John the Baptist in the case of Q 3:16), in light of recent criticisms levied against such an atomistic approach, it is interesting to note that together these traditions may attest to a recurrent motif in the Jesus tradition.[213] Even if some of the individual sayings cannot be demonstrated with

---

suggests that at least the first half of *Gos. Thom.* 82 may go back to John the Baptist (Ibid., 147-48).

212. See Ibid., 149.: "While decisive proof is unattainable, there is a strong possibility and a significant probability that Saying 82 of GThom presents an authentic saying of Jesus."

213. On these criticisms and recurrent attestation see pages 23-24 above.

a high degree of plausibility to have originated with Jesus, they may still bear witness to memories of the sort of thing Jesus pronounced. Taken together, Mark 9:49; Luke 12:49–50; 17:26–32; 23:31; *Gos. Thom.* 82 all, apparently independently, bear witness to a similar theme—one that is significantly taken up from the preaching of John the Baptist (Q 3:16)—that in the face of the coming of God's eschatological judgment all, including Jesus and his own disciples, must face a period of fiery testing or trial and that entry into the kingdom is in some sense contingent upon passing that ordeal.

CHAPTER 6

Saved through Fire

Introduction

IN OUR STUDY OF the fiery ordeal in New Testament eschatology, we have thus far considered the evidence found in the Gospels, giving special attention to the views of John the Baptist and the historical Jesus, while also considering the perspectives of the Gospel writers themselves. We now turn to the Pauline Epistles to see what "the apostle to the Gentiles" may have to add to our understanding of this subcategory of New Testament eschatology.

Only in three verses does Paul refer explicitly to fire—2 Thess 1:8; 1 Cor 3:13; 15. To these we might add this comment, which Paul addresses to the Romans: "if your enemies are hungry, feed them; if they are thirsty, give them something to drink; for by doing this you will heap burning coals on their heads" (Rom 12:20). The Romans text, however, strikes us as a metaphor for "the remorse and humiliation of the adversary" rather than as some means of eschatological judgment or punishment.[1] This reading is more congruent with the immediate context of Romans than is the interpretation that holds that eschatological punishment is in view for several reasons. Dunn notes four: 1) the passage contains several apparent allusions to the Sermon on the Mount, which is at odds with a wish for the destruction of one's enemy; 2) the verse quotes Prov 25:21–22, which may borrow imagery from an Egyptian repentance ritual, in which penitents carried bowls of hot coals on their heads to demonstrate that they were truly repentant; 3) the Targum of Prov 25:21–22 concludes ". . . and God will make him your friend" indicating that reconciliation with one's enemy, not his or her destruction, is the ultimate aim; and 4) in his quotation, Paul excludes the

---

1. Käsemann, *Commentary on Romans*, 349.

traditional conclusion of Prov 25:22, "and the Lord will reward you," which may indicate that Paul is intentionally removing the self-interested aspect of the verse in his belief that love of one's enemy does not lead to self justification but to the enemy's repentance and forgiveness.[2]

If Rom 12:21 is more concerned with repentance than with eschatological judgment, testing, and purification, we are thus left with the evidence from 2 Thess 1:8 and 1 Cor 3:13; 15. While the authenticity of 2 Thessalonians is often called into question, the fire accompanying the parousia in 1:8 appears to be quite similar to the fire that accompanies "the Day" in 1 Cor 3:13 and 15.[3] Just as the fire in 2 Thess 1:8 appears to reveal Jesus in his coming, the fire of 1 Cor 3:13 reveals either "the Day" or "the works" to be tested, and as in the former text the fire may play some role in "inflicting vengeance," in the latter it burns up the bad works. Nevertheless, despite these potential similarities, Paul (or Deutero-Paul) does not develop his discussion around the function of fire in 2 Thess 1:8, and we are left with very little to build upon in terms of its interpretation, so we can simply note its presence on the day of the parousia. We shall thus focus all of our attention on the passage in which the remaining references to fire occur: 1 Cor 3:10–15, giving special attention to the concluding, enigmatic verse, in which Paul states, "the builder will be saved, but only as through fire."

1 Corinthians 3:10–15

It was the opinion of Atto of Vercelli, the tenth-century theologian and bishop, that 1 Cor 3:15, in which Paul writes that certain "builders" of the Corinthian church "will be saved, but only as through fire" (σωθήσεται, οὕτως δὲ ὡς διὰ πυρός), numbered among those texts that the author of 2 Pet 3:16 had in mind when he wrote concerning Paul's letters that "[t]here are some things in them hard to understand."[4] The inherent difficulty of the text has been compounded by centuries of debate over whether or not the notion of being "saved through fire" ought to be taken as Pauline evidence for purgatory.[5] Indeed, in the West, this passage was historically read as a

---

2. For full discussion see Dunn, *Romans*, 750–51.

3. However, see the recent and compelling defense of the authenticity of 2 Thessalonians in Foster, "Who Wrote 2 Thessalonians?"

4. Cited in Robertson and Plummer, *Corinthians*, 65.

5. See especially the history of interpretation in Gnilka, *Ist 1 Kor. 3, 10-15 ein Schriftzeugnis für das Fegfeuer?* For further discussion see Bietenhard, "Kennt das Neue Testament die Vorstellung vom Fegfeuer?"; Michl, "Gerichtsfeuer und Purgatorium zu 1 Kor 3,12–15"; Townsend, "1 Corinthians 3:15 and the School of Shammai."

proof text for purgatory, an interpretation that was a point of contention far earlier than the Protestant Reformation.[6] This verse became a lightning rod in Catholic-Protestant debates, and until recently the rhetoric of Protestant exegetes commenting on this verse was clearly fuelled by controversies over the doctrine of purgatory.[7] While this sort of anti-papal rhetoric is thankfully absent from most modern commentaries, reticence regarding the potentially purifying function of fire persists. A few modern exegetes still maintain that this verse refers to post-mortem purification that leads to salvation;[8] however, the prevailing consensus insists that the fire of 1 Cor 3:13–15 plays no soteriological function in the purification of individuals whatsoever.

Three interrelated arguments typify the dominant position: (1) according to v. 13, the fire will test the *work* (τὸ ἔργον) of the builders, not the builders themselves; the fire therefore plays no role in the judgment of individuals;[9] (2) in the same verse, Paul states that the fire will *test* (δοκιμάσει), not that it will purify;[10] and (3) "saved, but only as through fire" (σωθήσεται, οὕτως δὲ ὡς διὰ πυρός) in v. 15 is a proverbial statement, meaning something akin to the expression found in Amos 4:11 and Zech 3:2: "you were like a brand snatched from the fire" (see Jude 23).[11] The expression is frequently paraphrased as "saved by the skin of one's teeth," indicating a narrow escape. A crucial assumption in this last argument is that the phrase διὰ πυρός must be taken in the local sense.[12] According to this reading, which is adopted by the majority of commentators, the fire in no way contributes to the builder's salvation.[13] Indeed, the builder is delivered

---

6. On the history of the interpretation of this verse and the patristic origins of belief in purgatory, see Gnilka, *1 Kor. 3, 10–15*; Le Goff, *Birth of Purgatory*, 52–95; Thiselton, *Corinthians*, 331–32.

7. See for instance, Matthew Henry, *Exposition*, 5:345.

8. Michl, "Gerichtsfeuer und Purgatorium," 395–401; Townsend, "1 Corinthians 3:15," 500–504.

9. Kuck, *Judgment*, 181: "The fire is not said to purify or punish the persons themselves."

10. Morris, *First Epistle of Paul to the Corinthians*, 66: "The fire is, of course, a fire of testing, not one of purifying." So also Barrett, *Corinthians*, 89; Conzelmann, *1 Corinthians*, 77, n. 84; Fee, *Corinthians*, 142; Yinger, *Paul, Judaism, and Judgment*, 217, n. 48.

11. See Bietenhard, "Kennt das Neue Testament?" 104; Fee, *Corinthians*, 144; Hays, *First Corinthians*, 56; Lang, *Die Briefe an die Korinther*, 55; Lietzmann, *An die Korinther*, 17; Thiselton, *Corinthians*, 315.

12. Barrett, *Corinthians*, 89; Konradt, *Gericht*, 269; Kuck, *Judgment*, 184; Robertson and Plummer, *Corinthians*, 65; Schrage, *Der erste Brief*, 304.

13. Donfried ("Justification and Last Judgment," 105) contends that "the verb σῴζω in v. 15 has nothing to do with Christology and is used here in an entirely secular sense of 'to rescue, to deliver from danger or harm.'"

despite having to pass through the danger the fire presents. On the other hand, if διὰ πυρός is taken in the instrumental sense, the verse would mean "the builder will be saved *by means of* fire," which would attribute a purifying—indeed, salvific—function to the fire itself.

Alexander N. Kirk has effectively dismantled the first of the above three presuppositions, disputing the widely held view that the building materials of v. 12 and the "work" (τὸ ἔργον) of vv. 14–15 refer to the *deeds* or *activities* of the builder. Kirk argues persuasively that the building materials and the work refer rather to human persons, the Christian believers who are joined to the church and who collectively comprise the temple of God.[14] Noting that immediately preceding and following the "building" metaphor are two related metaphors—"you are God's field" (v. 9) and "you are God's temple" (v. 16)—both of which refer to the Corinthian believers, he suggests that the building metaphor and the building materials likewise allude to the Corinthians.[15] Moreover, Paul's use of the singular "work" (τὸ ἔργον) in vv. 14–15 is similar to his use of the same noun in 9:11, where Paul inquires of the Corinthians "are you not my work [τὸ ἔργον] in the Lord?," unambiguously equating τὸ ἔργον with those whom he has converted.[16] The temple imagery employed in this passage also recalls other NT texts that refer to Christians being built up as a temple (see Eph 2:20–22; 1 Pet 2:4–8), while the expectation that Paul would receive reward or loss as determined by the fate of his "spiritual children" is consistent with his views regarding final rewards articulated elsewhere (see 1 Thess 3:5; Phil 2:16; 4:1; Gal 4:11; 2 Cor 1:14).[17] Moreover, this identification of the building materials and the work of the builder with converted persons finds support in a number of patristic exegetes.[18] Kirk also questions the consensus position that v. 15 is a proverb connoting a narrow escape, rightly pointing out how distant the purported parallels to this supposed proverb are.[19] However, he sidesteps the key argument in its favor, namely the assertion that the prepositional phrase διὰ πυρός must be taken in the local and not in the instrumental sense. Further, while he disputes the adequacy of reading v. 15 as a proverb, he offers no comment on how the fire functions if the saying is not taken proverbially. Thus, while Kirk's article presents a forceful and well-reasoned

---

14. Kirk, "Building with the Corinthians." Kirk is elaborating on a view introduced to modern scholarship by Schlatter, *Paulus der Bote Jesu*, 133.

15. Kirk, "Building with the Corinthians," 554.

16. Ibid., 557.

17. Ibid., 558–59.

18. Ibid., 560–62.

19. Ibid., 565–66.

challenge to one of the basic assumptions of the dominant interpretation of this pericope, the prevailing consensus requires further criticism.

The present argument, which dovetails with and expands upon Kirk's critique of the proverbial reading of 1 Cor 3:15, makes a case for the instrumental sense of διά in v. 15 and for the purifying and soteriological function of fire in vv. 13–15. After demonstrating Paul's use of temple imagery in this pericope's wider context, I shall argue that Paul is alluding to a specific temple text, LXX Malachi 3, where fire plays an explicitly refining function, and that Paul's allusion assumes this purificatory function of fire. I shall then proceed to a linguistic analysis of the Greek phrases διὰ πυρός and σῴζω + διά + genitive, contending that Paul employs the preposition διά instrumentally and thus assigns a soteriological function to the fire of eschatological judgment. While I have no wish to defend the view that this passage should be read as a proof text for purgatory, I do wish to argue that v. 15 is more than simply a proverb connoting a narrow escape.

A Refining Fire in God's Temple

First Corinthians 3:15 comes near the conclusion of a passage in which Paul has been urging upon his readers the necessity of church unity. Divisions have sprung up in Corinth, with some claiming allegiance to Apollos, others to Paul. In his appeal to unity, Paul employs two extended metaphors: one of planting (3:6–9) and one of building (3:10–15). The former analogy stresses that Paul and Apollos are "God's servants working together" (v. 9). Paul planted, Apollos watered, but both are working towards a common purpose. In continuity with the planting metaphor, the Corinthians are compared to "God's field" (v. 9). From here Paul transitions to another metaphor in which the Corinthians are compared to "God's building" (v. 9). It is in this context that he introduces the motif of fire:

> Κατὰ τὴν χάριν τοῦ θεοῦ τὴν δοθεῖσάν μοι ὡς σοφὸς ἀρχιτέκτων θεμέλιον ἔθηκα, ἄλλος δὲ ἐποικοδομεῖ. ἕκαστος δὲ βλεπέτω πῶς ἐποικοδομεῖ. θεμέλιον γὰρ ἄλλον οὐδεὶς δύναται θεῖναι παρὰ τὸν κείμενον, ὅς ἐστιν Ἰησοῦς Χριστός. εἰ δέ τις ἐποικοδομεῖ ἐπὶ τὸν θεμέλιον χρυσόν, ἄργυρον, λίθους τιμίους, ξύλα, χόρτον, καλάμην, ἑκάστου τὸ ἔργον φανερὸν γενήσεται, ἡ γὰρ ἡμέρα δηλώσει, ὅτι ἐν πυρὶ ἀποκαλύπτεται· καὶ ἑκάστου τὸ ἔργον ὁποῖόν ἐστιν τὸ πῦρ [αὐτὸ] δοκιμάσει. εἴ τινος τὸ ἔργον μενεῖ ὃ ἐποικοδόμησεν, μισθὸν λήμψεται· εἴ τινος τὸ ἔργον κατακαήσεται, ζημιωθήσεται, αὐτὸς δὲ σωθήσεται, οὕτως δὲ ὡς διὰ πυρός.

> According to the grace of God given to me, like a skilled master builder I laid a foundation, and someone else is building on it. Each builder must choose with care how to build on it. For no one can lay any foundation other than the one that has been laid; that foundation is Jesus Christ. Now if anyone builds on the foundation with gold, silver, precious stones, wood, hay, straw—the work of each builder will become visible, for the Day will disclose it, because it will be revealed with fire, and the fire will test what sort of work each has done. If what has been built on the foundation survives, the builder will receive a reward. If the work is burned up, the builder will suffer loss; the builder will be saved, but only as through fire. (1 Cor 3:10–15)

In this pericope, Paul has in mind the construction of a specific building—the temple. However, even those who detect this temple imagery rarely grasp its significance, particularly the importance of fire imagery in the temple. A fuller appreciation of this metaphor and its implications will contribute to our understanding of 1 Cor 3:15.

The Temple Metaphor

In vv. 16–17 Paul explicitly identifies the Corinthian church as the temple of God. Where he has previously stated, "you are God's field, God's building" (v. 9), he now affirms, "you are God's temple" (ναὸς θεοῦ) (v. 16). While scholars occasionally suggest that Paul is mixing his metaphors here with no discernible unity between them, he is in fact employing a coherent and consistent cluster of metaphors that extend from v. 5 to v. 16. While less obvious than the explicit statement "you are God's temple," the building metaphor likewise points to the temple. Paul describes the builders as making use of diverse materials: "gold, silver, precious stones, wood, hay, straw" (χρυσόν, ἄργυρον, λίθους τιμίους, ξύλα, χόρτον, καλάμην). Part of Paul's purpose is to indicate that just as the first three of these materials can endure fire and the latter three are combustible, the work of some will withstand the fire of judgment while the work of others will be burned up. Additionally, however, Paul has chosen these particular materials to evoke an identifiable image. As G. K. Beale observes, "The only other place in Scripture where a 'foundation' of a building is laid and 'gold', 'silver', and 'precious stones' are 'built' upon the foundation is Solomon's temple."[20] Several texts attest to the use of individual elements, such as stones (1 Kgs 5:17), gold (1 Kgs 6:20–21, 28, 30, 35), or silver (1 Chr 22:14, 16) in the construction of the temple;

---

20. Beale, *Temple*, 247. See Kuck, *Judgment*, 177–78; Wolff, *Der erste Brief*, 71–72.

however, it is to 1 Chr 29:2 in particular that Paul most clearly alludes. In this text David announces that he has provided the following materials for "the house of my God" (οἶκον θεοῦ μου): gold (χρυσίον), silver (ἀργύρον), wood (ξύλα), and precious stones (λίθον τίμιον). This allusion is reinforced by Paul's description of himself as a wise or skilled master builder (σοφὸς ἀρχιτέκτων) in 1 Cor 3:10, which recalls the description of the builder of the tabernacle in LXX Exod 35:31-33, who is filled with the spirit in "skill" or "wisdom" (σοφίας) and is said "to be a master builder [ἀρχιτεκτονεῖν] according to all the works of a master builder [ἀρχιτεκτονίας]." Significantly, this "master builder" also works with "gold" (χρυσίον), "silver" (ἀργύριον), "stone" (λίθον), and "wood" (ξύλα).[21] These intertextual links indicate that when Paul described the materials used by the builders and referred to himself as a "wise master builder," he was intentionally alluding to the construction of the temple.[22]

This temple metaphor, which is not made explicit until vv. 16 and 17, may extend all the way back to v. 5, where Paul compares the Corinthians to a field, for in the Hebrew Bible and other Jewish literature the temple is closely associated with garden imagery; indeed, the temple was filled with agricultural images (see 1 Kgs 6:18, 29, 32; 7:18-20, 22, 24-26; 7:42, 39-50).[23] Significantly, in certain texts from Qumran the community is described alternatively as a plantation and as a building in the context of referring to the community in terms of a temple.[24] 1QS VIII.5-8, for instance, states that the Community will be "an everlasting plantation, a holy house for Israel and the foundation of the holy of holies for Aaron."[25] Similarly, the juxtaposition of a planted vine and the temple occurs in Pseudo-Philo, *L.A.B.* 12.8-9: after the Israelites commit idolatry through their worship of the Golden Calf, Moses prays to the Lord, saying that God has "planted this vine" but that the vine "has not recognized its cultivator" and so the vine has been "burned up." Moses immediately shifts to a description of the temple: "you have adorned your house with precious stones and gold; and you have

---

21. Beale, *Temple*, 247.

22. Williams (*The Wisdom of the Wise*, 259) believes that the allusion is to Isa 3:3, where the phrase σοφὸν ἀρχιτέκτονα is used. However, the additional presence of gold, silver, stones, and wood in Exod 35:31-33 and the lack of any other shared vocabulary in Isa 3:3, indicate a greater resonance with the Exodus allusion, further suggesting a link to temple (or tabernacle) imagery. The LXX of Isa 3:3 diverges significantly from the MT, which, instead of "skilled builders," has והכם חרשים "skilled magicians."

23. Beale, *Temple*, 248. Beale is followed by Roy E. Ciampa and Brian S. Rosner, *Corinthians*, 151. See also Telford, *Barren Temple*, 208-12.

24. See Gärtner, *Temple and the Community*, 58.

25. See Hogeterp, *Paul and God's Temple*, 281.

sprinkled your house with perfumes and spices and balsam wood" before he returns to the image of the vine.[26]

Malachi 3 and the Refiner's Fire

In light of these various texts that similarly juxtapose the plantation metaphor with the temple metaphor, it is apparent that in vv. 6–17, Paul is not shifting haphazardly between disconnected metaphors, but very intentionally and consistently focusing on the image of the temple.[27] Paul's deliberate use of this metaphor significantly locates the testing fire of God's judgment in the context of the temple, not simply in some "fantastic, possibly apocalyptic buildings."[28] And this identification of the building as the temple is far from "inconsequential," as Joseph Fitzmyer suggests it is.[29] Indeed, the allusion to the "temple" (ναός) alongside the image of "fire" (πῦρ) that tests "silver" (χρυσόν) and "gold" (ἄργυρον) and which "will burn" (κατακαήσεται) the work that is like "stubble" (καλάμην) on "the day" (ἡμέρα) suggests an allusion to LXX Malachi 3, according to which the LORD, who is described as a "refiner's fire" (πῦρ χωνευτηρίου), comes into his "temple" (τὸν ναόν) on "the day" (ἡμέρα) of judgment and sits "as a refiner and purifier of silver and gold" (ὡς τὸ ἀργύριον καὶ ὡς τὸ χρυσίον) to "purify the descendants of Levi and refine them like gold and silver" (ὡς τὸ ἀργύριον καὶ ὡς τὸ χρυσίον) (vv. 2–3).[30] A few verses later, Malachi describes "the day" (ἡμέρα) as "burning [καιομένη] like an oven" and prophesies that "all the arrogant and all evildoers will be stubble" (καλάμη) and that the fire will "leave them neither root nor branch" (v. 19).[31] The verbal similarities between these two texts are strong and suggest an intertextual link.[32]

26. On *L.A.B.*'s depiction of Israel as a vineyard and the interpretation of God's *domus* as the temple, see Jacobson, *Pseudo-Philo's* Liber Antiquitatum Biblicarum, 1:497–500.

27. Peterson, "ἔργον," 440.

28. Conzelmann, *1 Corinthians*, 76.

29. Fitzmyer, *First Corinthians*, 197.

30. It may also be significant that both 1 Cor 3:13 and Mal 3:19 (MT) use "the day" (היום = ἡ . . . ἡμέρα) absolutely to refer to "the Day of the LORD"; see Delling, "ἡμέρα," 952. The LXX has ἡμέρα κυρίου.

31. The καλάμη ("stubble") of LXX Mal 3:19 explains its presence in 1 Cor 3:13 more adequately than does the thesis of Ford, "You are God's Sukkah." Ford argues that Paul's imagery derives from the construction of Sukkoth, for which hay and straw were used. She suggests they were adorned with tapestries embroidered with gold and silver thread and precious stones; however, such slender filaments of gold and silver thread fail to convey the sense of enduring strength that Paul seeks to communicate.

32. See Ciampa and Rosner, *Corinthians*, 153; J. Proctor, "Fire in God's House," esp.

	LXX Mal 3	1 Cor 3
temple	ναόν (v. 1)	ναός (v. 16)
day	ἡμέραν (v. 2); ἡμέρα (v. 19)	ἡμέρα (v. 13)
fire	πῦρ (v. 2)	πυρί, πῦρ (v. 13); πυρός (v. 15)
silver	ἀργύριον (v. 3)	ἄργυρον (v. 12)
gold	χρυσίον (v. 3)	χρυσόν (v. 12)
burn	καιομένη (v. 19)	κατακαήσεται (v. 15)
stubble/straw	καλάμη (v. 19)	καλάμην (v. 12)

More broadly, if we accept Kirk's thesis that the building materials and work of 1 Cor 3:12–15 represent not deeds but human persons, both LXX Mal 3:3, 19 and 1 Cor 3:12 contain two categories of materials that represent two sorts of people: in both texts the righteous are likened to precious metals while the wicked are compared to highly flammable substances.

The recognition that intertextual allusions and echoes influence readers' interpretations of texts has become a commonplace of contemporary scholarship. The influential work of Richard Hays on this subject has drawn the attention of New Testament scholars to the literary trope of *metalepsis* or *transumption*, which suggests that when one text alludes to another, "the figurative effect of the echo can lie in the unstated or suppressed (transumed) points of resonance between the two texts."[33] In light of the intertextual relationship between 1 Cor 3:10–15 and LXX Mal 3, we may thus suggest that the purificatory function of fire explicit in Malachi 3 is implicitly present in 1 Cor 3:10–15. In speaking of the testing of the work of the builders of the temple on "the day," Paul intended to recall Malachi's reference to the refining and purifying of the sons of Levi in the temple on "the day." Thus the testing of the work in 1 Cor 3:13 may carry connotations of purification and refinement.

11–14; Williams, *Wisdom*, 264–5; 269–72.

33. Hays, *Echoes of Scripture*, 20. See Allison, *The Intertextual Jesus*, 21: "An allusion is no end in itself but a suggestive element, a clue, an implied link to another text; it is a piece whose purpose is to summon what it has been subtracted from (see synecdoche). And it is up to readers to do the summoning."

Testing and Refining the Work of the Builders

The suggestion that 1 Cor 3:10–15 transumes the purifying function of fire of Malachi 3 is strengthened—not weakened, as is often assumed—by the testing function attributed to the fire in v. 12. As noted above, two qualities of materials are tested, those that will endure the fire and those that will not. According to v. 15, the flammable substances will be burned up (κατακαήσεται). This burning is the end result of the testing process for the inferior substances. We may assume a different result for those that endure the testing process. Significantly, in the Hebrew Bible and Septuagint, testing and refining are frequently mentioned in tandem (see Isa 48:10; Jer 9:6; Zech 13:9; Ps 66:10). Of particular relevance is the parallelism between the refining of silver and the testing of gold in Zech 13:9: "And I will put this third into the fire, and refine (πυρώσω) them as one refines silver, and test (δοκιμῶ) them as gold is tested." While technically speaking testing and refining were two distinct stages in the smelting process, the phrases are here placed in synonymous parallelism, and in the poetry of prophetic speech the two procedures are collapsed into a single metaphor.[34] Also noteworthy is Theodotian's translation of Dan 11:35. Where the MT states that in the tribulation, the wise ולברר וללבן . . . לצרוף "may be refined, purified, and cleansed," Theodotian makes explicit that this period is meant πυρῶσαι αὐτούς "to try them with fire." Notably, it was no great leap for the translator to equate this fiery test with the refining, purifying, and cleansing of human beings. Given other contextual clues, most notably the allusions to LXX Malachi 3, where the fire serves to refine and purify the sons of Levi as silver, it is very probable that Paul thought not only of the testing, but also of the refining, of the precious metals he mentions.[35] And as Kirk has demonstrated, these precious metals represent human persons. As fire purifies the sons of Levi in Malachi 3, the fire that attends "the day" in 1 Corinthians 3 will test the converts like silver and gold; what is imperfect will be burned away, and they will be refined and purified. Thus, arguments that emphasize the function of testing over against refining and the identity of the work as deeds over against individual believers are not as convincing as they may at first appear. We have seen that testing and refining were frequently linked together conceptually, and we have noted the plausibility of Kirk's argument that when Paul wrote of the work and materials of the builder, he had in mind not their teaching or their deeds, but those whom they had converted. The testing of the builder's work may therefore refer to the testing, refining,

---

34. See the excursus on "The Cupellation of Silver" in Holladay, *Jeremiah*, 1:230–32.
35. Kuck, *Judgment*, 181.

and purifying of individual believers by means of fire. This leads to the question of whether Paul assigns this fire any soteriological function, as may be indicated by his use of the verb σῴζω in v. 15.

Saved as through Fire

The crux of this passage, v. 15, which states that the builder "will be saved, but only as through fire" (σωθήσεται, οὕτως δὲ ὡς διὰ πυρός) has been identified as a proof text for purgatory by some, while the majority of interpreters support the assertion that "Paul is obviously borrowing from a common phrase, 'barely escaped from the fire.'"[36] Indeed, it is now an axiom of Pauline scholarship that we are dealing with a proverbial or quasi-proverbial saying. However, no exact parallel to the construction σῴζω + διὰ πυρός is known, and most of the purported parallels to this supposedly "common phrase" are quite distant. Certainly, none is sufficiently close to demonstrate the proverbial nature of our saying. It is my contention that we are dealing with neither proverb nor proof text, but with Paul's belief that on the day of the parousia—not in some post-mortem state of purgation—fire would play an instrumental role in the purification and salvation of some, just as it would play a destructive role in the judgment of others.

On the Ambiguity of διὰ πυρός

The lynchpin of the argument that 1 Cor 3:15 is a proverb alluding to a narrow escape achieved by passing through fire is the assertion that Paul employs the prepositional phrase διὰ πυρός in the local sense. To support this position, Walter Bauer cites a handful of Greco-Roman and Jewish texts that employ διὰ πυρός or διὰ φλογός in a local sense, where the image is of an individual rushing through a wall of fire to safety.[37] While Bauer's few examples do demonstrate the possibility of the local sense of this phrase, they do not exhaust the variety of meanings it could elicit, nor do they determine Paul's usage. Admittedly, the Septuagint does provide further examples of the local use (Num 31:23; 2 Chr 28:3; Isa 43:2; Zech 13:9; Ps 65:12 [66:12

---

36. Conzelmann, *1 Corinthians*, 77. Weiss (*Der erste Korintherbrief*, 83, n. 81) lists a number of texts that allegedly parallel v. 15: Strabo *Geog.* 3.5.11; Eur *Androm.* 487; *Elect.* 1182; Liv. 20,35; LXX Amos 4:11; Zech 3:2; Ps 65:12; Jude 23. Kirk ("Building with the Corinthians," 565–9) rightly notes how very distant these and other supposed parallels are.

37. BDAG, s.v. πῦρ. See Jos. *Ant.* 17.264; Diodorus Siculus 1.57.7–8; Cynic Epistles, *Crates* 6.

MT]; 4 Macc 18:14 [quoting Isa 43:2]). However, even in some texts that employ διὰ πυρός in the local sense, the phrase carries with it connotations of purification—particularly in texts that refer to precious metals. For instance, LXX Num 31:23, which follows a list of metals of various qualities, states, "everything that can withstand fire shall be clean ... but whatever cannot go through fire (διὰ πυρός), shall be passed through the water." The purifying function of the fire intensifies the cleansing function of water. Further, while I have already cited Zech 13:9 to demonstrate the parallelism between testing and refining, here I note the use of the phrase διὰ πυρός: "And I will put this third into the fire (διὰ πυρός), and refine them as one refines silver, and test them as gold is tested." Psalm 65:10–12 similarly states, "For you, O God, have tested us; you have tried us as silver is tried ... we went through fire (διὰ πυρός) and through water; yet you have brought us out to a spacious place." Thus, even when διὰ πυρός is used in the local sense, it may connote the purifying function of fire.

However, in addition to these examples of the local sense of διὰ πυρός, one could just as easily adduce several examples from the relevant literature where the phrase is used in the instrumental sense.[38] Philo, for instance, uses the instrumental sense in three of the four occurrences of διὰ πυρός in his corpus.[39] In *Abr.* 1 he writes of "terrible destructions which have taken place on earth by the agency of fire and water" (διὰ πυρός καὶ ὕδατος) while in *Mos.* 2.219 he refers to businesses "carried on by means of fire" (διὰ πυρός) and "instruments made by fire" (διὰ πυρός). *Letter of Aristeas* 87 likewise describes the burnt offering being consumed by fire (διὰ πυρός). Perhaps more significantly, we also encounter the instrumental usage in the Septuagint, the New Testament, and the Apostolic Fathers. The phrase frequently occurs in the instrumental sense in the Maccabean literature. 3 Maccabees 2:29 reports that Jews who were registered were to be "branded on their bodies with fire" (διὰ πυρός); 4 Macc 7:12 describes Eleazar's resolve in spite of being "consumed by the fire" (διὰ τοῦ πυρός); 4 Macc 9:9 alludes to "eternal torment by fire" (διὰ πυρός). Further, in the New Testament, 1 Pet 1:7 speaks of faith being "tested by fire" (διὰ πυρός). And the only two occurrences in the Apostolic Fathers are instrumental: *1 Clem* 11.1 describes how Sodom was "judged by fire" (διὰ πυρός) while *Herm. Vis.* 4 3.4 explains that gold is "tested by the fire" (διὰ τοῦ πυρός).

Testament of Abraham 13 (Rec A) is often brought into discussions of 1 Cor 3:10–15 due to the verbal and thematic links between the two texts.

38. Kuck (*Judgment*, 184, n. 179) gives the impression that the instrumental use of this phrase is much rarer than it actually is: "The only instrumental use of διὰ πυρός I could find is in 4 Macc 9:9 (eternal punishment by means of fire)."

39. The only exception is *Spec. Laws* 4.28.

While some have proposed that Paul was influenced by the *Testament of Abraham*,[40] and others have suggested that they were both drawing on a shared tradition,[41] the most plausible explanation for their strong similarities is that the *Testament of Abraham* was originally a Jewish document that Christian scribes have substantially redacted in light of 1 Cor 3:10–15.[42] One point of significant overlap is *T. Ab.* 13.11's echo of the διὰ πυρός from 1 Cor 3:15.[43] Significantly, in *T. Ab.* 13.11, the phrase is used instrumentally—δοκιμάζει τὰ τῶν ἀνθρώπων ἔργα διὰ πυρός—allowing us to conclude that whoever made use of Paul's saying here apparently also interpreted his phrase in the instrumental sense.

While all this evidence is highly suggestive, it does not point decisively in favor of the instrumental reading, for as I have demonstrated, the phrase could be used in either the local or instrumental sense. I have thus far merely sought to redress a lopsided emphasis on the local sense of this phrase in contemporary scholarship.

The Sense of σῴζω + διά + Genitive

In light of the flexible nature of the phrase διὰ πυρός, it is therefore instructive to consider the sense of the construction σῴζω + διά + genitive "to save by/through . . ." elsewhere in the Pauline corpus and more broadly in the contemporary literature. The construction occurs twice in the Septuagint, in eight other instances in the Greek New Testament (not counting 1 Cor 3:15), frequently in the Apostolic Fathers, and sparingly in Philo and Josephus. In the vast majority of these cases, the preposition takes the instrumental sense.

The sense of eight of the ten additional biblical occurrences of the phrase σῴζω + διά + genitive is uncontroversial:

1. ". . . so he [the LORD] saved them [Israel] by the hand [ἔσωσεν . . . διὰ χειρός] of Jeroboam son of Joash" (LXX 2 Kgs 14:27).
2. "The Lord has founded Zion, and the humble among the people will be saved through him" (δι' αὐτοῦ σωθήσονται) (LXX Isa 14:32b).

---

40. Fishburne, "1 Corinthians 3:10–15."
41. Hollander, "Testing by Fire."
42. Allison, *Testament of Abraham*, 291; Kuck, *Judgment*, 184.
43. See Allison, *Testament of Abraham*, 291.

3. "Indeed, God did not send the Son into the world to condemn the world, but in order that the world might be saved through him" (σωθῇ ... δι' αὐτοῦ) (John 3:17).

4. "But we believe that we are saved through the grace [διὰ τῆς χάριτος ... σωθῆναι] of the Lord Jesus" (Acts 15:11).

5. "Much more surely then, now that we have been justified by his blood, will we be saved through him [σωθησόμεθα δι' αὐτοῦ] from the wrath of God" (Rom 5:9).

6. "Now I would remind you, brothers and sisters, of the good news that I proclaimed to you ... through which also you are being saved" (δι' οὗ ... σῴζεσθε) (1 Cor 15:1–2).

7. "For by grace you have been saved through faith" (σεσῳσμένοι διὰ πίστεως) (Eph 2:8).

8. "He saved us through the water [ἔσωσεν ... διὰ λουτροῦ] of rebirth and renewal by the Holy Spirit" (Titus 3:5).

In each of the above passages—two of which are from undisputed Pauline epistles, and two of which are from Deutero-Pauline epistles—the phrase σῴζω + διά + genitive unambiguously takes the instrumental sense, and detailed discussion of each text need not detain us here. It is noteworthy that, with the obvious exception of the Septuagint texts, in each case the object of the preposition belongs to a cluster of theological themes related to the gospel: faith, grace, Jesus, the good news, or the waters of rebirth and the Holy Spirit. Indeed, one gathers the impression that resistance to the instrumental sense of "saved through fire" in 1 Cor 3:15 stems from the apparent inconsistency between being saved by means of fire, on the one hand, and being saved by grace, the gospel, Jesus, or faith in Christ, on the other hand. For this reason, the remaining two biblical occurrences of the phrase σῴζω + διά + genitive are of particular significance, despite the difficulties implicit in their own meanings.

According to the Deutero-Pauline 1 Tim 2:15, "she [woman] will be saved through childbearing" (σωθήσεται ... διὰ τῆς τεκνογονίας). The instrumental sense of this phrase is frequently contested, presumably due to the tension it creates with the conviction that one is justified by faith, not works. The local reading of the verse would suggest that the woman's life will be *preserved through* the dangerous process of giving birth.[44] Attractive as this reading may be to modern exegetes who find this verse overtly patriarchal, it does not adequately fit the context of 1 Tim 2:15, for in the

---

44. McCabe, *An Examination of the Isis Cult*, 77.

Pastoral Epistles σῴζω is always used soteriologically, and given this context, the instrumental sense of the preposition διά is most plausible.[45] The instrumental reading, moreover, coheres with the Jewish view that the travails of childbirth in some way overcome the curse of Eve (see Gen 3:16) and that women attain merit by fulfilling their duties as wives and mothers.[46]

We also have a rough parallel to this construction in 1 Pet 3:20, where we read that in the days of Noah, "eight persons were saved through water" (διεσώθησαν δι᾽ ὕδατος). At first glance, this verse appears to support the local reading, and some have taken it in this sense, for Noah and his family were preserved as they physically passed through the waters of the flood.[47] However, as v. 21 explains, "baptism, which this [the flood] prefigured, now saves [σῴζει] you." According to the allegorical logic of these verses, the water of the deluge typologically signifies the water of baptism and the verb διεσώθησαν stands parallel to σῴζει.[48] Admittedly, reading the *destructive* waters of the flood as a type signifying the *saving* waters of baptism is, as R. T. France puts it, "a little whimsical." However, as France himself goes on to observe, it is "certainly not beyond the imagination of a keen typologist."[49] Indeed, if we follow the logic of 1 Peter 3 itself, the sense of the verse seems to be that the flood waters were instrumental in saving Noah and his family by buoying them up in the ark while cleansing the world of wickedness, just as the waters of baptism are instrumental in the salvation and cleansing of the believer.[50] Thus, in our consideration of the construction σῴζω + διά + genitive in 1 Cor 3:15, it is significant that in every other biblical instance it is used in the instrumental sense.

This grammatical construction is also quite common in the Apostolic Fathers, and there too the instrumental sense dominates (see *1 Clem.* 9.4; 12.1; 58.2; *2 Clem.* 3.3; Pol. *Phil.* 1.3; *Herm. Vis.* 3.3.5; 3.8.3; 4.2.4; *Sim.* 9.12.3). One possible exception that may permit a local reading is *Herm. Sim.* 9.12.3. A dominant allegorical motif in the *Shepherd of Hermas* is the construction of a tower, and the stones used to build this tower are said to pass through a certain gate. In the explanation of the allegory in 9.12.3 the

---

45. Porter, "Saved by Childbirth," esp. 95–97.

46. See *Tg. Neof. 1* on Gen 3:16; *b. Ber.* 17a; *b. Sotah* 21a.

47. While the verb διασῴζω may indicate the local reading, meaning "to bring safely through," it may also mean "rescue without special feeling for the meaning of διά" (BDAG, s.v., διασῴζω). See *1 Clem.* 9.4: "through him [Noah] the Master saved (διέσωσεν) the living creatures that entered into the ark in harmony."

48. A cogent defense of the instrumental sense of διά in this passage is presented by Satta, "Baptism Doth Now Save Us," esp. 66–67.

49. France, "Exegesis in Practice," 273.

50. See *Herm. Vis.* 3.3.5.

Shepherd explains that some "will be saved through it [the gate]" (σῴζεσθαι δι' αὐτῆς). The very mention of a gate in conjunction with the preposition διά would appear to suggest a local reading, particularly since earlier in the text we read of individuals physically passing through this gate (see 9.3.4; 9.4.1). However, the Shepherd further explains that the gate is Christ, and since according to *Hermas's* soteriology Christ is the *means* of salvation, the instrumental sense is also possible. It may even be that the text is intentionally ambiguous, holding together the two possible readings with the local sense prevailing at the allegorical level and the instrumental sense coming into focus in its interpretation.

The evidence from Philo further buttresses our case, for he too uses the construction σῴζω + διά + genitive exclusively in the instrumental sense (*Leg.* 3.189; *Cherub.* 130; *Agr.* 1.13; *Abr.* 145). The most significant of these is in his discussion of the judgment of Sodom:

> Because of the five finest cities in it four were about to be destroyed by fire, and one was destined to be left unhurt and safe from every evil. For it was necessary that the calamities should be inflicted by the chastising power, and that the one which was to be saved should be saved by the beneficent power (σῴζεσθαι δὲ διὰ τῆς εὐεργέτιδος). (*Abr.* 145)

While the "beneficent power" that saved the one city is distinguished from the "chastising power" (i.e., the fire) that destroyed the four other cities, it is striking that the phrase σῴζεσθαι δὲ διά is used in the context of πῦρ where διά is used in the instrumental sense.

Finally, Josephus's use of the phrase σῴζω + διά + genitive admittedly presents something of an aberration, for he uses the construction on three occasions, and in each instance he employs διά in the local, not the instrumental, sense (*Ant.* 14.362; 17.275; *Life* 304). While this evidence indicates that one could in fact use the construction in the local sense, Josephus is the outlier. Further, Josephus's context must be taken into account: in each instance he refers not to eschatological salvation, but to physical preservation from danger in battle. This is quite different from Paul's use of the phrase, for which the biblical and early Christian evidence offers a far closer parallel. These three passages from Josephus notwithstanding, every other instance of the phrase σῴζω + διά + genitive that I have been able to find in the Greek literature most relevant to our study can or must assume the instrumental sense. This observation, alongside the number of examples of διὰ πυρός that allow that usage, points strongly in the direction of the instrumental sense of the phrase in 1 Cor 3:15.

We must, however, briefly pause to address the significance of the construction οὕτως ... ὡς to consider what, if any, impact it has upon our conclusion. C. K. Barrett cites two other occurrences of this construction in 1 Corinthians, one of which indicates the metaphorical nature of the subject (9:26), and the other of which indicates a statement of fact (4:1).[51] With no further explanation, he asserts that this phrase signals the metaphorical nature of our verse, rather than the factual one. This metaphorical reading is assumed by the common translation "but only as" and would seem to support the proverbial reading of v. 15. However, as Barrett himself concedes, the phrase may also be used make a factual statement.[52] To be sure, the precise phrase οὕτως δὲ ὡς occurs also in Aristotle (GA 734b 17-19) and Pseudo-Clement (Hom. 17.18), and in neither instance does it convey a metaphor or proverb.[53] Appeals to Paul's use of the construction οὕτως δὲ ὡς cannot therefore overturn the weight of the substantial evidence surveyed above.

The Soteriological Function of Fire

If the above linguistic evidence suggests that the διά in 1 Cor 3:15 is instrumental, in what sense is the fire of eschatological judgment instrumental in salvation, and more broadly speaking, how does this interpretation cohere with the soteriology Paul articulates elsewhere?

With Malachi 3 clearly in the background, the fire appears to save insofar as it purifies. Just as God refines inept priests in Mal 3:3, fire purifies incompetent builders in 1 Cor 3:15. The testing and refining function of fire in this text may be further illumined by similar motifs elsewhere in early Christian literature. For instance, according to Did. 16.5, "all humankind will come to the fiery test" (πύρωσιν τῆς δοκιμασίας) and "those who endure [ὑπομείναντες] in their faith will be saved by the curse itself" (σωθήσονται ὑπ' αὐτοῦ τοῦ καταθέματος). While the traditional reading of this text identifies "the curse" as Christ (see Gal 3:13), some recent interpreters have identified "the curse" that saves with "the fiery test."[54] If this reading is

---

51. Barrett, Corinthians, 89.
52. See Zeller, Der erste Brief an die Korinther, 164.
53. Aristotle GA 734b 17-19: "It is clear by now that there is something which fashions the parts of the embryo, but that this agent is not by way of being [οὕτως δὲ ὡς] a definite individual thing"; Pseudo-Clement, Hom. 17.18 (quoting Num 12:6-7): "If a prophet arise from amongst you, I shall make myself known to him through visions and dreams, but not so as [οὕτως δὲ ὡς] to my servant Moses."
54. The argument is presented most persuasively by Milavec, "Saving Efficacy of the Burning Process." See Draper, "Jesus Tradition," 282.

correct, *Did.* 16.5 closely parallels our reading of 1 Cor 3:15. A more certain parallel is *Herm. Vis.* 4.3.4: "just as gold is tested by fire [δοκιμάζεται διὰ τοῦ πυρός] and made useful, so also you who live in them are being tested [δοκιμάζεσθε]. Therefore those who endure [μείναντες] and pass through the flames will be purified by them" (ὑπ'αὐτῶν καθαρισθήσεσθε).[55] Here the fire that tests clearly has a purifying function. When Paul suggests that fire is instrumental to salvation, he appears to be thinking in similar terms.

This is not to question Paul's insistence that believers are justified by faith (see Galations 2–3 and Romans 3–4) and are ultimately saved through Christ (see 1 Cor 15:1–2; Rom 5:9). Indeed, in his building metaphor Paul assumes that the foundation is Christ (v. 12). Yet in Paul's soteriology the believer's justification and future salvation do not preclude the possibility of punishment for sins either in this life or at the last judgment (see Rom 14:10; 1 Cor 4:4–5; 5:5; 11:29–32; 2 Cor 5:10). The verb ζημιωθήσεται in v. 15a may suggest that some of the saved will be "lightly punished at the judgment, depending on their deeds,"[56] and in this instance, the builder's punishment may be the painful purification process of being saved διὰ πυρός.[57] Perhaps the most relevant Pauline parallel is 1 Cor 5:5: "hand this man over to Satan for the destruction of the flesh [εἰς ὄλεθρον τῆς σαρκός], so that [ἵνα] his spirit may be saved [σωθῇ] in the day [ἐν τῇ ἡμέρᾳ] of the Lord."[58] Here the circumstances of exclusion, suffering, and possible death contribute (ἵνα) to the individual's salvation on "the day."[59] Similarly, in 1 Cor 3:15, the fire of divine judgment on "the day" appears to be the circumstance through which the builder is purified of his sins and through which Christ saves.

---

55. Note the striking similarities between 1 Cor 3:13–15, *Did.* 16.5, and *Herm. Vis.* 4.3.4: there is a fiery test (πῦρ [αὐτὸ] δοκιμάσει//πύρωσιν τῆς δοκιμασίας// δοκιμάζεται διὰ τοῦ πυρός); some endure (μενεῖ//ὑπομείναντες//μείναντες); and they will be saved/purified by the fire (σωθήσεται . . . διὰ πυρός//σωθήσονται ὑπ' αὐτοῦ τοῦ καταθέματος//ὑπ' αὐτῶν καθαρισθήσεσθε).

56. E. P. Sanders, *Judaism*, 275; see BDAG, s.v., ζημιόω. Shanor ("Paul as Master Builder," esp. 469) notes that in some Greek temple inscriptions this term refers to the imposition of fines on incompetent builders. Some sort of penalty is in view beyond "the potential 'dis-honor' (i.e. shame) of lost reward on the Day of Judgment" (Herms, "Being Saved without Honor," 205).

57. So also Lietzmann, *Korinther*, 17: Gemeint ist wohl, daß er nach einiger Strafe (ζημιωθήσεται = ὡς διὰ πυρός) gerettet wird.

58. *Contra* Konradt (*Gericht*, 268), the significance of this parallel is not diminished by the fact that the punishment (11:32) and the destruction of the man's flesh (5:5) are not eschatological. As Sanders (*Paul, the Law and the Jewish People*, 108) suggests, "faults unpunished in this world will be punished at the judgment. Further, punishment at the judgment brings atonement, just as do punishment and death in this world."

59. See Barrett, *Corinthians*, 126–27.

Conclusion

The dominant reading of 1 Cor 3:10–15 rests upon the following three assertions: (1) Paul's statement in v. 13 that fire will test the work of the builder indicates that only *deeds* and not individuals will be subjected to the fire; (2) the reference to *testing* precludes us from attributing any purifying function to the fire; and (3) his assertion that the builder "will be saved, but only as through fire" (v. 15), is nothing more than a proverb connoting a narrow escape. Since Kirk has already called into serious question the first of the above premises, I have focused my attention on the remaining two.

In light of the probable allusion to Malachi 3, where fire plays a purifying and refining function, and given the frequent pairing of testing and refining in the Hebrew Bible, there is no obvious reason to exclude the potentially purificatory function of fire from Paul's statement that the fire will test the work of the builder. More substantially, since 1 Cor 3:15 has no precise parallel, it is difficult to accept the scholarly axiom that we are dealing with a common proverbial statement. Rather than accepting the consensus opinion, which is based on only very inexact parallels, I have considered this verse in its individual grammatical parts, for which there are ample parallels in the relevant Greek literature. The phrase διὰ πυρός can just as easily take the instrumental sense as the local sense, though scholars frequently draw attention only to the latter use. More significantly, in every other biblical occurrence of the construction σῴζω + διά + genitive it takes the instrumental sense. That this is the dominant usage is supported by its use in the related Greek literature. Further, since the phrase οὕτως δὲ ὡς may indicate either metaphor or fact, it presents no serious challenge to our reading.

Taken together with the recent contribution of Alexander Kirk, the arguments presented here require a significant paradigm shift in the interpretation of 1 Cor 3:10–15. The confidence with which so many modern commentators assert that διὰ πυρός must be taken in the local sense and that the fire cannot serve a purifying or soteriological function is unwarranted. It seems that those who have revolted against the use of 1 Cor 3:15 as a proof text for purgatory have overshot the mark in being overly skeptical regarding the possibility that fire may play a purifying function.[60] Indeed, the balance appears to be tipped in favor of the instrumental sense of διά and thus in favor of the soteriological function of fire in v. 15. The dominant reading, which has gone too long without critical scrutiny, cannot stand without further justification. Those wishing to defend it must give further consideration to Kirk's arguments and to the above intertextual and linguistic evidence.

---

60. See Héring, *Corinthians*, 23: "Yet it is not a question of purifying fire (no purgatory!), but of one which destroys worthless material."

CHAPTER 7

Local Persecutions and the Cosmic Conflagration

MOVING FROM THE PAULINE to the Petrine epistles, we enter into a body of literature fraught with difficulties, particularly with regard to dating and authorship. The Petrine literature contains several references to fire as a means of testing or destruction (1 Pet 1:7; 4:12; 2 Pet 3:10–15). However, as we shall observe, the motif is used inconsistently between the two letters, taking on greater, more cosmic proportions in the latter epistle. While questions of date, authorship, and provenance may seem peripheral to our investigation, an approximation of these will assist us in understanding the development of the fiery ordeal and the cosmic conflagration motifs in 1 and 2 Peter. Before proceeding to our exegesis of each text, therefore, questions related to their historical contexts will be addressed.

First Peter

Two passages in 1 Peter make explicit reference to fire as a metaphor for testing. In 1:6–7 we read the following:

> ἐν ᾧ ἀγαλλιᾶσθε, ὀλίγον ἄρτι εἰ δέον [ἐστὶν] λυπηθέντες ἐν ποικίλοις πειρασμοῖς, ἵνα τὸ δοκίμιον ὑμῶν τῆς πίστεως πολυτιμότερον χρυσίου τοῦ ἀπολλυμένου διὰ πυρὸς δὲ δοκιμαζομένου, εὑρεθῇ εἰς ἔπαινον καὶ δόξαν καὶ τιμὴν ἐν ἀποκαλύψει Ἰησοῦ Χριστοῦ·

> In this you rejoice, even if now for a little while you have had to suffer various trials, so that the genuineness of your faith—being more precious than gold that, though perishable, is tested by fire—may be found to result in praise and glory and honor when Jesus Christ is revealed.

The second text, 1 Pet 4:12-13, which falls near the conclusion of the letter, does not use the word πῦρ, but employs the more technical πύρωσις:

> Ἀγαπητοί, μὴ ξενίζεσθε τῇ ἐν ὑμῖν πυρώσει πρὸς πειρασμὸν ὑμῖν γινομένῃ ὡς ξένου ὑμῖν συμβαίνοντος, ἀλλὰ καθὸ κοινωνεῖτε τοῖς τοῦ Χριστοῦ παθήμασιν χαίρετε, ἵνα καὶ ἐν τῇ ἀποκαλύψει τῆς δόξης αὐτοῦ χαρῆτε ἀγαλλιώμενοι.
>
> Beloved, do not be surprised at the fiery ordeal that is taking place among you to test you, as though something strange were happening to you; But rejoice insofar as you are sharing Christ's sufferings, so that you may also be glad and shout for joy when his glory is revealed.

While it was once popular to posit that the original letter ended with the doxology in 4:11 and that 4:12ff was the contribution of a different, later author, most scholars now recognize the literary unity of the 1 Peter.[1] These two references to fire thus form a sort of *inclusio*, the first appearing in the opening of the letter and the second beginning its last major literary unit, indicating that the experiences of suffering referenced in the intervening material ought to be understood as a period of being tested by fire or a fiery ordeal.

Upon first reading, one can already ascertain that the fire motif has been taken in a somewhat different direction from that which we have seen in the Pauline Epistles. For instance, the fire that tests does not come in the future on "the Day," as in 1 Cor 3:10-15, but is already present in the ποικίλοις πειρασμοῖς that believers endure. This detail alone indicates that the fire must be taken metaphorically. It is no literal fire that attends the parousia, such as that we have seen in 1 Cor 3:10-15 and elsewhere, but a period of adversity through which the faith of individuals will be tested. An understanding of the provenance of 1 Peter will help to illumine the period of suffering to which the testing by fire metaphorically alludes.

Context and Date

As Feldmeier observes, "[w]ith the exception of Job, no biblical book deals so often and so extensively in relation to its length with a situation

---

1. Such partition theories were based on the assumption that the first letter (1:1—4:11) was written when persecution was only a potential reality that lay on the horizon while the second letter (4:12—5:14) was written in the midst of a present and very real persecution. See, for instance, Streeter, *The Primitive Church*, 128-31. In this case the "various trials" that test faith by fire in 1:6-7 would be distinct from the "fiery ordeal" of 4:12. Most modern commentators prefer the view that 1 Peter is a literary unity. For an excellent discussion, see Achtemeir, *1 Peter*, 58-62.

of suffering as 1 Peter does."[2] In attempting to identify this "situation of suffering," scholars have suggested a wide range of dates for the authorship of 1 Peter. Those who favor Petrine authorship, date it prior to 64 CE, the traditional date of Peter's execution, and thus tend to identify the suffering with the Neronic persecutions.[3] At the other end of the spectrum, some have suggested the period of Trajan in light of the correspondence between Pliny and Trajan regarding the proper protocol for dealing with Christians.[4] However, while the exhortation "Always be ready to make your defense [ἀπολογίαν] to anyone who demands from you an accounting [λόγον] for the hope that is in you" (1 Pet 3:15), employs terms such as ἀπολογίαν and λόγον, which may imply a judicial context, Donelson suggests,

> the language of "always" and "to anyone" suggests readiness for all kinds of situations and inquiries. Furthermore, since the content of the account is "the hope that is in you," the defence seems to be more a theological account, a statement of gospel, than a legal defence against a public accusation.[5]

In view of these observations, the ποικίλοις πειρασμοῖς are probably not any sort of systematic persecutions, whether under Nero or Trajan. Rather, the persecution implied by the letter itself appears to be at the hands not of government officials but of neighbors and former acquaintances such as those alluded to 1 Pet 4:4, who are surprised (ξενίζονται) that the Christ followers that 1 Peter addresses have renounced their former ways.[6]

---

2. Feldmeier, *First Letter of Peter*, 2.

3. This possibility receives consideration in Davids, *First Epistle of Peter*, 10. Davids, however, is noncommittal regarding date and authorship. I am not persuaded by the arguments in favor of Petrine authorship, or for that matter, the theory that Sylvanus acted as Peter's amanuensis. My thinking on this matter has been deeply influenced by the arguments of John Elliott, who argues for authorship by a Petrine school in Rome sometime between 72 and 96 CE. See his discussion of authorship and date of composition in Elliott *1 Peter*, 118–38.

4. See especially Knox, "Pliny and 1 Peter." Knox draws attention to the emphasis both Pliny and 1 Peter put on "the name of Christ."

5. Donelson, *I and II Peter and Jude*, 105.

6. The ξενίζονται here provides an ironic contrast to 4:12, where the author urges his readers "do not be surprised [μὴ ξενίζεσθε] at the fiery ordeal that is taking place among you." So Dubis, *Messianic Woes*, 63: "Although believers may be surprised by the new state of affairs, believers should not be." Dubis goes on to demonstrate convincingly that the believers' lack of surprise is most intelligible when read in light of the eschatological conviction that suffering must precede the end (note also the δεῖ in 1 Pet 1:6).

The Fiery Ordeal

Both 1:7, which speaks of "testing by fire" διὰ πυρὸς δὲ δοκιμαζομένου, and 4:12, which refers to the "fiery ordeal" that tests πυρώσει πρὸς πειρασμὸν, probably have in mind the persecution and suffering of Christians referred to in the body of the letter (2:12, 23; 3:14–17; 4:4, 14–17). First Peter 1:6–7 anticipates this suffering, and gives it meaning through a comparison with the testing of gold by fire. As Feldmeier observes, this passage argues "*a minore ad maius* from gold that passes away, to the so very much more valuable faith."[7] Thus, unlike Paul's use of a similar metaphor in 1 Cor 3:10–15, where the endurance of precious metals stands in contrast to the combustion of shoddy building materials easily consumed by fire, gold is here seen as something perishable, in contrast to faith, which is imperishable. The point is that even the purest gold that has endured the refining process will not withstand the fire of final judgment, but genuine faith will.[8] Those who remain faithful through the ποικίλοις πειρασμοῖς (v. 6) will prove the δοκίμιον ὑμῶν τῆς πίστεως "genuineness of [their] faith" (v. 7) just as the quality of gold is proved through the smelting process.

There is precedent for this sort of analogy in Greco-Roman thought. Seneca, for instance, writes, *Ignis aurum probat, miseria fortes viros* ("Fire tests gold, affliction, strong men; *Ep., On Prov.* 4:10). Yet many have noted that our author is here drawing on a long-established Wisdom tradition (see Ps 66:10 [LXX 65:10]; Isa 48:10; Prov 17:3).[9] It is Sirach 2:1–6 in particular that offers the most extensive verbal and conceptual links to our text:[10]

> τέκνον εἰ προσέρχῃ δουλεύειν κυρίῳ ἑτοίμασον τὴν ψυχήν σου εἰς <u>πειρασμόν</u> εὔθυνον τὴν καρδίαν σου καὶ καρτέρησον καὶ μὴ σπεύσῃς ἐν καιρῷ ἐπαγωγῆς κολλήθητι αὐτῷ καὶ μὴ ἀποστῇς ἵνα αὐξηθῇς ἐπ' ἐσχάτων σου πᾶν ὃ ἐὰν ἐπαχθῇ σοι δέξαι καὶ ἐν ἀλλάγμασιν ταπεινώσεώς σου μακροθύμησον ὅτι ἐν <u>πυρὶ δοκιμάζεται χρυσὸς</u> καὶ ἄνθρωποι δεκτοὶ ἐν καμίνῳ ταπεινώσεως πίστευσον αὐτῷ[11]

> My child, when you come to serve the Lord, prepare yourself for testing. Set your heart right and be steadfast, and do not be impetuous in time of calamity. Cling to him and do not depart,

---

7. Davids, *First Epistle of Peter*, 57; Feldmeier, *First Letter of Peter*, 83.
8. See Jobes, *1 Peter*, 95.
9. Feldmeier, *First Letter of Peter*, 84.
10. Jobes, *1 Peter*, 95.
11. Vocabulary shared by 1 Pet 1:6–7 and Sir 2:1–6 is underlined. For further discussion, see Liebengood, *Eschatology of 1 Peter*, 110–16.

so that you may be honored at the end of your life.¹² Accept whatever befalls you, and in times of humiliation be patient. For gold is tested in the fire, and those found acceptable, in the furnace of humiliation.

In addition to the shared vocabulary—πειρασμός, πῦρ, δοκιμάζω, χρυσός—both Sirach and 1 Peter envision a "time of calamity" and a reward at the end of one's life. Further, the "furnace of humiliation" metaphor may be echoed later in 1 Pet 4:12, which speaks of the "fiery ordeal" (πύρωσις).

We may also note the strong literary parallels between 1 Pet 1:6–7 and Wis 3:5–7:¹³

> καὶ <u>ὀλίγα</u> παιδευθέντες μεγάλα εὐεργετηθήσονται ὅτι ὁ θεὸς <u>ἐπείρασεν</u> αὐτοὺς καὶ <u>εὗρεν</u> αὐτοὺς ἀξίους ἑαυτοῦ ὡς <u>χρυσὸν</u> ἐν χωνευτηρίῳ <u>ἐδοκίμασεν</u> αὐτοὺς καὶ ὡς ὁλοκάρπωμα θυσίας προσεδέξατο αὐτούς καὶ ἐν καιρῷ ἐπισκοπῆς αὐτῶν ἀναλάμψουσιν καὶ ὡς σπινθῆρες ἐν καλάμῃ διαδραμοῦνται.

> Having been disciplined a little, they will receive great good, because God tested them and found them worthy of himself; like gold in the furnace he tried them, and like a sacrificial burnt offering he accepted them. In the time of their visitation they will shine forth, and will run like sparks through the stubble.

Here we encounter both verbal parallels with 1 Pet 1:6–7—ὀλίγος, πειράζω, εὑρίσκω, χρύσος, δοκιμάζω—and the notion that the righteous ought to maintain hope in the face of adversity, for those whose worth is revealed by the trials they presently endure will receive future reward, which is precisely what we encounter in 1 Peter's promise that the faith of believers will "be found to result in praise and glory and honor when Jesus Christ is revealed" (v. 7). Thus, in light of the fact that the metaphor of gold being tested by fire is amply attested in Jewish literature, we may conclude that while the Greco-Roman parallel is interesting, Peter's use of this motif probably derives from Israel's Wisdom tradition, and may draw specifically from Sir 2:1–6 and Wis 3:5–7.

This metallurgical analogy, moreover, which we have now encountered on several occasions, carries with it additional connotations of purification. As Davids observes, "in 1 Peter genuineness is the outcome of faith's being

---

12. I have followed the RSV at this point. The NRSV renders this clause, "so that the last days of your life may be prosperous."

13. Vocabulary shared between 1 Pet 1:6–7 and Wis 3:5–7 is underlined. See further Liebengood, *Eschatology of 1 Peter*, 112–13.

tested by fire, which is the same thing that fire reveals in gold."[14] Implicit in this metaphor is the notion of purification, for "as all the dross is separated by the melting of metal so that in the end only the pure precious metal remains, so suffering is understood as a process of separation in which faith is proven in that it is purified."[15] Or, as Goppelt puts it, "[a]ffliction should, like fire in which precious metal is purified, separate out what is foreign waste and test what is pure. The image illustrates not only the goal of the trial, but also its necessity."[16] In a similar line of thought, Donelson points to the dual outcome:

> Only in these trials does the true character of Christians emerge. This could be for good or ill. Thus there is a sense here of being tempted. Trials that come with being the elect can tempt one to cease being the elect. On the other hand, these trials have the capacity to improve, to refine. The trials become almost a good thing. As 1:7 will note, for gold to become gold it needs the fire.[17]

Thus in 1:7 we encounter a familiar motif, the use of fire for the testing and refining of precious metals as an analogy for the testing and purification of the righteous. Given the two possible outcomes, being consumed by the fire or being proved by the fire, it is appropriate to speak of the dual function of this fiery ordeal, and by implication to suggest that it may carry purificatory qualities.

Despite the rich allusiveness of the motif of faith tested by fire here in 1:7, a fuller understanding of Peter's use of the metaphor depends upon the exegesis of 4:12–19. We have already seen repeated occurrences of the motif of fire that tests, particularly in our discussion of 1 Cor 3:10–15. It is in 1 Pet 4:12, however, that we first encounter the term πύρωσις, a word which literally means "burning," but which can carry several additional connotations; most significantly, it can mean a "burning ordeal" or "severe suffering."[18]

The term πύρωσις is quite rare in the Septuagint and Greek New Testament, where it occurs only in Amos 4:9, Prov 27:21a, and Rev 18:9, 18. The occurrence in Proverbs—"The crucible is for silver, and the furnace is for gold"[19]—has been especially significant in recent discussion, for in that

---

14. Davids, *First Epistle of Peter*, 57.
15. Feldmeier, *First Letter of Peter*, 83.
16. Goppelt, *Commentary on I Peter*, 90.
17. Donelson, *I & II Peter and Jude*, 33.
18. BDAG s.v. πύρωσις 2.
19. The second half of the proverb indicates, "a person is tested by being praised." The Proverb itself thus appears to lack eschatological significance.

text πύρωσις stands parallel to δοκίμιον, which translates the Hebrew מצרף "crucible":

LXX: δοκίμιον ἀργύρῳ καὶ χρυσῷ πύρωσις

MT: מצרף לכסף וצור לזהב

This connection between πύρωσις and מצרף has been exploited in an influential thesis by Emilie T. Sander, who argued that the word מצרף "crucible" functions as a technical term in the Dead Sea Scrolls, where it refers to "THE trial of the end-time, the eschatological ordeal or test. It is the time or situation of the testing which the faithful members of God's elect apocalyptc [sic] community must undergo before they are vindicated and the dominion of Belial comes to an end" (see 1QS I 17; VII 4; 1QM XVI 15; XVII 1, 8–9; 1QHa IV 16; 1QpPs XXXVII 2, 19; 4QFlor II 1; CD XX 27).[20] She further suggests that "the Qumran usage of מצרף provides the specific context in which πύρωσις is used in 1P 4:12."[21] While some of the particulars of Sander's thesis remain unconvincing, the general observation that there are strong similarities between the use of מצרף in the Dead Sea Scrolls to refer to an eschatological period of testing and the term πύρωσις in 1 Pet 4:12 is suggestive and compelling.[22] It is most persuasive, as noted by Michaels, in those texts where "מצרף, or 'crucible'... seems to have embraced both the testing of the righteous and the final punishment of the wicked."[23] Sander's claim fits the eschatological scenario of 1 Peter: "The end of all things is at hand" (4:7). Moreover, Dubis notes the use of the term by Josephus in his description of the destruction of Sodom and Gomorrah by fire (*Antiq.* 1.203) and in Rev 18:9, 18, where the fate of Babylon is described in terms of πύρωσις, thus providing further confirmation of the term's eschatological connotations. Perhaps most significant, however, is the striking parallel in *Did.* 16:5. In an eschatological discourse describing the last days, the author writes, "Then human creation will come to the fire of testing [τὴν πύρωσιν τῆς δοκιμασίας], and many will fall away and perish, but those who endure in their faith will be saved by the curse itself" (trans, Ehrman, LCL).

20. Sander ΠΥΡΩΣΙΣ, 43–44. For our discussion of some of the texts employing the term מצרף ("crucible") see pages 97, 102–3, 107, 109–10, 116 above. See the additional discussion in Dubis, *Messianic Woes*, 76–85.

21. Sander ΠΥΡΩΣΙΣ, 43.

22. See the critical yet appreciative analyses of Michaels, *1 Peter*, 260–61; Dubis, *Messianic Woes*, 76–85. Dubis, in particular, calls into question several of Sander's "subsidiary conclusions" while ultimately affirming, and even strengthening the argument that πύρωσις was a technical term for eschatological tribulation (82). Achtemeier deems her argument "ingenious if somewhat unconvincing" (*1 Peter*, 306, n. 25).

23. Michaels, *1 Peter*, 260.

Just as 1 Pet 4:12 follows the term πύρωσις with a qualifier pointing to the testing nature of the fire, πυρώσει πρὸς πειρασμὸν, *Did.* 16:5 offers a similar qualifier: τὴν πύρωσιν τῆς δοκιμασίας. It was the view of Sander that the need for the qualifying appositional phrase in 1 Pet 4:12, πρὸς πειρασμὸν, indicated that πύρωσις had lost its technical sense by the time 1 Peter reached its final form. However, Dubis has convincingly argued that the qualifier in 1 Peter makes excellent sense in its context: whereas at Qumran the term מצרף was used with reference to a period of testing for all, both the elect and the wicked, "First Peter 4:12 . . . is focusing upon the reason the πύρωσις comes to the *righteous*, namely, to test them (πρὸς πειρασμὸν) with the goal of "approving" them (see 1:6–7)."[24] Dubis's interpretation finds substantial support in the use of the qualifier in *Did.* 16:5. Following Sander and Dubis, therefore, we may conclude that 1 Pet 4:12 envisions a period of eschatological testing that precedes the end, a period which entails the refining of an elect community, resulting in the removal of those who are unfit as well as the purification of the righteous.

The Temple Motif, Once Again

When discussing 1 Cor 3:10–15 above, I noted that Paul employs an extended metaphor in which he likens the community of believers to the temple and suggests that this community will be tested with fire, perhaps alluding to the coming of God's fiery judgment into the temple which is prophesied in Mal 3:1–5; 4:1. In light of this temple imagery and potential allusion to Malachi, it is significant that Dennis E. Johnson has discovered similar temple imagery in 1 Peter, culminating in 4:12–19, with a potential allusion to Mal 3:1–5; 4:1.[25] As in 1 Cor 3:10–15, the testing of Christians is likened to the testing of precious metals in 1 Pet 1:7, thus recalling the metallurgical motif in Mal 3:1–5. Later, in 1 Pet 2:5–6, believers are identified as "living stones" (λίθοι ζῶντες) that are to be built into a "spiritual house" (οἶκος πνευματικός) which has Christ as its "cornerstone" (ἀκρογωνιαῖον) (2:6), once again picking up the architectural imagery we saw in 1 Cor 3:10–15. The suggestion that this spiritual house is in fact a temple is supported by the admonition to the believers that they are also to be a "holy priesthood" (ἱεράτευμα ἅγιον), offering "spiritual sacrifices" (πνευματικὰς θυσίας), which also calls to mind how the purified descendants of Levi will present to God "offerings in righteousness" (θυσίαν ἐν δικαιοσύνῃ) (LXX Mal 3:3). The motif culminates in 4:17–18:

24. Dubis, *Messianic Woes*, 84.
25. Johnson, "Fire in God's House."

> For the time has come for judgment to begin with the house of God (τοῦ οἴκου τοῦ θεοῦ);[26] if it begins with us, what will be the end for those who do not obey the gospel of God? And "If it is hard for the righteous to be saved, what will become of the ungodly and the sinners?"[27]

If, following Johnson, we are correct in identifying this οἴκου τοῦ θεοῦ as the temple, we have an image of the fire of judgment beginning in the temple of God and moving outwards.

As Johnson suggests, "[a]lthough Ezek 9:6 may well have influenced 1 Pet 4:17 verbally (*archomai apo*), conceptually it is Mal 3:1–5; 4:1 that provide the pattern for the escalation of eschatological judgment as it moves out from the house of God to those outside the covenant."[28] This reading is supported by the links to Mal 3:1–5; 4:1 noted above. Like the Levites of Mal 3:1–5, the house of God will be refined and purified by God's fiery presence while those "who do not obey the gospel of God" will experience the Day as a furnace that burns up the arrogant and evildoers like stubble as in Mal 4:1. Thus we may have another allusion to Malachi, wherein fire plays a dual function, simultaneously refining the elect while consuming the wicked.

A Metaphor for Present Suffering

More clearly than any other text we have considered, 1 Peter makes use of the fire motif in a purely metaphorical sense. Whereas the proclamation of John the Baptist, the words of Jesus, or the passage from Paul could entail an actual, physical fire that accompanies the Day of the Lord, in 1 Pet 1:7; 4:12 the fire is quite clearly metaphorical. This may indicate a new direction in

---

26. I have here departed from the NRSV and followed Johnson in interpreting τοῦ οἴκου τοῦ θεοῦ as "the house of God" rather than as "the household of God."

27. It is interesting to note how the question "if it begins with us, what will be the end for those who do not follow the gospel of God?" parallels Jesus' question in Luke 23:31: "For if they do this when the wood is green, what will happen when it is dry?" While no significant vocabulary is shared between the two (only εἰ . . . τί) the thought patterns are nearly identical, and they share a context of fiery judgment: both could be similarly paraphrased: "if this experience of suffering, which is likened to fire in some sense, happens to us/me, what will happen to the wicked?" Both also imply that the fiery judgment of others is to be far greater.

28. Johnson, "Fire in God's House." Spicq (*Les Épîtres de Saint Pierre*, 160) asserts that the allusions to Ezek 9:6 and Mal 3:1–5 are "sans doute." Elliot, moreover, notes that "[t]he idea of judgment's commencing with God's own people is traditional," citing Amos 3:1–2, 14; Jer 1:13–16; 25:29 (32:29 LXX); Ezek 9:5–6; 21:7–22; Mal 3:1–18; 2 *Bar.* 13:9–10; *T. Benj.* 10:8.

the use of the motif of testing by fire in early Christianity. As we shall presently see, 2 Peter takes the motif in an entirely different direction.

Second Peter

As in 1 Peter, the situation implied in 2 Peter is one of persecution; however, the persecution in the second epistle centers on the problem of scoffers ridiculing believers in view of the delay of the parousia. Using the genre of a last testament, the author has Peter address a purportedly future generation with these words: "First of all you must understand this, that in the last days scoffers will come, scoffing and indulging their own lusts and saying, 'Where is the promise of his coming? For ever since our ancestors died, all things continue as they were from the beginning of creation!'" (2 Pet 3:3-4). The scoffers maintain an "eschatological skepticism" which results in an ethical permissiveness, and the link between eschatology and ethics is 2 Peter's primary concern.[29] It is this critique of the eschatological orientation of Christians in light of the apparent non-fulfillment of the second coming that 2 Peter 3 seeks to address.

Context and Date

Second Peter has a good claim to be the latest composition in the New Testament. While some maintain Petrine authorship and thus date the epistle prior to 64 CE,[30] others have argued for a date in the late first century.[31] However, 2 Peter is widely regarded as pseudepigraphical and is most frequently dated to sometime in the early to middle second century.[32] I accept this majority opinion. As we shall see, this later date coheres with the more developed eschatology with regard to the function of fire we see in 2 Peter, for whereas all of the texts we have considered thus far envision fire playing a role in the eschatological judgment on the Day of the Lord by testing and perhaps purifying individuals, 2 Peter radically expands the scope of this fire, giving it cosmic proportions that are more in keeping with Stoic cosmology than with Jewish eschatology.[33]

29. Bauckham, *Jude, 2 Peter*, 51; Stephens, *Annihilation or Renewal?* 125.
30. For a fairly recent defense of Petrine authorship see Kruger, "Authenticity."
31. See especially Bauckham, *Jude, 2 Peter*, 158.
32. See Kelly, *Epistles of Peter*; Donelson, *I & II Peter and Jude*. Donelson's comments capture the uncertainty involved: "Sometime between 120 and 150 C.E. would be a good guess, but it is only a guess" (209).
33. On the Stoic and wider Hellenestic context of the cosmic conflagration in

Ekpyrosis in Stoic Thought

Before turning to the text of 2 Peter, we would do well to consider the functions of fire in Stoic thought and the doctrine of *ekpyrosis*. Unlike Plato, who held to a dualistic worldview, the early Stoics, beginning in the early third century BCE, advocated a monistic cosmology, according to which the world was providentially guided by a divine reason. As did Heraclitus, they identified this reason with fire.[34] For the Stoics, fire was not only the guiding principle that animated the universe; it was also the ultimate source from which the universe derives and at the same time a destructive force that would ultimately consume the world. In this regard they were also unlike Plato, who believed the universe to be eternal. For Stoic cosmology, moreover, fire plays a central role at every stage, from generation of the universe (cosmogony), to its governance by divine providence, to its conflagration (eschatology), each stage of which requires our consideration.

Regarding Stoic cosmogony, as Michael Lapidge, one of the few to write on the subject, observes, "[t]he Stoics could not believe in a creation *ex nihilo*, and therefore they were obliged to posit an eternally existing substance (*ousia*) out of which universes arose and into which they dissolved."[35] Moreover, while the early Stoics were theoretically monistic, they accepted the Platonic and Aristotelian principle that "genesis could only take place from the interaction of opposite forces";[36] thus, it was believed that "there were two ἀρχαί, one of which was passive (πάσχον) and was called ὕλη, the other of which was active (ποιοῦν) and was called θεός."[37] This latter, active principle, which was called θεός (god), was defined as μέγα πῦρ καὶ συνεχές

---

2 Peter, see especially Adams, *Stars Will Fall*, 114–24; Glasson, *Greek Influence*, 74–80; Harrill "Stoic Physics"; Pearson, "Indo-European Eschatology"; van der Horst, *Hellenism—Judaism—Christianity,* 271–92. The influence of Stoicism is sometimes debated due to the fact that 2 Peter envisions one final conflagration and not an infinite cycle. However, as Harrill notes, "This protest contains the unexamined presupposition that all possible allusions to Stoic ἐκπύρωσις must refer precisely to every detail of a set of theoretical speculations" (127). He demonstrates, to the contrary, that Roman allusions to ἐκπύρωσις often appeal to select Stoic topoi rather than embracing a technical and stystematically worked out doctrine. Further, offering a helpful analogy, he compares 2 Peter's cultural (rather than technical) familiarity with Stoicism to "how a modern person might know and use, more or less, some Freudian terms (ego, superego, the subconscious, a 'Feudian slip'), picked up from the common currency and authority of such discourse in modern society and not from any actual education in the psychoanalytic theories of Sigmund Freud" (128).

34. Cancik, "End of the World," 84–125.
35. Lapidge, "Stoic Cosmology," 161–85; 163.
36. Ibid., 163.
37. Lapidge, "ἀρχαί and στοιχεῖα," 241.

"a great and sustaining fire" (SVF 2.1045). Several other fragments clearly indicate the central role played by fire in the generation of the cosmos. Eusebius comments, "[a]ccording to the Stoics, who say that the fiery and hot substance is the command center of the universe, and that god is corporeal and is the creative force itself, (being) none other than the energy of fire" (SVF 2.1032).[38] Another ancient source adds, "the Stoics defined god as endowed with mind, a creative fire, going methodically about the business of creation of the universe" (SVF 2.1027; see 1.171; 2.1133–34).[39] For the Stoics, then, fire was a divine, creative, active and intelligent force from which all else derived.

The principle, referred to as ὕλη (matter, stuff), represents the passive aspect of the originative substance, and for the major early Stoics, namely Zeno, Cleanthes, and Chrysippus, this principle is represented by water or "pre-cosmic moisture."[40] It is from the creative fire's interaction with the passive moisture that all matter—earth, water, air, and fire—was believed to generate. The fire that is a result of this creative process is apparently qualitatively different from the creative fire and should not be confused with it.[41] While Chrysippus appears to have believed that the next stage of universal existence was dominated by a cosmic *pneuma* (wind or breath), which animated the living universe, for Zeno and Cleanthes, this role was played by the creative fire of the sun, which continued to feed upon the vapors of moisture emanating from the sea. Indeed, Cleanthes states, "the sun consists of fire and is nurtured by the vapours from the Ocean, because no fire could continue to exist without some sort of food" (SVF 1.504).[42] Though it is not clear how he reconciled this with his theory of the *pneuma*, Chrysippus similarly affirmed, "the heavenly bodies along with the sun are kindled from the sea" (SVF 2.579).

This creative fire continues to feed upon the vapors of moisture that are exhaled from the sea until all of the moisture of the earth is consumed. As van der Horst puts it, the sun

> sends these vapours back again after having consumed their moisture, so that eventually the whole cosmos will be set on fire,

38. Quoted in Lapidge, "Stoic Cosmology," 164.

39. Ibid.

40. Ibid., 166.

41. Lapidge ("Stoic Cosmology," 167) warns, "There is a tendency among commentators of Stoicism—ancient and modern—to assume that these two fires are really one. The Stoics themselves, however, distinguished carefully between them: fire the *archē*, that is, the *theos*, was described as 'creative fire' (*pyr technikon*), whereas fire the element (*stoicheion*) was described as 'destructive fire' (*pyr atechnon*) (SVF 1. 120)."

42. Cited in Mansfield, "Providence and the Destruction of the Universe," 150.

> because inevitably there must come a moment when celestial fire, which feeds off terrestrial moisture, will dry up the earth and so consume it.... So it is when the earth receives back from the sun its own accumulated, dehydrated, and heated vapours, that it catches fire and is totally destroyed. Then the divine, creative fire of this heavenly body turns into the destructive fire that converts its sustenance into itself.[43]

Cicero, when describing this element of Stoic cosmology, observes, "when the moisture has been used up [*umore consumpto*] neither can the earth be nourished nor will the air continue to flow, being unable to rise upward after it's drunk up all the water; thus nothing will remain but fire" (SVF 2.593).[44] Thus, once the fire has consumed all of the moisture in the world, the fire consumes everything in a cosmic conflagration: "Common Nature [understood as fire] becomes greater and more, and finally dries up everything and takes it back into itself" (SVF 2.599).

Some of the early Stoics went to great lengths to differentiate between the creative fire that plays the active role in generating matter and the destructive fire that ultimately consumes the universe. As Mansfield notes, Cleanthes says there are two kinds of fire: "burning and destructive fire on the one hand, benevolent, sustaining and vital fire on the other. The sun and the other heavenly bodies, since they are causes of life on earth, are said to consist of this vital fire, which is also present within ordinary living beings" (SVF 1.504).[45] Similarly, Zeno avers, "there are two kinds of fire, the one nontechnical (uncreative), converting its sustenance into itself, the other technical (craftsmanlike, creative); the latter variety is found within plants and animals, and is the substance of the heavenly bodies" (SVF 1.120).[46] There can be no doubt that the creative, craftsmanlike fire is to be distinguished from the created, terrestrial fire;[47] however, this prompts us to ask the important question of to what degree the creative ought to be differentiated from the destructive fire that functions to convert moisture into itself and ultimately bring about the cosmic conflagration. Mansfield stipulates that the "distinction between the two kinds of fire postulated by Zeno and Cleanthes does not, apparently, preclude their fundamental unity,"[48] citing Cleanthes to indicate that both the creative and destructive functions are

---

43. Van der Horst, *Hellenism—Judaism—Christianity*, 274.
44. Ibid.
45. Mansfield, "Providence and the Destruction of the Universe," 151.
46. Ibid., 152.
47. See Lapidge, "Stoic Cosmology," 167.
48. Mansfield, "Providence and the Destruction of the Universe," 155.

attributed to the sun: "the contact [of its rays] is of such a nature that it not only warms, but often actually burns."[49] In considering the manifold functions of fire in the distinct phases of cosmology, Mansfield observes that "the effective action of fire is, during cosmogony, wholly benevolent in the sense described; that, within the generated and ordered universe itself, its action is for the most part benevolent, but also to a slight extent apparently non-benevolent; and that, in the long run, the latter capacity comes to predominate throughout." He concludes that, "at least for Zeno and Cleanthes, there is a double cycle of fire in its opposite aspects, in which the craftsmanlike fire and the destructive fire dominate at opposite ends of the cycle, whereas during the stretches in between each in turn slowly gains the upper hand."[50]

Even the destructive function of fire, however, should not be viewed in a negative light. Indeed, according to Mansfield, "Chrysippus speaks of total conflagration in wholly positive terms."[51] To be sure, when speaking of *ekpyrosis*, Chrysippus insists that the cosmos turns into ἀθγή, which can be translated as "brightness" or "light" (SVF 2.611).[52] This positive interpretation of *ekpyrosis* is confirmed by fragments that indicate that in the cosmic conflagration evil is destroyed (2.606) and suggest that the fire plays a purifying function (2.617, 630).[53] Moreover, van der Horst insists that *ekpyrosis* was seen as a joyous occasion, and was "a wholly positive event." Thus, he concludes, "[t]he final conflagration is—in a sense—not a destruction, it is an act of god in his benevolent providence," emphasizing that "any element of divine judgment or punishment is completely lacking in this Stoic concept."[54]

This positive pronouncement regarding the function of *ekpyrosis* can be maintained for several reasons. First, the consumption of matter by divine fire entails the union of all that is created with the creative force itself, which can be construed quite positively. In the words of Mansfield again:

> All things have become reunited into this one pure god: the unity or "sympathy" of things is far better realized during *ekpyrosis* than either before or after. Simultaneously, whatever vestiges of grimness total conflagration might still have possessed

49. Ibid.
50. Ibid., 156.
51. Ibid., 176.
52. Ibid.
53. Lapidge ("Stoic Cosmology," 180) notes, "[e]ach of the early Stoics taught that the universe, after certain definite periods of time (whose duration is never specified), dissolves into fire and so 'purifies' itself. This process of purification was apparently called *katharsis* by the Stoics (SVF 2, 598); it was a purification in the sense that, at *ekpyrosis*, the universal substance consisted in nothing but pure fire."
54. van der Horst, *Hellenism—Judaism—Christianity*, 275.

> in the theories of Zeno (who has all things return to "fire" . . .) and Cleanthes (who has all things return to "flame" . . .) is abolished when this event is spoken of in terms of "brightness"; the absorption of all things into light is hardly to be regarded as something even imaginatively to be feared.[55]

Secondly, after destroying the created world, the fire that consumes and purifies the universe resumes its original creative function and the process repeats itself: "at certain fated times the whole universe will be converted into fire; next, it is again made into an ordered universe. The primal fire is so to speak a kind of seed containing the *logoi* of all things that have become, do become, and will become" (SVF 1.98).[56] Thus nothing was lost in the process of *ekpyrosis*; the whole of the universe simply returned to its original state only to be recreated once again through the same process. "What is surprising, perhaps," as Lapidge notes, "is that each newly generated universe was identical in every detail to the one(s) that had preceded it."[57]

Thus for the Stoics fire was the "element par excellence" (SVF 2.413).[58] It played the active role in the creation of the universe and was understood as an intelligent, providential *theos*. Likewise, through the sun it provided beneficial, creative fire and at the same time destructive, consumptive fire throughout the existence of the material universe as it emanated warmth and light while consuming the moisture exhaled from the sea. Lastly, and perhaps most significantly for our study, it played the lead role in the cosmic conflagration in which the entirety of created matter was consumed by the destructive fire. This process of creation, sustenance, and destruction was an infinitely repeating cycle that went on creating and destroying an infinite number of universes. As we consider the function of fire in 2 Peter 3, we shall note both the similarities to and departures from the Stoic doctrine of *ekpyrosis*.

Reserved for Fire

In responding to "the scoffers" who insist that Christ is not returning—"For ever since our ancestors died, all things continue as they were from the beginning of creation!" (2 Pet 3:4)—the author begins by appealing to biblical testimony of God's active involvement in the world, first in creation and

---

55. Mansfield, "Providence and the Destruction of the Universe," 177.
56. Cited in Ibid., 145.
57. Lapidge, "Stoic Cosmology," 180.
58. Cited in Mansfield, "Providence and the Destruction of the Universe," 145.

subsequently in God's dramatic intervention in sending the Flood on Noah's generation. The author writes:

> λανθάνει γὰρ αὐτοὺς τοῦτο θέλοντας ὅτι οὐρανοὶ ἦσαν ἔκπαλαι καὶ γῆ ἐξ ὕδατος καὶ δι' ὕδατος συνεστῶσα τῷ τοῦ θεοῦ λόγῳ, δι' ὧν ὁ τότε κόσμος ὕδατι κατακλυσθεὶς ἀπώλετο· οἱ δὲ νῦν οὐρανοὶ καὶ ἡ γῆ τῷ αὐτῷ λόγῳ τεθησαυρισμένοι εἰσὶν πυρὶ τηρούμενοι εἰς ἡμέραν κρίσεως καὶ ἀπωλείας τῶν ἀσεβῶν ἀνθρώπων.

> They deliberately ignore this fact, that by the word of God heavens existed long ago and an earth was formed out of water and by means of water, through which the world of that time was deluged with water and perished. But by the same word the present heavens and earth have been reserved for fire, being kept until the day of judgment and destruction of the godless. (2 Pet 3:5–7)

In this first section of the apology for the delay of the parousia, the author appeals to biblical history in response to his opponents' implied position that the cosmos is indestructible and that God does not intervene in history. In his response we encounter a typology in which the Flood of Noah's generation foreshadows the destructive fire of the day of judgment. Before addressing the significance of that typology, however, we must consider the claim that "an earth was formed out of water and by means of water, through which the world of that time was deluged with water and perished" (vv. 5–6).

Verses 5 and 6 consist of three prepositional phrases regarding the relation of the water to the creation and destruction of the earth. The earth was formed ἐξ ὕδατος (from water) and δι' ὕδατος (by means of, or through water), δι' ὧν (by which [water]) the earth was deluged and destroyed. The meanings of the first and third phrases are fairly uncontroversial. That the earth was created from or out of water probably alludes to the watery chaos of Genesis 1 out of which God created order by separating the waters and the earth, and most commentators agree on this point. Even clearer is the reference to the deluge of water by which the earth was destroyed during the time of Noah. Less evident and more controversial, however, is the meaning of the phrase δι' ὕδατος. The translators of the NRSV, which is quoted above, interpret this phrase in the instrumental sense. In what sense the earth was created *by means of water*, however, is unclear. Those who accept the instrumental sense of the preposition in this passage suggest that it refers to the nourishing rainfall that caused the plants of the earth to grow. Yet this reading is far from obvious.

More recently Edward Adams has argued that the phrases ἐξ ὕδατος and δι' ὕδατος may allude "to a particular Greek or Hellenistic cosmological tradition, one in which water is specifically identified as the substance out of which the world was made."[59] While admitting the presence of an allusion to Genesis 1, Adams suggests that the particular cosmological tradition that most adequately contextualizes the double construction "from water and through water" is Stoicism. In Stoic thought, the cosmos proceeds through a series of stages from fire through water, earth, air, and back again. Thus, as Adams argues:

> On the basis of the Stoic account of cosmic origins, it would be quite correct to say that the cosmos was formed "out of" water, since water, though not the archetypal element, was nevertheless the immediate substance out of which the cosmos was made, the malleable, corporeal stuff which the divine craftsman shaped and adapted into an ordered world. It would be equally correct to say that the heavens and the earth were formed "through" water, since water was not the original state of things but one of the material alterations experienced by the universe on its way to becoming a fully formed structure.[60]

While Adams confesses that no exact verbal parallel to δι' ὕδατος exists in Stoic literature to indicate that the cosmos passes "through water," he is able to cite Diogenes 7.136, 142, which employs the phrase δι' ἄερος to indicate that the elements function as transitional steps between rarefied fire and life sustaining cosmos.[61]

As is often noted, the author of 2 Peter envisages three epochs, punctuated by two cataclysmic judgments: that from the creation of the world to the flood; the present period between the flood and the judgment by fire; and the future new creation.[62] However, if Adams is correct that Stoic cosmology forms the backdrop for the phrases "from water and through water,"

---

59. Adams, *Stars Will Fall*, 212.

60. Ibid., 212–13. Adams notes that most commentators reject the Stoic interpretation, suggesting Old Testament or intertestamental Jewish influences. In his discussion of Jewish texts, however, Adams demonstrates that destruction of the world by fire is very rare and only evident in the *Sibylline Oracles*, where Stoic influence is present. Nonetheless, it is quite evident that in this passage Stoic cosmology coincides with Jewish eschatology in an early Christian syncretism, for while the passage of earth through water and its ultimate return to fire may reflect Stoic cosmology, God's judgment of the created order through the flood of Genesis typologically points to God's eschatological judgment by fire on the day of Christ's return.

61. Ibid., 213.

62. Bauckham, *Jude, 2 Peter*, 299.

we may infer that in the author's view of cosmic origins there existed a state of pure fire prior to the watery chaos of Gen 1:2.[63] Just as the earth was formed from water and returned to a watery chaos in the Noahic deluge, on a much larger scope the cosmos ultimately derived from a state of pure fire and is thus "reserved for fire," for it will return to a state of pure fire.

With regard to the phrase "reserved (τεθησαυρισμένοι) for fire" (v. 7), it is Bauckham's estimation that the word θησαυρίζω "to store up" draws upon a common notion that the rewards of the righteous and the punishments of the wicked are "stored up" in heaven until the day of judgment. Bauckham cites several Jewish and Christian texts to support this reading, but the most compelling is Pseudo-Sophocles (*apud* Clem. Alex. *Strom.* 5.14.121.4), who argues that at the eschatological judgment the air "will open the storehouse full of fire" (πυρὸς γέμοντα θησαυρόν; see *Pss. Sol.* 9:5; *4 Ezra* 7:77, 83–84; *Frg. Tg.* Deut 32:34; Rom 2:5; *Clem. Hom.* 16:20).[64] This raises the question of whether by saying that the cosmos is "reserved for fire" on the day of judgment the author is suggesting that the heavens and the earth will be destroyed in the sense of total annihilation and thus no longer in existence, or whether we ought to think of this judgment by fire in some other sense.

The former appears to be the view of Stephens, who writes, "it is within 2 Peter that we find the clearest representative of an early Christian belief in cosmic annihilation."[65] 2 Peter's language certainly does sound destructive. Bauckham, however, argues as follows:

> The idea of destruction of the antediluvian world need not be taken to mean total annihilation. Rather, just as it was created by being brought out of the primeval ocean, so it was destroyed when it was once again submerged in the primeval ocean. The ordered world (κόσμος) reverted to chaos.[66]

Similarly, while Adams speaks in terms of destruction, he does so in a qualified sense. He is essentially in agreement with Bauckham, suggesting that the cosmos will be destroyed in fire just as the earth was destroyed by water in the Noahic flood when it returned to a state of watery chaos. Thus, while the return to a state of pure fire constitutes a destruction of the world, it would be inappropriate to refer to a state of annihilation or non-existence. Notably, while the heavens and earth are reserved for fire, the day of judgment is a day of destruction (ἀπώλετο) for the wicked in the sense that the earth

---

63. Adams, *Stars Will Fall*, 213.
64. Cited in Bauckham, *Jude, 2 Peter*, 300.
65. Stephens, *Annihilation or Renewal?* 124. Stephens in turn cites Kelly, *Epistles of Peter and Jude*, 224. Kelly likewise speaks in terms of annihilation.
66. Bauckham, *Jude, 2 Peter*, 299.

was destroyed (ἀπωλείας) in the deluge. The parallel being made, therefore, between the destruction of the earth by water and the destruction of the heavens and earth by fire would apparently not point in the direction of annihilation of the cosmos but towards a return to a previous state; whereas in the flood it returned to a state of watery chaos, on the day of judgment it will return to a state of pure fire.[67] The notion is therefore one of renewal through fire—note the "new heavens and new earth" of v. 13 below—rather than annihilation and removal from existence.[68] We will return to this issue below in our discussion of the verb εὑρεθήσεται in v. 10.

The Elements Will Be Dissolved with Fire

The author of 2 Peter briefly redirects his argument to expound on the relativity of time in relation to God (v. 8) and to argue that the purported delay is accounted for by appealing to divine forbearance (v. 9) before he returns to the fire motif (v. 10–13). When he does return to the subject of judgment by fire, the mingling of Stoic cosmology and Jewish Christian eschatology is even more apparent. In 2 Pet 3:10–13 we read:

> Ἥξει δὲ ἡμέρα κυρίου ὡς κλέπτης, ἐν ᾗ οἱ οὐρανοὶ ῥοιζηδὸν παρελεύσονται στοιχεῖα δὲ καυσούμενα λυθήσεται καὶ γῆ καὶ τὰ ἐν αὐτῇ ἔργα [οὐκ][69] εὑρεθήσεται. τούτων οὕτως πάντων λυομένων ποταποὺς δεῖ ὑπάρχειν [ὑμᾶς] ἐν ἁγίαις ἀναστροφαῖς καὶ εὐσεβείαις, προσδοκῶντας καὶ σπεύδοντας τὴν παρουσίαν τῆς τοῦ θεοῦ ἡμέρας δι' ἣν οὐρανοὶ πυρούμενοι λυθήσονται καὶ στοιχεῖα καυσούμενα τήκεται. καινοὺς δὲ οὐρανοὺς καὶ γῆν καινὴν κατὰ τὸ ἐπάγγελμα αὐτοῦ προσδοκῶμεν, ἐν οἷς δικαιοσύνη κατοικεῖ.

> But the day of the Lord will come like a thief, and then the heavens will pass away with a loud noise, and the elements will be dissolved with fire, and the earth and everything that is done on

---

67. Noah's flood is frequently conjoined with the cosmic conflagration. Just as the earth was once destroyed by water, it will again be consumed by fire. The expectation of destruction by fire may result from God's promise never again to flood the earth with water. For the pairing of flood and fire in apocalyptic discourse, see especially Philo, *Mos.* 2.48.263; Jos. *Ant.* 1.70–71; *L.A.E.* 49.3; *Sib. Or.* 3.689–90. See also Chaine, "Cosmogonie Aquatique"; Schlosser, "Les Jours."

68. Thiede, "Pagan Reader," 53; Wolters, "Worldview and Textual Criticism," 408; Davids, *Letters of 2 Peter and Jude*, 271.

69. I have bracketed the οὐκ because it appears in the main text of NA28 (without brackets) but not in NA27, where it is noted only as a textual variant. For reasons given below, I favor the reading represented by NA27. The English translation "will be disclosed" follows NA27.

> it will be disclosed. Since all these things are to be dissolved in this way, what sort of persons ought you to be in leading lives of holiness and godliness, waiting for and hastening the coming of the day of God, because of which the heavens will be set ablaze and dissolved, and the elements will melt with fire? But, in accordance with his promise, we wait for new heavens and a new earth, where righteousness is at home.

Here we see the common motif of fire accompanying the day of the Lord (see 1 Cor 3:10–15) and some apparent allusions to the Jesus tradition. The day is said to come "like a thief," which is a potential reference to Matt 24:42–44 (see 1 Thess 5:3; Rev 3:3; 16:15), just as the verb παρελεύσονται "will pass away" may allude to Matt 5:18; 24:35; Mark 13:31; Luke 21:33. Unique to our text's description of the day of the Lord, however, is the use of the adverb ῥοιζηδόν "with a loud noise," which is a *hapax* in the New Testament. The word itself is onomatopoeic and may allude to the sound of a burning fire, which roars and crackles.[70]

The identification of the στοιχεῖα has proved problematic, with at least three main options open for consideration. First, some have suggested that στοιχεῖα is a reference to angelic beings. This parallels the interpretation of Paul's use of the term in reference to hostile spiritual beings (Gal 4:3; Col 2:8, 20). For reasons stated below regarding the use of the term in vv. 12 and 13, however, I find this proposal improbable. Much more plausible is the proposal that the author had in mind celestial bodies, such as the sun, moon, and stars (see Isa 13:10; Ezek 32:7–8; Mark 13:24–27; Rev 6:13).[71] Bauckham notes that this interpretation is popular among most commentators and finds support in the literature of the second century CE (Theophilus, *Ad Autol.* 1.4–6; 2.15, 35; Justin 2 *Apol.* 5.2; *Dial.* 23:3; Polycrates, *apud* Eusebius, *Hist. Eccl.* 3.31.2; Tatian, *Oratio* 9–10).[72] Moreover, this interpretation is suggested in part by the apparent location of the στοιχεῖα between the heavens and the earth in v. 10. Given that elsewhere heavenly bodies such as stars and moons are referred to as στοιχεῖα, interpreters have understandably found this proposal compelling.

The other dominant solution is that στοιχεῖα refers to the basic elements that comprise all of creation, typically water, air, fire, earth.[73] This

---

70. A variety of translations are cited in Strange, *Exegetical Summary*, 269. Note the similar detail in the great conflagration described in *Sib. Or.* 4.175, which is accompanied by μύκημα καὶ ὄμβριμον ἦχον "bellowing noise and mighty sound." That text, like this one, bears the marks of Stoic influence.

71. Kraftchick, *Jude, 2 Peter*, 163.

72. Bauckham, *Jude, 2 Peter*, 316. See Delling, "στοιχεῖα," 681–82.

73. Reicke, *Epistles of James, Peter, and Jude*, 180; Adams, *Stars Will Fall*, 223;

meaning is widely attested (see *Herm. Vis.* 3:13:3; Aristides, *Apol.* 3–7; *Sib. Or.* 3:80–92), and it fits the Hellenistic and Stoic context we have already observed, particularly in light of the fact that the elements are consumed by fire in the cosmic conflagration, which was a common feature of Stoic cosmology.[74] Its strength is that it makes better sense of vv. 12–13 than the identification with angelic beings or astral bodies, both of which readings are problematized by verse 12, where the οὐρανοί "will be set ablaze and dissolved" and the στοιχεῖα "will melt with fire" with no explicit reference to the earth at all.[75] Indeed, the dissolution of heavens and the melting of the elements in v. 12 appears to parallel the "heavens and the earth are stored up for fire" in verse 7 and the resultant new heavens and new earth in verse 13 so that στοιχεῖα seem to be synonymous with earth (γῆ). If the elements that make up the earth are in view, this parallelism between the elements and the earth makes perfect sense. Thus, while compelling arguments can be made for the identification with celestial bodies, we are left favoring the interpretation that it is the basic physical elements that comprise the cosmos that are in view.

According to v. 10, the στοιχεῖα "will be dissolved with fire" (καυσούμενα λυθήσεται), and later, v. 12 states that "the heavens will be set ablaze and dissolved [λυθήσονται], and the elements will melt [τήκεται] with fire." According to Stephens "the main verb here, λυθήσονται, is best translated in the sense of 'dissolved,' seeming to suggest either the annihilation of the στοιχεῖα or at least their disintegration into their constituent parts."[76] The verb is the future passive of λύω, which here means "to reduce something by violence into its components," which would advise us against Stephens's suggestion that the στοιχεῖα may be annihilated.[77] Moreover, in vv 11–12, λυθήσονται stands in parallel to τήκεται, and the former verb occurs elsewhere with reference to the "elements," particularly in Philo, *Aet. Mund.* 110, where he refers to the changes that occur from one element to another: "Therefore the steep road begins with the earth; for when it is wasted away [τηκομένη] it endures a change to water, and the water when it has evaporated is changed into air, and the air when rarefied is changed into fire" (see Isa 34:4; *1 En.* 1:6; *T. Levi* 4:1). The use of the verb by a Jewish near contemporary of 2 Peter who in the immediate context goes on to speak of the cosmic confla-

---

Stephens, *Annihilation or Renewal?* 132. Delling, "στοιχεῖα," 686.

74. It may be possible to conclude that the term is inclusive of both earth and the heavenly bodies, as does Thiede, "Pagan Reader," 82.

75. This juxtaposition of "the heavens and the elements" similarly makes the interpretation of στοιχεῖα as angelic beings improbable.

76. Stephens, *Annihilation or Renewal?* 131.

77. BDAG s.v. στοιχεῖα.

gration suggests that what is in view here is not annihilation but alteration into a more basic elemental state.

Lastly, the meaning of the phrase γῆ καὶ τὰ ἐν αὐτῇ ἔργα εὑρεθήσεται, literally, "the earth and all the works on it *will be found*," has been deemed a *crux interpretum* and has provoked textual critics, commentators, and translators alike. The source of confusion is the final verb εὑρεθήσεται, which is well attested (א B K P 424c 1175 1739txt 1852 syrph, hmg arm Origen). There are three general approaches to this problematic text. Some argue in favor of variant readings; others emend the text, suggesting solutions for which there is no textual support; and others still have sought to make sense of the verb εὑρεθήσεται as it stands. The extant textual variants include the following:

1. ... οὐκ εὑρεθήσεται ("... will not be found"; sa Harclaean Syriac).
2. ... εὑρεθήσεται λυόμενα ("... will be found dissolved"; P72).
3. ... ἀφανισθήσονται ("... will vanish"; C).
4. ... κατακαήσεται ("... will be burned up"; A 048 049 056 0142 33 614 Byz Lect syrh copbo eth *al*).
5. some omit the phrase altogether (Ψ vg Pelagius *al*).

However, in addition to being supported by a strong textual tradition, εὑρεθήσεται is the *lectio difficilior* and thus is more likely to be original.[78] One can imagine how variant (1) and (2) attempt to make sense of the verb while variants (3) and (4) try to interpret it. Lastly, (5) fails to explain adequately why this phrase would ever be added. In addition to the above variants, some modern scholars have supposed that even the earliest Greek text is corrupt and have proposed numerous textual emendations, seeking to mitigate the tension that exists with the dominant reading.[79] Since we believe a strong case can be made for the earliest reading, εὑρεθήσεται, we need not list all of the speculative reconstructions, which are too numerous to include here.

What then might be meant by εὑρεθήσεται? While some have deemed the present text "devoid of meaning"[80] or "incomprehensible,"[81] others have

---

78. Reicke, *Epistles of James, Peter, and John*, 185.

79. These are conveniently collected in Bauckham, *Jude, 2 Peter*, 317–18; Metzger, *Textual Commnetary* 636–37.

80. Metzger, *Textual Commentary*, 636.

81. Van den Heever, "In Purifying Fire," 107.

identified it as a technical verb relating to a context of judgment.[82] Danker, for instance, has observed a similar usage in the Psalms of Solomon 17:8:[83]

κατὰ τὰ ἁμαρτήματα αὐτῶν ἀποδώσεις αὐτοῖς ὁ θεός
εὑρεθῆναι αὐτοῖς κατὰ τὰ ἔργα αὐτῶν

According to their sins you repaid them, O God;
they were found according to their deeds.[84]

The parallelism between the two parts in this passage, with sins being repaid in the first clause and deeds being "found" in the second, indicates that εὑρεθῆναι is here used in a judicial sense.[85] And it is this connotation of judgment that, in Danker's view, makes 2 Pet 3:10 intelligible.

Al Wolters accepts that the verb suggests a context of judgment. However, he argues that 2 Pet 3:10 envisions this judgment as involving a smelting process that results in the purification of the world. Accepting the authenticity of εὑρεθήσεται, Wolters proposes the verb is here used in a technical sense to indicate "*eschatological survival . . .* meaning 'to have survived,' 'to have stood the test,' 'to have proved genuine.'"[86] To support this claim, Wolters appeals to three texts where, so he argues, the verb appears in its passive form and in an absolute sense, as it is used here: 2 Pet 3:14; 1 Pet 1:7; and *Ep. Barn.* 21:6. The first of these occurs just four verses after 2 Pet 3:10. Thus, in the immediate context of the verse under investigation, the author writes, "Therefore, beloved, while you are waiting for these things, strive to be found [εὑρεθῆναι] by him at peace, without spot or blemish" (2 Pet 3:14). Here the verb is used to indicate the result of a juridical process, and as Danker observes, "if εὑρεθήσεται in vs. 10 was the original reading, then the echo in vs. 14 is artistically designed, and in keeping with the writer's trend of thought."[87] Similarly, in 1 Pet 1:7, the author exhorts his readers to endure their various trials "so that the genuineness of your faith—being more precious than gold that, though perishable, is tested by fire—may be found [εὑρεθῇ] to result in praise and glory and honor when Jesus Christ

---

82. See Lenhard, "Noch einmal," who notes that Esther 2:23; Zeph 3:12; and Jer 50:20 use the language being "found" in contexts of judgment. More recently, Juza, "Echoes of Sodom and Gomorrah," has observed that the allusion may be to those who would be "found" to be righteous in LXX Gen 18:26–32, where the verb εὑρίσκω appears seven times. Juza discerns numerous other verbal links to the story of Sodom and Gomorrah, and there is, of course, the explicit reference in 2 Pet 2:6.

83. Danker, "2 Peter 3:10," 85.

84. My translation.

85. Danker, "2 Peter 3:10," 85.

86. Wolters, "Worldview and Textual Criticism," 410.

87. Danker, "2 Peter 3:10," 84. See Wilson, "Εὑρεθήσεται in 2 Peter 3:10," 44.

is revealed." As Wolters observes, in this verse "the passive of *heuriskō* describes, again in an eschatological context, the surviving of a purifying fire."[88] Wolters then goes on to assert, "*heurethē* is here used absolutely, without predicate, exactly comparable to *heurethēsetai* in 2 Pet 3:10."[89] The third text to which Wolters appeals is *Ep. Barn.* 21:6: "Be instructed by God, seeking out what the Lord seeks from you and then doing it, in order that you may be found [εὕρητε] in the day of judgment." Regarding this passage Wolters writes, "[t]he parallel with 2 Pet 3:10 is so close (the same absolute use of *heuriskesthai*, the same eschatological context, the same link with eschatological exhortation) that it looks like an explicit verbal echo."[90]

Wolters situates this particular use of the verb in the context of fiery judgment on the Day of the Lord. He argues that the complete destruction of the world is not envisaged here, for the verb καίομαι ("to burn up") is never used. Rather, 2 Peter employs the verb πυρόομαι, which "is regularly used of metals being heated in a smelting furnace" (see LXX Zech 13:9; Rev 1:15; 3:18).[91] This combination of smelting imagery and the Day of the Lord suggests to Wolters an allusion to Mal 3:2–4; however, here "it is the entire cosmos, not just the Israelite priesthood, that is to be refined in the crucible of judgement on the great day of God's appearance."[92] This reading finds modest support in 2 *Clem.* 16:3, which appears to be a very early interpretation of 2 Pet 3:10:[93]

> γινώσκετε δέ ὅτι ἔρχεται ἤδη ἡ ἡμέρα τῆς κρίσεως ὡς κλίβανος καιόμενος καὶ τακήσονταί τινες τῶν οὐρανῶν καὶ πᾶσα ἡ γῆ ὡς μόλιβος ἐπὶ πυρὶ τηκόμενος καὶ τότε φανήσεται τὰ κρύφια καὶ φανερὰ ἔργα τῶν ἀνθρώπων.
>
> But you know that the day of judgment is already coming as a blazing furnace, and some of the heavens will dissolve, and the whole earth will be like lead melting in a fire, and then everyone's works, the secret and the public, will be revealed.

Here, in the phrase "as a blazing furnace," the allusion to Malachi 3 is made explicit. In addition, the verb εὑρεθήσεται is replaced by the more comprehensible φανήσεται ("will be revealed"), indicating perhaps that a very early

---

88. Wolters, "Worldview and Textual Criticism," 410.
89. Ibid.
90. Ibid., 411.
91. Ibid., 409.
92. Ibid.
93. Bauckham posits a Jewish apocalyptic source common to both 2 Peter and 2 *Clement*. This is, however, unnecessary; the far simpler solution of literary dependence is more satisfactory.

reader of 2 Peter understood this obscure verb to suggest that the fire plays a revelatory and not a destructive role. This revelatory function would cohere with the context of smelting—a testing process which reveals the quality of precious metals—as Wolters infers.

However, while these observations are indeed significant, Wolters's argument has been subjected to some criticism. Van den Heever, for instance, notes that in the parallels Wolters cites, εὑρεθήσεται/εὑρίσκω is not used in the same unqualified way that it is 2 Pt 3:10.[94] In 1 Pet 1:7, the believers' faith will be "found *to result in praise and glory and honor*"; in 2 Pet 3:14, those addressed should "strive to be found by him *at peace, without spot or blemish*"; and in *Ep. Barn.* 21:6, those who do the Lord's bidding will "be found *in the day of judgment*."[95] However, while Wolters has perhaps slightly overstated his case, and his suggestion that εὑρεθήσεται carries a technical metallurgical sense lacks sufficient evidence, the occurrence of the passive form of the same verb in combination with metallurgical imagery in 1 Pet 1:7 (and Wis 3:5), which may have inspired 2 Peter's usage, and the verse's interpretation in *2 Clem.* 16:3, which reads it as an allusion to the refining process envisioned in Mal 3:2–4, is indeed suggestive. We may still conclude that the author envisages a purgative or purifying fire that consumes everything, thus resulting in a new heavens and a new earth, even if we remain uncertain as to whether he was thinking in terms of metallurgy with specific reference to Mal 3:2–4. Thus, while Wolters's thesis has justifiably received some criticism, he has made an important contribution and insights may be gleaned from his discussion even if his argument is not accepted wholesale.

Drawing upon the Stoic doctrine of *ekpyrosis*, 2 Peter therefore appears to suggest that on the last day the heavens and the earth will return to a state of pure fire—a state similar to that from which it originated. Unlike Stoic thought, however, which posits an infinite cycle of conflagrations and returns, the thought of 2 Peter appears more linear. Like the refining fire of the smelter in Malachi 3, the cosmic conflagration melts the physical elements, thus resulting in a new, purified heaven and earth with no indication of further conflagrations. The Stoic notion of a cosmic conflagration has been superimposed on the framework of Jewish eschatology.

---

94. Van den Heever, "In Purifying Fire," 109.

95. While van den Hever is technically correct that none of these examples employs εὑρεθήσεται/εὑρίσκω in an unqualified sense, *Ep. Barn.* 21:6 in particular remains remarkably germane, for the prepositional phrase "in the day of judgment" only qualifies *when* the finding will take place, not *how* the believer will be found.

Conclusion

One of the more interesting observations to be made with regard to the fiery ordeal in the Petrine Epistles is that whereas 1 Peter employs the testing by fire motif in a purely metaphorical manner to refer to local occurrences, 2 Peter radically expands the scope of the apparently literal fire to give it cosmic dimensions. Despite this significant difference, both letters appear to makes use of the motif of testing by fire by drawing upon the image of the smelting process, which tests and purifies precious metals. Whereas 1 Peter frames the image in the context of well-known Wisdom traditions, 2 Peter situates it in the context of Stoic cosmology. One could easily highlight the distinctiveness of each letter's use of the motif, but perhaps equally as striking is what they hold in common—drawing upon the image of the smelter's furnace, they suggest that the fire that tests also purifies the faithful.

CHAPTER 8

Concluding Remarks

Summary

IN THE FOREGOING PAGES we have set out examine instances in the New Testament where *everyone*, both the righteous elect and the recalcitrant sinners, are subjected to eschatological judgment by fire. Previous scholarship on fire in the Bible, surveyed in chapter 1, has tended to take one of two approaches: on the one hand, there are those who have offered surveys of all of the different functions of fire in the Old Testament (Laughlin) or the New Testament (Lang, "πῦρ," *TDNT*) or both (Lang, *Feuer*); and, on the other hand, there are those who have focused entirely on a single pericope (Gnilka; 1 Cor 3:10-15) or verse (Dunn; Luke 3:16//Matt 3:11). Rather than following this general pattern, we have sought to examine a distinct motif that recurs throughout the New Testament.[1] This motif, according to which all are subjected to fire, may best be understood as a fiery ordeal or test that results in the destruction of the wicked and the preservation or even purification of the righteous. While this motif does not dominate in any single New Testament text, it is a recurring theme that spans from some the New Testament's earliest documents (or sources)—1 Corinthians, Mark and Q—to our latest—2 Peter.

In chapter 1 we offered an introduction to and a historiographical context for our subject, a survey of relevant literature, and a discussion of methodology. Chapter 2 surveyed Old Testament texts related to fire in order to demonstrate the multiple functions of fire in the biblical worldview and to draw attention to the much overlooked purificatory function of fire.

---

1. There is some similarity between our approach of singling out a particular motif and the approach of Mayer, who, in searching for Zoroastrian influence in the Hebrew Bible, also focuses on the motif of the fiery ordeal.

In chapter 3 we turned to the apocalyptic literature of Second Temple Judaism. Here we focused specifically on the fire of judgment and particularly on the testing function of fire. These chapters served as a contextual framework for situating our discussion of the fiery ordeal in New Testament eschatology. The upshot of this survey of Jewish literature was that fire could serve a punitive or a refining function depending upon the status of the individual subjected to the fire. While the wicked are punished and ultimately destroyed by the fire, the upright are either preserved from the dangers of the blazing fire, pass through it safely, or are positively transformed by it.

Chapter 4 focused on fiery ordeal in the Gospels, focusing on John the Baptist's proclamation of a coming one who would baptize with Holy Spirit and fire. John's proclamation of the Coming One who would baptize in Holy Spirit and fire helped situate sayings of Jesus, discussed in chapter 5, such as his cryptic utterance concerning salting with fire in Mark 9:49 and his mention of fire and baptism in Luke 12:49–50, these verses in turn suggested to us that the original form of the Baptist's proclamation may not have included reference to the Holy Spirit. The remainder of chapter 5 considered certain enigmatic sayings of Jesus which may allude to a period of testing or tribulation through fire for the righteous, including Luke 17:26–32, which alludes to the days of Noah and the days of Lot; Luke 23:31, in which Jesus invokes the parable of the green wood and the dry; and *Gos. Thom.* 82, according to which those who are near Jesus are also near the fire, which we suggested was the fire of tribulation or testing. According to each of these traditions (with the possible exception of Luke 17:26–32), Jesus envisioned a final period of fiery judgment, from which he did not exclude his disciples. If even they were expected to endure this judgment by fire, we may understand this as a sort of fiery ordeal, which he expected the righteous to pass.

Chapter 6 considered the manifestation of this motif in the Pauline and epistles, particularly 1 Cor 3:10–15, where we observed a probable allusion to Mal 3:1–6; 4:1. We suggested that the imagery of fire testing the work of the builders alongside the temple imagery in 1 Cor 3:10–15 draws upon the very similar imagery of fire coming into the temple in Malachi 3 to purify the priests of Levi, thus suggesting the purificatory role of fire in this passage. Moreover, through an appeal to linguistic evidence pertaining to the use of the phrase "saved through" in Jewish and Christian writings, we argued that the concluding phrase "saved through fire" ought to be taken in the instrumental sense, not in the local sense as is so often assumed. The fire of this text appears to be a literal fire that attends the Day of the Lord and plays a decisive role in the testing, refining, and salvation of individuals.

Chapter 7 turned to the Petrine Epistles. In our examination of 1 Peter 1:7; 4:12 we encountered a purely metaphorical interpretation of the fiery ordeal. The former draws upon metallurgical imagery to compare the testing of faith with the testing of gold by fire. We concluded that just as gold is tested and refined through the smelting process, so is faith understood to be tested and purified through strife and persecution. Similarly in 4:12 we noted the similar interpretation of the "fiery ordeal" where it was applied, much as in the Dead Sea Scrolls' use of the term מצרף "crucible," to a present period of suffering, and thus the fire is a purely metaphorical description of a present period of persecution, and not, as in 1 Corinthians, an expected feature of the future day of the Lord. Nonetheless, the metaphorical fire of testing still appears to carry a purificatory function. Lastly, we considered 2 Pet 3:10–13 and the depiction of the cosmic conflagration therein. We observed the author's appropriation of Stoic physics, but noted how he combined it with Jewish eschatology and incorporated both into a Christian synthesis. The author expected that at the last day, the στοιχεῖα or the physical elements that comprise heaven and earth, would melt and return to a state of pure fire from which a new heavens and a new earth would emerge. The verb εὑρεθήσεται "will be found," may allude to finding (i.e., judging) the quality and content of the earth and all who are on it after a period of testing by fire, and may thus recall the smelting process, as Wolters has argued.

Contributions and Implications

Notably, many of the New Testament texts we have considered are frequently identified as *cruces interpretum*. Consider, for instance, the numerous textual variants and proposed emendations for Mark 9:49 and 2 Pet 3:10, or the doctrinal disputes that have raged over the interpretation of 1 Cor 3:10–15. It is our hope that identifying these passages as belonging to the general motif of the fiery ordeal or test will illumine their meanings and contribute to a greater understanding of them both collectively and individually. Further, the history of Catholic-Protestant relations is marked by doctrinal disputes about purgatory. While I have refrained from exploring the development of the doctrine of purgatory, it is apparent that the fiery ordeal in which fire may play a purificatory function may have implications for theologians concerned with the origins of this doctrine. First Corinthians 3:10–15, in fact, was in the West long considered a proof text for purgatory. While our exegesis in no way confirms the doctrine of purgatory, it may open avenues for discussion of the origins of this doctrine and facilitate a Protestant revaluation of the purificatory function of fire in New Testament eschatology. Lastly, while we have observed a sustained motif through the Gospels, Paul, and the Petrine Epistles, it is quite evident that at least in the last of these the emphasis has shifted. 1 and 2 Peter appear to take the same motif in opposite directions. Whereas 1 Peter, in a sense, demythologizes the apocalyptic language and applies it metaphorically to his readers' context, 2 Peter expands the motif, giving it ever more cosmic proportions. This recognition may have implications for the understanding of the early church's reception of other apocalyptic themes in the Gospels and Paul.

Areas for Further Study

At several points our discussion of New Testament texts was illumined by related traditions in the Apostolic Fathers. Given our focus on New Testament eschatology, we were unable to thoroughly examine those traditions in their own right. In at least three passages, all of which were given brief mention above, the fiery ordeal may be featured, namely 2 *Clem.* 16:3; *Did.* 16:5; *Herm. Vis.* 4:3:4 (see also *Mart. Pol.* 15:2). Further consideration of these texts, as well as some early Christian apocalypses, such as the *Apocalypse of Peter* and the *Apocalypse of Paul*, would help to round out the discussion of this motif in Early Christian eschatology.

In addition, in chapters 6 and 7 we observed that Paul, 1 Peter, and 2 Peter (or at least 2 *Clement's* interpretation of it), appeal to imagery from Malachi 3. Given the influence of this passage on John the Baptist in his proclamation of the coming Day of the Lord and his fashioning himself after Elijah, as prophesied in Malachi 4, it may prove instructive to examine the history of interpretation of Malachi 3–4 in Jewish intertestamental literature and trace its influence on New Testament Eschatology.

Bibliography

Achtemeier, Paul J. *1 Peter: A Commentary on First Peter*. Minneapolis: Fortress, 1996.
Adams, Edward. *The Stars Will Fall from Heaven: Cosmic Catastrophe in the New Testament and its World*. LNTS 347. London: T. & T. Clark, 2007.
Allen, Leslie C. *Jeremiah: A Commentary*. OTL. Louisville: Westminster John Knox, 2008.
Allison, Dale C. *Constructing Jesus: Memory, Imagination, and History*. Grand Rapids: Baker Academic, 2010.
———. "The Continuity between John and Jesus." *JSHJ* 1 (2003) 16–22.
———. "Elijah Must Come First." *JBL* 103 (1984) 256–58.
———. *The End of the Ages Has Come: An Early Interpretation of the Passion and Resurrection of Jesus*. Philadelphia: Fortress, 1985.
———. "How to Marginalize the Traditional Criteria of Authenticity." In *Handbook for the Study of the Historical Jesus*, edited by Tom Holmén and Stanley E. Porter, 4 vols., 1:3–30. Leiden: Brill, 2011.
———. "Jesus & the Victory of Apocalyptic." In *Jesus & the Restoration of Israel*, edited by Carey C. Newman, 126–41. Downers Grove, IL: IVP, 1999.
———. *Resurrecting Jesus: The Earliest Christian Tradition and its Interpreters*. London: T. & T. Clark, 2005.
———. *Testament of Abraham*. CEJL. Berlin: de Gruyter, 2003.
Alter, Rorbert. *The Five Books of Moses: A Translation with Commentary*. New York: Norton, 2004.
Anrich, G. "Clemens Und Origenes als Begründer der Lehre vom Fegfeuer." In *Theologisch Abhandlungen für H. J. Holtymann*, edited by W. Nowack, 95–120. Tübingen: Mohr, 1902.
Arens, Eduardo. *The ΗΛΘΟΝ-Sayings in the Synoptic Tradition: A Historico-Critical Investigation*. OBO 10. Gottingen: Vandenhoeck & Ruprecht, 1976.
Baarda, T. J. "Mark IX.49." *NTS* 5 (1959) 318–21.
Barrett, C. K. *A Commentary on the First Epistle to the Corinthians*. BNTC. Peabody, MA: Hendrickson, 1993.
———. *The Holy Spirit and the Gospel Tradition*. London: SPCK, 1947.
Bauckham, Richard. *The Fate of the Dead: Studies on Jewish and Christian Apocalypses*. SNT 93. Leiden: Brill, 1998.
———. *Jude, 2 Peter*. WBC 50. Waco, TX: Word, 1983.

Bauer, Johannes B. "Echte Jesuswort." In *Evangelien aus dem Nilsand*, edited by W. C. v. Unnik, 108–50. Frankfurt am Main: Heinrich Scheffler, 1960.

———. "Das Jesuswort 'Wer mir nahe ist.'" *TZ* 15 (1959) 446–50.

Bautch, Kelly C. *A Study of the Geography of 1 Enoch 17–19: "No One Has Seen What I Have Seen."* JSJSup 81. Leiden: Brill, 2003.

Beale, G. K. *The Temple and the Church's Mission: A Biblical Theology of the Dwelling Place of God.* NSBT. Leicester, UK: Apollos, 2004.

Beasley-Murray, George R. *Baptism in the New Testament.* London: Paternoster, 1962.

———. *Jesus and the Kingdom of God.* Grand Rapids: Eerdmans, 1986.

Becker, Jürgen. *Jesus of Nazareth.* Berlin: de Gruyter, 1998.

———. *Johannes der Täufer und Jesus von Nazareth.* Neukirchen-Vluyn: Neukirchener Verlag, 1972.

Bengel, Johann Albrecht. *Gnomon of the New Testament.* Vol 2. Edinburgh: T. & T. Clark, 1858.

Bernstein, Alan E. *The Formation of Hell: Death and Retribution in the Ancient and Early Christian Worlds.* Ithaca, NY: Cornell University Press, 1993.

Best, Ernst. "Mark's Preservation of the Tradition." In *The Interpretation of Mark*, edited by William R. Telford, 153–68. Edinburgh: T. & T. Clark, 1995.

———. "Spirit-Baptism." *NovT* 4 (1960) 236–43.

Beuken, Willem A. M. *Isaiah II/2: Isaiah 28-39.* HCOT. Leuven: Peeters 2000.

Bietenhard, Hans. "Kennt das Neue Testament die Vorstellung vom Fegefeuer?" *TZ* 3 (1947) 101–22.

Black, Matthew, James C. Vanderkam, et al. *The Book of Enoch or I Enoch: A New English Edition with Commentary and Textual Notes.* SVTP 7. Leiden: Brill, 1985.

Blackman, Aylward M., and H. W. Fairman. "The Myth of Horus at Edfu: II. C. The Triumph of Horus over His Enemies: A Sacred Drama (Continued)." *JEA* 29 (1943) 2–36.

Blenkinsopp, Joseph. *A History of Prophecy in Israel.* Louisville, KY: Westminster John Knox, 1996.

———. *Isaiah 1–39: A New Translation with Introduction and Commentary.* AB 19. New York: Doubleday, 2000.

Block, Daniel I. *The Book of Ezekiel: Chapters 1–24.* NICOT. Grand Rapids: Eerdmans, 1997.

Boccaccini, Gabrielle. *Enoch and the Messiah Son of Man: Revisiting the Book of Parables.* Grand Rapids: Eerdmans, 2007.

Bock, Darrell L. *Luke.* 2 Vols. BECNT. Grand Rapids: Baker, 1994.

Boring, M. Eugene. *Mark: A Commentary.* NTL. Louisville, KY: Westminster John Knox, 2006.

Bovon, François. *Luke 1: A Commentary on the Gospel of Luke 1:1–9:50.* Hermeneia: A Critical and Historical Commentary on the Bible. Minneapolis: Fortress, 2002.

Box, G. H. *The Ezra-Apocalypse: Being Chapters 3–14 of the Book Commonly Known as 4 Ezra (or II Esdras).* London: Pitman & Sons, 1912.

Bretscher, Paul. G. "'Whose Sandals'? (Matt 3:11)." *JBL* 86 (1967) 81–87.

Broadhead, Edwin K. "An Authentic Saying of Jesus in the Gospel of Thomas?" *NTS* 46 (2000) 132–49.

Bronner, Leah. *The Stories of Elijah and Elisha as Polemics against Baal Worship.* POS 6. Leiden: Brill, 1968.

Brown, Raymond. "John the Baptist in the Gospel of John." In *New Testament Essays*, 132-40. Milwaukee: Bruce, 1965.
Brownlee, William H. "John the Baptist in the New Light of Ancient Scrolls." In *The Scrolls and the New Testament*, edited by K. Stendahl, 33-53. New York: Harper, 1957.
———. *The Midrash Pesher of Habakkuk*. SBLMS 24. Missoula, MT: Scholars, 1979.
Budge, E. A. Wallis, Ed. *Coptic Texts*. New York: AMS, 1977.
Buitenwerf, Reiuwerf. *Book III of the Sibylline Oracles and its Social Setting*. SVTP 17. Leiden: Brill, 2003.
Bulhart, W. "Ignis sapiens." *SE* 13 (1967) 72-103.
Bultmann, Rudolf. *The History of the Synoptic Tradition*. New York: Harper & Row, 1976.
Burkett, Delbert R. *The Son of Man Debate: A History and Evaluation*. SNTSMS 107. Cambridge: Cambridge University Press, 1999.
Burney, C. F. *The Poetry of Our Lord: An Examination of the Formal Elements of Hebrew Poetry in the Discourses of Jesus Christ*. Oxford: Clarendon, 1925.
Cadbury, Henry J. *The Style and Literary Method of Luke*. Cambridge: Harvard University Press, 1920.
Cadoux, Cecil John. *The Historic Mission of Jesus: A Constructive Re-Examination of the Eschatological Teaching in the Synoptic Gospels*. London: Lutterworth, 1941.
Caird, George B. *The Gospel of St. Luke*. New York: Seabury, 1964.
Callow, J. "Some Initial Thoughts on the Passive in New Testament Greek." *Selected Technical Articles Related to Translation* 15 (1986) 32-37.
Calvin, John. *A Harmony of the Gospels, Matthew, Mark and Luke*. Grand Rapids: Eerdmans, 1972.
Cancik, Hubert. "The End of the World, of History, and of the Individual in Greek and Roman Antiquity." In *The Encyclopedia of Apocalypticism: The Origins of Apocalypticism in Judaism and Christianity*, edited by Bernard McGinn, John J. Collins and Stephen J. Stein, 3 vols., 1:84-125. New York: Continuum, 1998.
Casey, Maurice. *The Solution to the "Son of Man" Problem*. LNTS 343. London: T. & T. Clark, 2007.
Cazelles, Henri. "Connexions et Structure de Gen XV." *RB* 69 (1962) 321-49.
Chajes, Hirsch Peretz. *Markus-Studien*. Berlin: Schwetschke, 1899.
Charles, R. H. *The Book of Enoch, or 1 Enoch*. 2nd ed. Oxford, Clarendon, 1912.
Charlesworth, James H., ed. *The Old Testament Pseudepigrapha*. Garden City, NY: Doubleday, 1983.
Ciampa, Roy E., and Brian S. Rosner. *The First Letter to the Corinthians*. PNTC. Grand Rapids: Eerdmans, 2010.
Cogan, Mordechai, and Hayim Tadmor. *I Kings: A New Translation with Introduction and Commentary*. AB 10. Garden City, NY: Doubleday, 2001.
Cogan, Mordechai, and Hayim Tadmor. *II Kings: A New Translation*. AB 11. Garden City, NY: Doubleday, 1988.
Cohn, Norman. *Cosmos, Chaos, and the World to Come: The Ancient Roots of Apocalyptic Faith*. 2nd ed. New Haven: Yale University Press, 2001.
Coleman, N. D. "Note on Mark ix 49, 50: A New Meaning for ἅλας." *JTS* 24 (1923) 387-96.
———. "'Salt' and 'Salted' in Mark ix. 49, 50." *ExpT* 48 (1937) 360-62.

Collins, Adela Yarbro. *Cosmology and Eschatology in Jewish and Christian Apocalypticism.* JSJSup 50. Leiden: Brill, 1996.
———. *Mark: A Commentary.* Hermeneia: A Critical and Historical Commentary on the Bible. Minneapolis: Fortress, 2007.
Collins, John J. *Daniel: A Commentary on the Book of Daniel.* Hermeneia: A Critical and Historical Commentary on the Bible. Minneapolis: Fortress, 1993.
———. *Seers, Sybils, and Sages in Hellenistic-Roman Judaism.* Leiden: Brill, 1997.
———. *The Sibylline Oracles of Egyptian Judaism.* SBLDS 13. Missoula, MT: Scholars, 1974.
———. "The Works of the Messiah." *DSD* 1 (1994) 98–112.
Conzelmann, Hans. *1 Corinthians: A Commentary on the First Epistle to the Corinthians.* Hermenia: A Critical and Historical Commentary on the Bibile. Philadelphia: Fortress, 1975.
———. *The Theology of St. Luke.* Philadelphia: Fortress, 1961.
Couchoud, Paul-Louis. "Notes de Critique Verbale sur St Marc et St Matthieu." *JTS* 34 (1933) 113–38.
———. "Was the Gospel of Mark Written in Latin?" *CrQ* 5 (1928) 35–79.
Cramer, John Anthony, Ed. *Catenae Graecorum Patrum in Novum Testamentum, I, Catenae in Ev. S. Matthaei et S. Marci.* Hildesheim: Georg Olms Verlagsbuchhandlung, 1967.
Cranfield, C. E. B. *The Gospel according to Saint Mark: An Introduction and Commentary.* CGTC. Cambridge: Cambridge University Press, 1977.
Creed, John Martin. *The Gospel according to St. Luke: The Greek Text with Introduction, Notes and Indices.* London: MacMillan, 1953.
Cross, Frank Moore. *Canaanite Myth and Hebrew Epic: Essays in the History of the Religion of Israel.* Cambridge: Harvard University Press, 1973.
Crossan, John Dominic. *The Historical Jesus: The Life of a Mediterranean Jewish Peasant.* San Francisco: HarperSanFrancisco, 1991.
Comont, Franz. *Lux Perpetua.* Paris: Librairie Orientaliste Paul Geuthner, 1949.
Court, John. *The Book of Revelation and the Johannine Apocalyptic Tradition.* JSNTSup 190. Sheffield, UK: Sheffield Academic Press, 2000.
Cunningham, Scot. *'Through Many Tribulations': The Theology of Persecution in Luke-Acts.* JSNTSup 142. Sheffield, UK: Sheffield Academic Press, 1997.
Dahl, Nils A. "The Origin of Baptism." In *Interpretationes ad Vetus Testamentum Pertinentes Sigmundo Mowinckel: Septuagenario Missae,* edited by Nils A. Dahl and Arvid S. Kapelrud, 36–52. Oslo: Forlaget Land Og Kirke, 1955.
Danker, F. W. "2 Peter 3:10 and Psalm of Solomon 17:10." *ZNW* 53 (1962) 82–86.
Dapaah, Daniel S. *The Relationship between John the Baptist and Jesus of Nazareth: A Critical Study.* Lanham, MD: University Press of America, 2005.
Davids, Peter H. *The First Epistle of Peter.* NICNT. Grand Rapids: Eerdmans, 1990.
———. *The Letters of 2 Peter and Jude.* PNTC. Grand Rapids: Eerdmans, 2006.
Davies, Philip R. *IQM, The War Scroll from Qumran: Its Structure and History.* Rome: Biblical Institute, 1977.
Davies, Stevan L. *Jesus the Healer: Possession, Trance, and the Origins of Christianity.* New York: Continuum, 1995.
Davies, W. D., and D. C. Allison. *A Critical and Exegetical Commentary on the Gospel according to Saint Matthew.* 3 vols. ICC. Edinburgh: T. & T. Clark, 1988–97.

Davies, W. D., and D. David Daube, eds. *The Background of the New Testament and Its Eschatology: Studies in Honour of C. H. Dodd.* Cambridge: Cambridge University Press, 1956.
De Vries, Simon J. *1 Kings.* WBC 12. Waco, TX: Word, 1985.
DeConick, April D. *Recovering the Original Gospel of Thomas: A History of the Gospel and its Growth.* LNTS 286. London: T. & T. Clark, 2005.
―――. *The Original Gospel of Thomas in Translation: With a Commentary and New English Translation of the Complete Gospel.* LNTS 287. London: T. & T. Clark, 2006.
―――. *Seek to See Him: Ascent and Vision Mysticism in the Gospel of Thomas.* VCSup 33. Leiden: Brill, 1996.
Delling, G. "βάπτισμα βαπτισθῆναι." *NovT* 2 (1957) 92–115.
Derrett, J. D. M. "Salted with Fire: Studies in Texts: Mark 9:42–50." *Theo* 76 (1973) 364–68.
Dillmann, August. *Das Buch Henoch.* Leipzig: Vogel, 1853.
Dodd, C. H. *Parables of the Kingdom.* New York: Scribner, 1961.
Donelson, Lewis R. *I & II Peter and Jude: A Commentary.* NTL. Louisville, KY: Westminster John Knox, 2010.
Donfried, Karl Paul. "Justification and Last Judgment in Paul. *ZNW* 67 (1976) 90–110.
Douglas, Mary. *Leviticus as Literature.* Oxford: Oxford University Press, 1999.
Draper, Jonathan. "The Jesus Tradition in the Didache." In *Gospel Perspectives 5*, edited by David Wenham, 269–87. Sheffield, UK: JSOT, 1984.
Dubis, Mark. *Messianic Woes in First Peter: Suffering and Eschatology in 1 Peter 4:12–19.* SiBL 33. New York: Lang, 2002.
Duchesne-Guillemin, J. *The Hymns of Zarathustra.* London: Butler & Tanner, 1952.
Dunn, James D. G. *Baptism in the Holy Spirit: A Re-Examination of the New Testament Teaching on the Gift of the Spirit in Relation to Pentecostalism Today.* SBT 2/15. Naperville, IL: Allenson, 1970.
―――. "Birth of a Metaphor: Baptized in the Spirit." *ExpT* 89 (1978) 134–38, 173–75.
―――. *Christology in the Making: A New Testament Inquiry into the Origins of the Doctrine of the Incarnation.* 2nd ed. Grand Rapids: Eerdmans, 1989.
―――. *The Epistles to the Colossians and to Philemon: A Commentary on the Greek Text.* NIGTC. Exeter, UK: Paternoster, 1996.
―――. *Jesus and the Spirit: A Study of the Religious and Charismatic Experience of Jesus and the First Christians as Reflected in the New Testament.* Grand Rapids: Eerdmans, 1997.
―――. *Jesus Remembered, I, Christianity in the Making.* Grand Rapids, Eerdmans, 2003.
―――. *Romans 9–16.* WBC 38B. Dallas, TX: Word, 1988.
―――. "Spirit-and-Fire Baptism." *NovT* 14 (1972) 81–92.
Dupont-Sommer, André. *The Essene Writings from Qumran.* Oxford: Blackwell, 1961.
Edsman, Carl-Martin. *Le Baptême de feu.* ASNU. Uppsala: Almqvist & Wiksell, 1940.
Edwards, Richard A. *The Sign of Jonah in the Theology of the Evangelists and Q.* SBT 2/18. London: SCM, 1971.
Eichrodt, Walther. *Ezekiel: A Commentary.* OTL. Philadelphia: Westminster, 1970.
Eisenman, Robert H. *James the Just in the Habakkuk Pesher.* SPB 35. Leiden: Brill, 1986.
Elliott, John H. *1 Peter: A New Translation with Introduction and Commentary.* AB 37B. New York: Doubleday, 2000.

Elliott, John K. "An Eclectic Textual Commentary on the Greek Text of Mark's Gospel." In *The Language and style of the Gospel of Mark*, 189–211. Leiden: Brill, 1993.

Ellis, Earle E. *The Gospel of Luke*. NCB. London: Thomas Nelson, 1966.

Emmel, Stephen. "The Recently Published Gospel of the Savior ('Unbekanntes Berliner Evangelium'): Righting the Order of Pages and Events." *HTR* 95 (2002) 45–72.

Eno, Robert B. "The Fathers and the Cleansing Fire." *ITQ* 53 (1987) 184–202.

Eve, Eric. "Meier, Miracle and Multiple Attestation." *JSHJ* 3 (2005) 23–45.

Faierstein, Morris M. "Why Do the Scribes Say that Elijah Must Come First." *JBL* 100 (1981) 75–86.

Fee, Gordon D. *The First Epistle to the Corinthians*. NICNT. Grand Rapids: Eerdmans, 1987.

Feldmeier, Reinhard. *The First Letter of Peter: A Commentary on the Greek Text*. Waco, TX: Baylor University Press, 2008.

Fieger, Michael. *Das Thomasevangelium: Einleitung, Kommentar und Systematik*. NTAbh 22. Münster: Aschendorff, 1991.

Fields, Weston. W. "'Everyone will be Salted with Fire' (Mark 9:49)." *GTJ* 6 (1985) 299–304.

———. *Sodom and Gomorrah: History and Motif in Biblical Narrative*. JSOTSup 231. Sheffield, UK: Sheffield Academic Press, 1997.

Fishburne, C. W. "1 Corinthians 3:10–15 and the Testament of Abraham." *NTS* 17 (1970) 109–15.

Fitzmyer, Joseph A. *First Corinthians: A New Translation with Introduction and Commentary*. AYB 32. New Haven: Yale University Press, 2008.

———. *The Gospel according to Luke: Introduction, Translation, and Notes*. AB 2–28A. Garden City, NY: Doubleday, 1981–85.

———. "More about Elijah Coming First." *JBL* 104 (1985) 295–96.

Fleddermann, Harry. "The Discipleship Discourse (Mark 9:33–50)." *CBQ* 43 (1981) 57–75.

Fleddermann, Harry T. *Mark and Q: A Study of the Overlap Texts*. Leuven: Peeters, 1995.

Flusser, David. "The Baptism of John and the Dead Sea Sect." In *Essays on the Dead Sea Scrolls: In Memory of E. L. Sukenik*, edited by Y. Yadin and C. Rabin. Jerusalem: H. Ha-Sefer, 1961.

Ford, J. Massyngberde. "You are God's Sukkah (I Cor. III. 10–17)." *NTS* 21 (1974) 139–42.

Fossum, Jarl E. *The Name of God and the Angel of the Lord: Samaritan and Jewish Concepts of Intermediation and the Origin of Gnosticism*. WUNT 1/36. Tübingen: Mohr Siebeck, 1985.

France, R. T. "Exegesis in Practice: Two Examples." In *New Testament Interpretation: Essays on Principles and Methods*, edited by I. H. Marshall, 252–81. Exeter: Paternoster, 1977.

———. *The Gospel of Mark: A Commentary on the Greek Text*. NIGTC. Grand Rapids, Eerdmans, 2002.

Frayer-Griggs, Daniel. "'Everyone Will Be Baptized in Fire': Mark 9.49, Q 3.16, and the Baptism of the Coming One." *JSHJ* 7 (2009) 254–85.

———. "Neither Proof Text nor Proverb: The Instrumental Sense of διά and the Soteriological Function of Fire in 1 Corinthians 3.15." *NTS* 59 (2013) 517–34.

Freedman, David Noel. "Burning Bush." *Bibl* 50 (1969) 245–46.

Fuller, Reginald H. *The Mission and Achievement of Jesus: An Examination of the Presuppositions of New Testament Theology*. SBT 12. London: SCM, 1954.
Funk, Robert. W., Roy W. Hoover, et al., eds. *The Five Gospels: The Search for the Authentic Words of Jesus*. San Francisco, Harper Collins, 1993.
García Martínez, F. and A. S. van der Woude. "A 'Groningen' Hypothesis of Qumran Origins and Early History." *RQ* 14 (1990) 521–41.
Garnet, Paul. *Salvation and Atonement in the Qumran Scrolls*. WUNT 2. Tübingen: Mohr Siebeck, 1977.
Gärtner, Bertil E. *The Temple and the Community in Qumran and the New Testament: A Comparative Study in the Temple Symbolism of the Qumran Texts and the New Testament*. SNTSMS 1. Cambridge: Cambridge University Press, 1965.
Gaster, Theodor H. *Myth, Legend, and Custom in the Old Testament: A Comparative Study with Chapters from Sir James G. Frazer's Folklore in the Old Testament*. New York: Harper & Row, 1969.
Gathercole, Simon J. *The Gospel of Thomas: Introduction and Commentary*. TENTS 11. Leiden: Brill 2014.
Geiger, Ruthild. *Die lukanischen Endzeitreden. Studien zur Eschatologie des Lukas-Evangeliums*. EH 23/16. Frankfurt: Lang, 1973.
Gerstenberger, Erhard S. *Leviticus: A Commentary*. OTL. Louisville: Westminster John Knox, 1996.
Gibson, Margaret Dunlop, ed. *The Commentaries of Isho'dad of Merv, Bishop of Hadatha (c. 850 A.D.) in Syriac and English*. HSem. Cambridge: Cambridge University Press, 1911.
Gill, John. *Gill's Commentary, V, Matthew-Acts*. Grand Rapids: Baker, 1980.
Glasson, T. F. *Greek Influence in Jewish Eschatology: With Special Reference to the Apocalypses and Pseudepigraphs*. London: SPCK, 1961.
———. *The Second Advent: The Origin of the New Testament Doctrine*. London: Epworth, 1963.
Glazier-McDonald, Beth. *Malachi: The Divine Messenger*. SBLDS 98. Atlanta: Scholars, 1987.
Gnilka, Joachim. *Das Evangelium nach Markus*. 2 vols. EKKNT 2. Zürich: Neukirchen-Vluyn, 1979.
———. *Ist 1 Kor. 3, 10–15 ein Schriftzeugnis für das Fegfeuer? Eine exegetisch-historische Untersuchung*. Düsseldorf: Triltsch, 1955.
———. *Jesus of Nazareth: Message and History*. Peabody, MA: Hendrickson, 1997.
Gould, E. P. *A Critical and Exegetical Commentary on the Gospel according to St. Mark*. ICC. Edinburgh: T. & T. Clark, 1992.
Grant, Robert M., and David Noel Freedman. *The Secret Sayings of Jesus*. Garden City, NY: Doubleday, 1960.
Green, Joel B. *The Gospel of Luke*. NICNT. Grand Rapids: Eerdmans, 1997.
Greenberg, Moshe. *Ezekiel 1–37: A New Translation with Introduction and Commentary*. AB 22. Garden City, NY: Doubleday, 1983.
Gregg, Brian Han. *The Historical Jesus and the Final Judgment Sayings in Q*. WUNT 2/207. Tübingen: Mohr Siebeck, 2006.
Grotius, Hugo. *Operum Theologorum, II.I, Annotationes in Quatuor Euangelia & Acta Apostolorum*. Amsterdam: Blaeu, 1679.
Guenther, Heinz O. "A Fair Face is Half the Portion: The Lot Saying in Luke 17:28–29." *Forum* 6 (1990) 56–66.

Gundry, Robert H. *Mark: A Commentary on his Apology for the Cross*. Grand Rapids: Eerdmans, 1993.

Ha, John. *Genesis 15: A Theological Compendium of Pentateuchal History*. BZAW 181. Berlin: de Gruyter, 1989.

Halperin, David J. *The Faces of the Chariot: Early Jewish Responses to Ezekiel's Vision*. TSAJ 16. Tübingen: Mohr Siebeck, 1988.

Hamerton-Kelly, R. G. *Pre-Existence, Wisdom, and the Son of Man: A Study of the Idea of Pre-Existence in the New Testament*. SNTSMS 21. Cambridge: Cambridge University Press, 1973.

Harnack, Adolf. *New Testament Studies, II, The Sayings of Jesus: The Second Source of St. Matthew and St. Luke*. CTL 23. New York: Putnam's Sons, 1908.

Harrill, J. Albert. "Stoic Physics, the Universal Conflagration, and the Eschatological Destruction of the 'Ignorant and Unstable' in 2 Peter." In *Stoicism in Early Christianity*, edited by Tuomas Rasimus, Troels Engberg-Pedersen and Ismo. Dunderberg, 115–40. Grand Rapids: Baker Academic, 2010.

Harris, J. G. *The Qumran Commentary on Habakkuk*. London: Mowbray, 1966.

Hauck, F. "ἅλας." TDNT 1 (1964) 228–29.

Haupt, Paul. "The Burning Bush and the Origin of Judaism." *APSP* 48 (1909) 354–69.

Hays, Richard B. *Echoes of Scripture in the Letters of Paul*. New Haven: Yale University Press, 1989.

———. *First Corinthians*. IBC. Louisville, KY: John Knox, 1997.

Hedrick, Charles W., and Paul A. Mirecki. *Gospel of the Savior: A New Ancient Gospel*. Santa Rosa, CA: Polebridge, 1999.

Henderson, Ian H. "'Salted with Fire' (Mark 9.42–50): Style, Oracles and (Socio) Rhetorical Gospel Criticism." *JSNT* 80 (2000) 44–65.

Hengel, Martin. *Judaism and Hellenism: Studies in Their Encounter in Palestine during the Early Hellenistic Period*. Philadelphia: Fortress, 1974.

Héring, Jean. *The First Epistle of Saint Paul to the Corinthians*. London: Epworth, 1962.

Hiers, Richard H. *Jesus and the Future: Unresolved Questions for Understanding and Faith*. Atlanta: John Knox, 1981.

Hinnells, John R. "The Zoroastrain Doctrine of Salvation in the Roman World: A Study of the Oracle of Hystaspes." In *Man and His Salvation: Studies in Memory of S. G. F. Brandon*, edited by Eric J. Sharpe and John R. Hinnells, 125–48. Manchester: Manchester University Press, 1973.

Hofius, Otfried. "Unknown Sayings of Jesus." In *The Gospel and the Gospels*, edited by Peter Stuhlmacher, 336–60. Grand Rapids: Eerdmans, 1990.

Hogeterp, Albert L. A. *Paul and God's Temple: A Historical Interpretation of Cultic Imagery in the Corinthian Correspondence*. BTS 2. Leuven: Peeters, 2006.

Holladay, William Lee. *Jeremiah 1: A Commentary on the Book of the Prophet Jeremiah Chapters 1–25*. Hermeneia: A Critical and Historical Commentary on the Bible. Philadelphia: Fortress, 1986.

Hollander, Harm W. "The Testing by Fire of the Builders' Works: 1 Corinthians 3:10–15." *NTS* 40 (1994) 89–104.

Hollenbach, Paul. "The Conversion of Jesus: From Jesus the Baptizer to Jesus the Healer." *ANRW* II/25.1 (1982) 196–219.

Holm-Nielsen, S. *Hodayot: Psalms from Qumran*. ATDan 2. Aarhus: Universitetsforlaget, 1960.

Holmes, Michael W., ed. The *Apostolic Fathers: Greek Texts and English Translations*. Grand Rapids: Baker Academic, 2007.
Holtzmann, Heinrich Julius. *Die synoptischen Evangelien: ihr Ursprung und geschichtlicher Charakter*. Leipzig: Engelmann, 1863.
Hooker, Morna. *The Gospel according to Saint Mark*. BNTC. Peabody, MA: Hendrickson, 1991.
Horgan, Maurya P. *Pesharim: Qumran Interpretations of Biblical Books*. CBQMS 8. Washington, DC: Catholic Biblical Association of America, 1979.
Hughes, John H. "John the Baptist: The Forerunner of God Himself." *NovT* 14 (1972) 191–218.
Jacobson, Howard. *A Commenttary on Pseudo-Philo's* Liber Antiquitatum Biblicarum, *with Latin Text and English Translation*. 2 vols. Leiden: Brill, 1996.
Jastrow, Marcus. *A Dictionary of the Targumim, the Talmud Babli and Yerushalmi, and the Midrashic Literature*. Peabody, MA: Hendrickson, 2005.
Jeremias, Joachim. *Die Sprache des Lukasevangeliums: Redaktion und Tradition im Nicht-Markusstoff des dritten Evangeliums*. KEK. Göttingen: Vandenhoeck und Ruprecht, 1980.
———. "γέεννα." *TDNT* 1 (1964) 657–58.
———. *New Testament Theology: The Proclamation of Jesus*. New York: Scribner, 1971.
———. *The Parables of Jesus*. New York: Scribner, 1955.
———. *Unknown Sayings of Jesus*. London: SPCK, 1964.
Jeremias, Jörg. *Theophanie: Die Geschichte einer Alttestamentlichen Gattung*. Neukirchen-Vluyn: Neukirchener Verlag, 1965.
Jobes, Karen H. *1 Peter*. BECNT. Grand Rapids: Baker Academic, 2005.
Johnson, Dennis E. "Fire in God's House: Imagery from Malachi 3 in Peter's Theology of Suffering (1 Pet 4:12–19)." *JETS* 29 (1986) 285–94.
Juza, Ryan. "Echoes of Sodom and Gomorrah on the Day of the Lord: Intertextuality and Tradition in 2 Peter 3:1–13." *BBR* 24 (2014) 227–45.
Kaiser, Otto. *Isaiah 1–12: A Commentary*. OTL. Philadelphia: Westminster, 1983.
Käsemann, Ernst. *Commentary on Romans*. Grand Rapids: Eerdmans, 1980.
Keck, Leander E. *A Future for the Historical Jesus: The Place of Jesus in Preaching and Theology*. Nashville: Abingdon, 1971.
Kelly, J. N. D. *A Commentary on the Epistles of Peter and of Jude*. BNTC. New York: Harper & Row, 1969.
Kirk, Alexander N. "Building with the Corinthians: Human Persons as the Building Materials of 1 Corinthians 3.12 and the 'Work' of 3.13–15." *NTS* 58 (2012), 549–70.
Kirschner, Robert. "The Rabbinic and Philonic Exegeses of the Nadab and Abihu Incident (Lev. 10:1–6)." *JQR* 73 (1983) 375–93.
Kloppenborg, John S. *The Formation of Q: Trajectories in Ancient Wisdom Collections*. Harrisburg, PA: Trinity, 1987.
———. *Q Parallels: Synopsis, Critical Notes & Concordance*. Sonoma, CA: Polebridge, 1988.
Knibb, Michael A. *The Ethiopic Book of Enoch: A New Edition in the Light of the Aramaic Dead Sea Fragments*. Oxford: Oxford University Press, 1978.
Knox, John. "Pliny and I Peter: A Note on I Pet 4:14–16 and 3:15." *JBL* 72 (1953) 187–89.

Konradt, Matthias. *Gericht und Gemeinde: Eine Studie zur Bedeutung und Funktion von Gerichtsaussagen im Rahmen der paulinischen Ekklesiologie und Ethik im 1 Thess und 1 Kor*. Berlin: de Gruyter, 2003.
Köster, Helmut. "συνέχω." *TDNT* 7 (1964) 885.
Kraeling, Carl H. *John the Baptist*. New York: Scribner, 1951.
Kraftchick, Steven J. *Jude, 2 Peter*. ANTC. Nashville, TN: Abingdon, 2002.
Kruger, Michael J. "The Authenticity of 2 Peter." *JETS* 42 (1999) 645–71.
Kuck, David W. *Judgment and Community Conflict: Paul's use of Apocalyptic Judgment Language in 1 Corinthians 3:5—4:5*. NovTSup 66. Leiden: Brill, 1992.
Kugel, James L. *The Bible As It Was*. Cambridge: Harvard University Press, 1997.
Kugler, R. "Priesthood at Qumran." *The Dead Sea Scrolls after Fifty Years: A Comprehensive Assessment*, edited by P. W. Flint and J. C. VanderKam, 93–116. Leiden: Brill, 1998.
Kümmel, Werner Georg. *Introduction to the New Testament*. Nashville, TN: Abingdon, 1966.
———. *Promise and Fulfillment: The Eschatological Message of Jesus*. SBT 23. Naperville, IL: Allenson, 1957.
Lang, Friedrich. *Die Briefe an die Korinther*. NTD 7. Göttingen, Vandenhoeck & Ruprecht, 1968.
———. "Das Feuer im Sprachgebrauch der Bibel: Dargestellt auf dem Hintergrund der Feuervorstellungen in der Umwelt." PhD diss., Tübingen, 1950.
———. πῦρ, κτλ. *TDNT* 6:928–52 Grand Rapids: Eerdmans, 1968.
Lapidge, Michael. "ἀρχαί and στοιχεῖα: A Problem in Stoic Cosmology." *Phronesis* 18 (1973) 240–78.
———. "Stoic Cosmology." In *The Stoics*, edited by J. M. Rist, 161–85. Berkeley, CA: University of California Press, 1978.
Latham, James E. *The Religious Symbolism of Salt*. ThH 64. Paris: Beauchesne, 1982.
Laughlin, John H. "'The 'Strange Fire' of Nadab and Abihu." *JBL* 95 (1976) 559–65.
———. "A Study of the Motif of Holy Fire in the Old Testament." Phd diss., Southern Baptist Theological Seminary, 1975.
Le Goff, Jacques. *The Birth of Purgatory* Chicago: University of Chicago Press, 1984.
Leaney, A. R. C. *The Rule of Qumran and its Meaning*. NTL. Philadelphia: Westminster, 1966.
Lelyveld, Margaretha. *Les Logia de la Vie dans l'Evangile selon Thomas: à la Recherche d'une Tradition et d'une Rédaction*. NHS 34. Leiden: Brill, 1987.
Levine, Baruch A. *Leviticus: The Traditional Hebrew Text with the New JPS Translation*. JPSTC. Philadelphia: Jewish Publication Society, 1989.
Levine, Baruch A. *Numbers: A New Translation with Introduction and Commentary*. AB 4. New York: Doubleday, 1993.
Liebengood, Kelly D. *The Eschatology of 1 Peter: Considering the Influence of Zechariah 9–14*. SNTSMS 157. Cambridge: Cambridge University Press, 2014.
Lietzmann, Hans. *An die Korinther I–II*. HNT 9. Tübingen: Mohr Siebeck, 1949.
Lightfoot, John. *A Commentary on the New Testament from the Talmud and Hebraica, II, Matthew-Mark*. Grand Rapids: Baker, 1979.
Lohmeyer, Ernst. *Das Evangelium des Markus*. KEK. Göttingen: Vandenhoeck & Ruprecht, 1953.

Lohse, Eduard. *Colossians and Philemon: A Commentary on the Epistles to the Colossians and to Philemon*. Hermeneia: A Critical and Historical Commentary on the Bible. Philadelphia: Fortress, 1971.
Loisy, Alfred. *The Birth of the Christian Religion*. New Hyde Park, NY: University Books, 1962.
Lövestam, Evald. *Jesus and 'This Generation': A New Testament Study*. ConBNT 25. Stockholm: Almqvist & Wiksell, 1995.
Lüdemann, Gerd. *Jesus after Two Thousand Years: What He Really Said and Did*. Amherst, NY: Prometheus, 2000.
Lührmann, Dieter. *Die Redaktion der Logienquelle*. WMANT 33. Neukirchen-Vluyn: Neukirchener, 1969.
Luz, Ulrich. *Matthew 1–7: A Commentary*. Hermeneia. Minneapolis, Fortress, 2007.
Malchow, Bruce V. "The Messenger of the Covenant in Mal 3:1." *JBL* 103 (1984) 252–55.
Mansfield, J. "Providence and the Destruction of the Universe in Early Stoic Thought: With Some Remarks on the 'Mysteries of Philosophy.'" In *Studies in Hellenistic Religions*, edited by M. J. Vermaseren, 129–88. Leiden: Brill, 1979.
Manson, T. W. *The Sayings of Jesus*. Grand Rapids: Eerdmans, 1957.
Mansoor, Menahem. *The Thanksgiving Hymns*. Grand Rapids, Eerdmans, 1961.
Marcus, Joel. *Mark 9–16*. AYB 27A. New Haven: Yale University Press, 2009.
Marjanen, Antti. "Is *Thomas* a Gnostic Gospel?" *Thomas at the Crossroads: Essays on the Gospel of Thomas*, edited by R. Uro, 107–39. Edinburgh: T. & T. Clark, 1998.
Marshall, I. H. *The Gospel of Luke: A Commentary on the Greek Text*. NIGTC. Exeter, UK: Paternoster, 1978.
Martin, Dale. "Jesus in Jerusalem: Armed and Not Dangerous." *JSNT* 37 (2014) 3–24.
März, C.-P. " 'Feuer auf die Erde zu werfen, bin ich gekommen': Zum Verständnis und zur Entstehung von Lk 12,49." In *À cause de l'Évangile: Mélanges offerts à Dom Jacques Dupont*, edited by F. Refoulé, 479–511. Paris: Cerf, 1985.
Mattill, A. J. *Luke and the Last Things: A Perspective for the Understanding of Lukan Thought*. Dillsboro, NC: Western North Carolina, 1979.
Mayer, Rudolf. *Die biblische Vorstellung vom Weltenbrand: eine Untersuchung über die Beziehungen zwischen Parsismus und Judentum* Bonn: Selbstverlag des Orientalischen Seminars der Universität Bonn, 1956.
McCabe, Elizabeth. *An Examination of the Isis Cult with Preliminary Exploration into New Testament Studies*. Lanham, MD: University Press of America, 2008.
McKnight, Scot. *Jesus and his Death: Historiography, the Historical Jesus, and Atonement Theory*. Waco, TX: Baylor University Press, 2005.
Meier, John P. *A Marginal Jew: Rethinking the Historical Jesus, Volume 1, The Roots of the Problem and the Person*. New York: Doubleday, 1991.
———. *A Marginal Jew: Rethinking the Historical Jesus, Volume 2, Mentor, Message and Miracles*. New York, Doubleday, 1994.
Metzger, Bruce M. *A Textual Commentary on the Greek New Testament*. Stuttgart: Deutsche Biblegesellschaft, 1994.
Meyer, Ben F. *The Aims of Jesus*. London: SCM, 1979.
Meyer, August Wilhelm. *Critical and Exegetical Hand-Book to the Gospels of Mark and Luke*. New York: Funk and Wagnalls, 1884.
Meyers, Carol L., and Eric M. Meyers. *Zechariah 9–14: A New Translation with Introduction and Commentary*. AB 25C. New York: Doubleday, 1993.
Michaels, J. Ramsey. *1 Peter*. WBC 49. Waco, TX: Word, 1988.

Michl, Johannes. "Gerichtsfeuer und Purgatorium zu 1 Kor 3,12–15." In *Studiorum Paulinorum Congressus Internationalis Catholicus*, 395–401. AnBib 17–18, vol. 1. Rome: Pontimicio Instituto Biblico, 1963.

Milavec, Aaron. *The Didache: Faith, Hope, & Life of the Earliest Christian Communities, 50–70 CE*. New York: Newman, 2003.

———. "The Saving Efficacy of The Burning Process in *Didache* 16:5." In *The Didache in Context: Essays on its Text, History, and Transmission*, edited by Clayton N. Jefford, 131–55. NovTSup 77. Leiden: Brill, 1995.

Milgrom, Jacob. *Leviticus 1–16: A New Translation with Introduction and Commentary*. AB 3. New York: Doubleday, 1991.

———. *Numbers: The Traditional Hebrew Text with the New JPS Translation*. JPSTC. Philadelphia: Jewish Publication Society, 2003.

Milik, Józef T., and Matthew Black. *The Books of Enoch: Aramaic fragments of Qumrân Cave 4*. Oxford: Clarendon, 1976.

Montefiore, C. G. *The Synoptic Gospels*. London, MacMillan, 1909.

Morgenstern, Julian. *The Fire upon the Altar*. Chicago: Quadrangle, 1963.

Morison, James. *A Practical Commentary on the Gospel according to St. Mark*. London: Hodder and Stoughton, 1882.

Morris, Leon. *The First Epistle of Paul to the Corinthians: An Introduction and Commentary*. Grand Rapids: Eerdmans, 1958.

Müller, F. Max, W. E. West, et al. *The Sacred Books of the East*. 50 vols. Oxford: Oxford University Press, 1879–1910.

Müller, Mogens. *The Expression 'Son of Man' and the Development of Christology: A History of Interpretation*. London: Equinox, 2008.

Nauck, Wolfgang. "Salt as a Metaphor in Instructions for Discipleship." ST 6 (1952) 165–78.

Neirynck, Frans. "The Tradition of the Sayings of Jesus: Mark 9, 33–50." In *The Dynamism of Biblical Tradition*, edited by P. Benoit, 62–74. New York: Paulist, 1967.

Nickelsburg, George W. E. *1 Enoch 1: A Commentary on the Book of 1 Enoch*. Hermeneia: A Critical and Historical Commentary on the Bibiel. Minneapolis: Fortress, 2001.

———. "Enoch, First Book of." *ADB*, edited by David N. Freedman, 2:508–16. New York: Doubleday, 1992.

———. *Jewish Literature between the Bible and the Mishnah: A Historical and Literary Introduction*. Minneapolis: Fortress, 2005.

Nickelsburg, George W. E., and James C. VanderKam. *1 Enoch: A New Translation Based on the Hermeneia Commentary*. Minneapolis: Fortress, 2004.

Niederwimmer, Kurt. *The Didache: A Commentary*. Hermenia. Minneapolis: Fortress, 1998.

Nineham, Dennis E. *The Gospel of St. Mark*. New York: Seabury, 1963.

Nolland, John. *Luke*. WBC 35. 3 vols. Dallas, TX: Word, 1989–93.

Noth, Martin. *Exodus: A Commentary*. OTL. Philadelphia: Westminster, 1962.

Oepke A. "βάπτω, κτλ." *TDNT* 1 (1964) 529–46.

Oesterley, W. O. E. *II Esdras (The Ezra Apocalypse)*. WC. London: Methuen, 1933.

Öhler, Markus. "The Expectation of Elijah and the Presence of the Kingdom of God." *JBL* 118 (1999) 461–76.

Origen. *Homilies 1–14 on Ezekiel*. Edited by Thomas P. Scheck. ACW 62. New York: Newman, 2010.

———. *Homilies on Luke: Fragments on Luke*. Edited by Joseph T. Lienhard. FC 94. Washington, DC: Catholic University of America Press, 1996.

———. *Prayer; Exhortation to Martyrdom*. Edited by John J. O'Meara. ACW 19. Westminster, MD: Newman, 1954.

Oster, Richard. *1 Corinthians*. Joplin, MO: College Press, 1995.

Otto, Rudolf. *The Kingdom of God and the Son of Man: A Study in the History of Religion*. London: Lutterworth, 1943.

Owen, Paul L., and Larry W. Hurtado, eds. *'Who is this Son of Man?': The Latest Scholarship on a Puzzling Expression of the Historical Jesus*: LNTS 390. London: T. & T. Clark, 2011.

Pallis, Alexander. *A Few Notes on the Gospels according to St. Mark and St. Matthew Based Chiefly on Modern Greek*. Liverpool: Liverpool Booksellers, 1903

Pardee, Nancy. "The Curse that Saves (DIDACHE 16:5)." In *The Didache in Context: Essays on its Text, History, and Transmission*, edited by Clayton N. Jefford, 156–76. NovTSup 77. Leiden: Brill, 1995.

Parker, Pierson. "The Posteriority of Mark." In *New Synoptic Studies*, edited by W. Farmer, 67–142. Macon, GA: Mercer University Press, 1983.

Patterson, Stephen J. "Fire and Dissension: Ipsissima Vox Jesu in Q 12:49, 51–53?" *Forum* 5 (1989) 121–39.

———. *The Gospel of Thomas and Jesus*. Sonoma, CA: Polebridge, 1993.

Perdue, Leo G. *Proverbs*. Louisville, KY: John Knox, 2000.

Pesch, Rudolf. *Das Markus-Evangelium*. 2 vols. HTKNT. Darmstadt: Wissenschaftliche Buchgesellschaft, 1976–77.

Peterson, Erik. "ἔργον in der Bedeutung 'Bau' bei Paulus." *Bib* 22 (1941) 439–41.

Pitre, Brant James. "Blessing the Barren and Warning the Fecund." *JSNT* 23 (2001) 59–80.

Plummer, Alfred. *A Critical and Exegetical Commentary on the Gospel according to St. Luke*. New York: Scribner's, 1896.

Polag, A. *Fragmenta Q: Textheft zur Logienquelle*. Neukirchen-Vluyn: Neukirchener Verlag, 1982.

Poole, Matthew. *A Commentary on the Whole Bible, III, Matthew–Revelation*. McLean, VA: MacDonald, 1962.

Porter, Stanley E. "What Does it Mean to be "Saved by Childbirth" (1 Timothy 2:15)." *JSNT* 49 (1993) 87–102.

Powys, David J. *'Hell': A Hard Look at a Hard Question: The Fate of the Unrighteous in New Testament Thought*. PBM. Carlisle, UK: Paternoster, 1998.

Proctor, John. "Fire in God's House: Influence of Malachi 3 in the NT." *JETS* 36 (1993) 9–14.

Ramelli, Ilaria. "Origen's Interpretation of Violence in the Apocalypse: Destruction of Evil and Purification of Sinners." In *Ancient Christian Interpretations of "Violent Texts" in the Apocalypse,* edited by Joseph Verheyden, Tobias Nicklas, and Andreas Merkt, 46–62. SUNT 92. Göttingen: Vandenhoeck & Ruprecht, 2011.

———. "Stromateis VII and Clement's Hints at the Theory of *Apokatastasis*." In *The Seventh Book of the* Stromateis: *Proceedings of the Colloquium on Clement of Alexandria*, edited by Matyáš, Vít Havrda, Vít Hušek, Jana Plátová, 239–60. SVC 17. Leiden: Brill, 2012.

Reicke, Bo. *The Epistles of James, Peter, and Jude*. AB 37. Garden City, NY: Doubleday, 1964.

Reid, John. "The Baptism of Water and the Baptism of Fire." *ExpT* 25 (1914) 306–7.
Reinhart, Karl. "Heraklitis Lehre vom Feuer." *Hermes* 77 (1942) 1–27.
Reiser, Marius. *Jesus and Judgment: The Eschatological Proclamation in its Jewish Context*. Minneapolis: Fortress, 1997.
Ringgren, Helmer. *The Faith of Qumran: Theology of the Dead Sea Scrolls*. New York: Crossroad, 1995.
Robertson, Archibald, and Alfred Plummer. *A Critical and Exegetical Commentary on the First Epistle of St. Paul to the Corinthians*. Edinburgh: T. & T. Clark, 1929.
Robinson, John A. T. "Elijah, John and Jesus: An Essay in Detection." In *Twelve New Testament Studies*, 28–52. London: SCM, 1962.
Rothschild, Clare K. *Baptist Traditions and Q*. WUNT 1/190. Tübingen: Mohr Siebeck, 2005.
Sander, Emilie T. "ΠΥΡΟΣΙΣ and the First Epistle of Peter 4:12." Phd diss., Harvard Divinity School, 1966.
Sanders, E. P. *Jesus and Judaism*. Philadelphia: Fortress, 1985.
———. *Paul, the Law and the Jewish People*. Minneapolis: Fortress, 1983.
Satta, Ronald F. "'Baptism Doth Now Save Us': An Exegetical Investigation of 1 Peter 3:20–21." *EvJ* 25 (2007) 65–72.
Scaliger, Joseph. "Notae." In *ΤΗΣ ΚΑΙΝΗΣ ΔΙΑΘΗΚΗΣ ΑΠΑΝΤΑ. Novi Testamenti: Libri Omnes, Recens nunc Editi: cum Notis & Animaduersionibus Doctissimorum, Praesertim Verò Roberti Stephani, Iosephi Scaligeri, Isaaci Casauboni*. London: Richard Whittaker, 1633.
Schäfers, Joseph. *Eine altsyrische antimarkionitische Erklärung von Parabeln des Herrn und zwei andere altsyrische Abhandlungen zu Texten des Evangeliums*. NTAbh 6. Münster: Aschendorff, 1917.
Schaltter, Adolf. *Paulus der Bote Jesu: eine Deutung seiner Briefe an die Korinther*. Stuttgart: Calwer, 1934.
Schlosser, Jacques. "Les jours de Noe et de Lot. A propos de Luc xvii, 26–30." *RB* 80 (1973) 13–36.
Schmidt, Karl Ludwig. (1969). *Der Rahmen der Geschichte Jesu*. Damstadt, Wissenschaftlich Buchgesselschaft.
Schnackenburg, Rudolf. "Der Eschatologische Abschnitt Lk 17, 20–37." In *Mélanges bibliques en hommage au R. P. Béda Rigaux*, edited by A. L. Descamps and R. P. A. Halleux, 213–34. Gembloux: Duculot, 1970.
Schneemelcher, Wilhelm, Edgar Hennecke, and Robert McLachlan Wilson. *New Testament Apocrypha*. 2 vols. Louisville, KY: Westminster/John Knox, 1991.
Schoedel, William R. *Ignatius of Antioch: A Commentary on the Letters of Ignatius of Antioch*. Hermeneia. Philadelphia, Fortress, 1985.
Schrage, Wofgang. *Der erste Brief an an die Korinther (1 Kor 1,6–6,11)*. EKKNT 7. Neukirchen-Vlyun: Neukirchener, 1991.
Schwartz, Howard. *Tree of Souls: The Mythology of Judaism*. New York: Oxford University, 2004.
Schwarz, Günther. "πας πυρι αλισθησεται (Markus 9,49)." *BN* 11 (1980) 45.
Schweitzer, Albert. *The Quest of the Historical Jesus*. Minneapolis: Fortress, 2001.
Schweizer, Eduard. *The Good News according to Luke*. Atlanta: John Knox, 1984.
———. *The Good News according to Mark*. Richmond, VA: John Knox, 1970.
———. "πνεῦμα, πνευματικός." *TDNT* 6 (1968) 332–451.
Scobie, Charles H. H. *John the Baptist*. London: SCM, 1964.

Scriba, Albrecht. *Die Geschichte des Motivkomplexes Theophanie: Seine Elemente, Einbindung in Geschehensabläufe und Verwendungsweisen in altisraelitischer, frühjüdischer und frühchristlicher Literatur*. FRLANT 167. Göttingen: Vandenhoeck & Ruprecht, 1995.

Seitz, Oscar J. F. "What Do These Stones Mean?" *JBL* 79 (1960) 247–54.

Sharman, Henry Burton. *The Teaching of Jesus about the Future according to the Synoptic Gospels*. Chicago: University of Chicago Press, 1909.

Sidebottom, E. M. "So-Called Divine Passive in the Gospel Tradition." *ExpT* 87 (1976) 200–204.

Spicq, Celsas. *Les Épîtres de Saint Pierre*. SB. Paris: Lecoffre, 1966.

Stephens, M. B. *Annihilation or Renewal? The Meaning and Function of New Creation in the Book of Revelation*. WUNT 2/307. Tübingen: Mohr Siebeck, 2011.

Stone, Michael E. *Fourth Ezra: A Commentary on the Book of Fourth Ezra*. Hermeneia. Minneapolis: Fortress, 1990.

Strack, Hermann L. and Paul Billerbeck. *Kommentar zum Neuen Testament aus Talmud und Midrasch*. 4 vols. München: Beck, 1922–28.

Strange, David. *An Exegetical Summary of 2 Peter*. Dallas, TX: SIL International, 2003.

Stuckenbruck, Loren T. *1 Enoch 91–108*. CEJL. Berlin: de Gruyter, 2007.

———. *The Book of Giants from Qumran: Texts, Translation, and Commentary*. TSAJ 63. Tübingen: Mohr Siebeck, 1997.

Taylor, Joan E. *The Immerser: John the Baptist within Second Temple Judaism*. SHJ. Grand Rapids: Eerdmans, 1997.

Taylor, Vincent. *The Formation of the Gospel Tradition*. London: MacMillan, 1953.

———. *The Gospel according to St. Mark*. London: MacMillan, 1953.

Telford, William R. *The Barren Temple and the Withered Tree: A Redaction-Critical Analysis of the Cursing of the Fig-Tree Pericope in Mark's Gospel and its Relation to the Cleansing of the Temple Tradition*. JSNTSup 1. Sheffield, UK: JSOT, 1980.

Theissen, Gerd, and Dagmar Winter. *The Quest for the Plausible Jesus: The Question of Criteria*. Louisville, KY: Westminster John Knox, 2002.

Thiede, Carsten P. "A Pagan Reader of 2 Peter: Cosmic Conflagration in 2 Peter 3 and the Octavius of Minucius Felix." *JSNT* 26 (1986) 79–96.

Thiselton, Anthony C. *The First Epistle to the Corinthians: A Commentary on the Greek Text*. NIGTC. Grand Rapids, Eerdmans, 2000.

Thomas, D. Winton. "A Consideration of Some Unusual Ways of Expressing the Superlative in Hebrew." *VT* 3 (1953) 209–24.

Thomas, J. *Le mouvement baptiste en Palestine et Syrie (150 av. J.-C.–300 ap. J.-C.)*. Gembloux: Duculot, 1935.

Tiller, P. A. *A Commentary on the Animal Apocalypse of I Enoch*. Atlanta: Scholars, 1993.

Tödt, Heinz Eduard. *The Son of Man in the Synoptic Tradition*. NTL. London: SCM, 1965.

Torrey, Charles Cutler. *The Four Gospels: A New Translation*. New York: Harper, 1933.

———. *Our Translated Gospels: Some of the Evidence*. New York: Harper, 1936.

Townsend, John T. "1 Corinthians 3:15 and the School of Shammai." *HTR* 61 (1968) 500–504.

Trapp, John. *A Commentary on the Old and New Testaments*. 5 vols. Eureka, CA: Tanski, 1997.

Trumbower, Jeffrey. "The Role of Malachi in the Career of John the Baptist." In *The Gospels and the Scriptures of Israel*, edited by C. A. Evans and W. R. Stegner, 28-41. Sheffield, UK: Sheffield Academic Press, 1994.
Tuckett, Christopher M. *Q and the History of Early Christianity: Studies on Q*. Edinburgh: T. & T. Clark, 1966.
———. "Thomas and the Synoptics." *NovT* 30 (1988) 132-57.
Tuell, Steven S. *Ezekiel*. Peabody, MA: Hendrickson, 2009.
Turner, Alice K. *The History of Hell*. New York: Harcourt Brace, 1993.
Valantasis, Richard. *The Gospel of Thomas*. London: Routledge, 1997.
van den Heever, G. A. "In Purifying Fire: World View and 2 Peter 3:10." *Neot* 27 (1993) 107-18.
van der Horst, Pieter W. *Hellenism-Judaism-Christianity: Essays on their Interaction*. Leuven: Peeters, 1998.
VanderKam, James C. *The Dead Sea Scrolls Today*. Grand Rapids: Eerdmans, 1994.
———. *An Introduction to Early Judaism*. Grand Rapids: Eerdmans, 2001.
Van Unnik, W. C. "The 'Wise Fire' in a Gnostic Eschatological Vision." In *Kyriakon: Festschrift für Johannes Quasten*, edited by P. Granfield and J. A. Jungmann, 277-88. Münster: Aschendorff, 1970.
Verhoef, Pieter A. *The Books of Haggai and Malachi*. NICOT. Grand Rapids: Eerdmans, 1987.
Vermès, Geza. *The Complete Dead Sea Scrolls in English*. London: Penguin, 2004.
Voltaire, François-Marie. *Philosophical Dictionary*. New York: Coventry House, 1932.
Volz, Paul. *Die Eschatologie der Jüdischen Gemeinde im Neutestamentlichen Zeitalter: Nach den Quellen der Rabbinischen, Apokalytischen und Apokryphen Literatur*. Hildesheim: Georg Olms Verlagsbuchhandlung, 1966.
von Rad, Gerhard. *Genesis: A Commentary*. OTL Philadelphia: Westminster, 1972.
Von Wahlde, Urban C. "Mark 9:33-50: Discipleship: The Authority that Serves." *BZ* 29 (1985) 49-67.
Wacholder, Ben Zion. (2007). *The New Damascus Document: The Midrash on the Eschatological Torah of the Dead Sea Scrolls: Reconstruction, Translation and Commentary*. STDJ 56. Leiden: Brill.
Webb, Robert L. "The Historical Enterprise and Historical Jesus Research." In *Key Events in the Life of the Historical Jesus: A Collaborative Exploration of Context and Coherence*, edited by Darrell L. Bock and Robert L. Webb, 9-93. Grand Rapids: Eerdmans, 2009.
———. "Jesus' Baptism by John: Its Historicity and Significance." In *Key Events in the Life of the Historical Jesus: A Collaborative Exploration of Context and Coherence*, edited by Darrell L. Bock and Robert L. Webb, 95-150. Grand Rapids: Eerdmans, 2009.
———. "John the Baptist and his Relationship to Jesus." In *Studying the Historical Jesus: Evaluation of the State of Current Research*, edited by Bruce Chilton and Craig A. Evans, 179-229. Leiden: Brill, 1994.
———. *John the Baptizer and Prophet: A Socio-Historical Study*. JSNTSup 62. Sheffield, UK: JSOT, 1991.
Weiss, Johannes. *Der erste Korintherbrief*. KEK 5. Göttingen: Vandenhoeck & Ruprecht, 1910.
———. *Jesus' Proclamation of the Kingdom of God*. Philadelphia: Fortress, 1971.
Wellhausen, Julius. *Das Evangelium Marci*. Berlin: Reimer, 1903.

Wesley, John. *Commentary on the Whole Bible: A One-Volume Condensation of His Explanatory Notes*. Grand Rapids: Zondervan, 1990.
Westermann, Claus. *Genesis 12-36*. CC. Minneapolis: Fortress, 1985.
———. *Isaiah 40-66: A Commentary*. OTL. Philadelphia, Westminster, 1969.
Wildberger, Hans.. *Isaiah 1-12*. CC. Minneapolis: Fortress, 1991.
———. *Isaiah 28-39*. CC. Minneapolis: Fortress, 2002.
Williams, H. H. Drake. *The Wisdom of the Wise: The Presence and Function of Scripture within 1 Cor. 1:18—3:23*. AGJU 49. Leiden: Brill, 2001.
Wilson, Ian. *Out of the Midst of the Fire: Divine Presence in Deuteronomy*. SBLDS 151. Atlanta: Scholars, 1995.
Wilson, W. E. "Εὑρεθήσεται in 2 Pet. iii. 10." *ExpT* 32 (1920) 44–45.
Wink, Walter. "Jesus' Reply to John: Matt 11:2–6/Luke 7:18–23." *Forum* 5 (1989) 121–28.
Winston, David. "The Iranian Component in the Bible, Apocrypha, and Qumran: A Review of the Evidence." *HR* 5 (1966) 183–216.
Wolff, Christian. *Der erste Brief des Paulus an die Korinther*. 2 vols. THKNT 7. Leipzig: Evangelische, 1966.
Wolters, Al. "Worldview and Textual Criticism in 2 Peter 3:10." *WTJ* 49 (1987) 405–13.
Wright, N. T. *The New Testament and the People of God*. COQG. Minneapolis: Fortress, 1992.
Yadin, Yigael. *The Scroll of the War of the Sons of Light against the Sons of Darkness*. London: Oxford University Press, 1962.
Zeller, Dieter. *Der erste Brief an die Korinther*. KEK 5. Göttingen: Vandenhoeck & Ruprecht, 2010.
Zimmerli, Walther. *Ezekiel: A Commentary on the Book of the Prophet Ezekiel*. 2 vols. Hermeneia Philadelphia: Fortress, 1979.
Zimmermann, Heinrich. "'Mit Feuer gesalzen werden': Eine Studie zu Mk. 9, 49." *TQ* 139 (1957) 28–39.
Zmijewski, Josef. *Die Eschatologiereden des Lukas-Evangeliums; eine traditions-und redaktionsgeschichtliche Untersuchung zu Lk 21, 5-36 und Lk 17, 20-37*. BBB 40. Bonn: Hanstein, 1972.
Zöckler, Thomas. *Jesu Lehren im Thomasevangelium*. NHMS 47. Leiden: Brill, 1999.

Ancient Document Index

Old Testament/Hebrew Bible

Genesis
1	123, 234
1:2	235
3:16	213
6:1–4	71, 182
9:18–27	42
15	27–28, 33n23
15:17	10, 108
18:17	40
18:20	42
18:20—19:28	40–43
18:23	40, 42
18:26–32	240n82
19	33n23
19:15b	43
19:24	1, 10, 22, 40, 48, 58, 75, 95, 122
19:24–26	155
19:25	43, 58
19:29b	43
19:31–32	41

Exodus
3	33n23
3:1–6	28–29
3:2	10, 34, 51
3:2–4	75
9:24	10, 36
13:21	10, 28, 75
14:24	10, 28
15:7	87
15:10	87
19	10, 33n23, 35
19:16–18	30–32, 83
19:18	31–32, 34, 51
23:20	67
24:17	10, 30–31, 34, 45, 51
25:22	45
32	46
35:31–33	205, 205n22

Leviticus
1:7	45
2:13	153
2:13b	154
3:2	10
6:12–13	45
9:23	10
9:24	34, 44–45, 48, 51
9:24a	44
10:2	44, 48

Numbers
3:4	22, 110
5:11–31	141
12:6–7	215n53
14:14	28

Numbers (continued)

16	48
16:16–17	49
16:20	48
16:35	22, 47–49
16:36–39	49
18:19	155
31:22	35
31:23	66, 142, 209–10
31:23a	49

Deuteronomy

4:12	30, 30n16, 31, 36
4:15	30n16, 36
4:33	30n16, 36
4:24	10
5:4	30n16, 36
5:22	30n16, 36
5:23	31
5:24	36
5:26	30n16, 36
9:3	10
9:10	30n16, 36
10:4	30n16, 36
32:22	12
32:41	73
39:23	155

Joshua

4:1–9	144

Judges

6	33n23
6:21	10
13	33n23

2 Samuel

22:8–16	31, 31n21
22:9	31n21

1 Kings

1:9–12	33
5:17	204
6:18	205
6:20–21	204
6:28	204
6:29	205
6:30	204
6:32	205
6:35	204
7:18–20	205
7:22	205
7:24–26	205
7:42	205
7:39–50	
18	33, 51
18:30–39	10
18:38	10, 44
19:12	34

1 Chronicles

21:26	10, 34, 44, 51
22:14	204
22:16	204
29:2	205

2 Chronicles

1:7	10
7:3	34, 44, 51
13:5	155
28:3	209

Nehemiah

9:12	10
9:19	10

Esther

2:23	240

Job

18:16–18	59
39:23	98n107

Psalms

11:6	75

18:4	99	30:27–28	53, 68, 137
18:7–15	31, 39n38	33	17
18:8	32n21	33:10–12	55–56
18:9	12	33:10–16	68
60:8	134	33:12	79n42
65:10–12	210	33:14	10
65:12	209, 209n36	33:14b	67
66:10	208, 221	33:14–16	55–56
66:12b	127	34:4	238
75:9	169	34:5	120
80:1	45	34:8–10	58, 68, 68n90
89:47	12	34:9	58
97:3	38	39:23	98
97:5	88	43:2	56, 209–211
99:1	45	43:2a	137
104	38	48:10	62, 202, 221
107:33–34	155	50:11	93
108:9	134	51:17	169
144:5–6	38	66:2	108
		66:15	133
		66:15–16	60
		66:22–24	60
		66:24	87

Proverbs

17:3	17, 107, 109n149
25:21–22	199
25:22	200
26:20–21	167
27:21	107, 115–116
27:21a	223

Jeremiah

1:13–16	227n28
4:11–12	
6:27–30	63, 68, 112–14
6:29	17
9:6	208
15:14	12
17:6	155
21:12	12
21:14	57
25:21	60
25:28	170
34:18–19	27
49:17–18	58
50:20	240n82

Isaiah

1:21–26	63n83
1:24–26	68
1:24	17
1:25b	112
1:25–26	26, 62
3:3	205n22
4:4	137
5:24	53, 68, 87
6	34
6:1–7	35
10:17	100n119
11:4	129
13:10	237
14:13	75
19:1	109
29:5–6	53, 68
30:27	15

Ezekiel

1:4–5	35–36
1:13–14	36
1:15–21	37
1:26	37
1:27	37

Ezekiel (continued)

1:28	10
5:2	120
9:5–6	227n28
9:6	226, 227n28
15:18	57
16:4	155
17:24	99
20:45–47	57–58, 183, 185–188
20:46	188n167
22:18–22	63n83, 64, 68
22:20	17
22:31	12
23:31–34	170
24:1–13	64–65, 68
24:3–13	113
24:12	160
24:12–13	114
25:16	119
28:14	75
28:16	75
30:5	119
30:13	109
32:7–8	237
38:22	75, 133
39:6	133
43:24	154

Daniel

2:34–35	129
3	56
6:23	116
7	10, 16, 37, 52, 54, 82, 129–30, 141, 178
7:1	
7:9–10	36
7:10	3, 14, 73, 98, 129, 137
7:13	129
11:35	116–17, 208
11:40—12:3	108
12:1	117

Hosea

7:4	59
9:16	184n150

Joel

2:28–29	137
2:30	133

Amos

1:4	33n23, 99
1:7	33n23
1:10	33n23
1:12	33n23
2:1b	55, 79n42
2:2	33n23
2:5	33n23
3:1–2	226n28
4:2	115
4:9	223
4:9a	115
4:11	201, 209n36
4:11b	115, 116n173
7:4	99
9:1–4	98n112
10b	115

Obadiah

18	87

Micah

1:4	88, 128

Nahum

1:5	88
3:3	97n107

Habakkuk

1–2	94
2:16	170

ANCIENT DOCUMENT INDEX   271

Zephaniah

1:18	2
2:5-6	119
2:9	155
3:2	240

Zechariah

2:5	34n23, 120
2:9	120
3:2	201, 209n36
12:6	28, 108
13:1	66
13:8-9	65-66, 66
13:9	17, 208-10, 241
13:9a	66

Malachi

3	6, 206-208, 215, 217, 241, 245, 247
3-4	144, 247
3:1-4	66-68
3:1-5	225-25
3:1-6	245
3:1-18	226n28
3:2	2, 34n23, 102, 107
3:2-3	17-18, 26
3:2-4	241-42
3:3	102, 207, 215, 225
3:19	10, 26, 28, 133, 206n30 206n31, 207
4	247
4:1	2, 18, 59, 67, 87, 144n48, 225-26, 245
4:5	133

Apocrypha

1 Esdras

6:24	154
6:30	154

2 Esdras

13-14	126-30

Tobit

14:6	80

Judith

16:17	10

Wisdom

3:5	242
3:5-7	222
10:4-6	179n131
14:11	80

Sirach

2:1-6	221-22
2:5	17
21:9	10

Baruch

48:39	12

3 Maccabees

2:4-5	179n131
2:29	120

4 Maccabees

5:32	79
6:24	79
7:12	79, 210
9:9	79, 210, 210n38
9:19	79
9:22	79

4 Maccabees (continued)

10:14	79
11:19	79
11:26	79
12:12	79
14:9–10	79
15:15	79
18:12–20	79
18:14	210

Old Testament Pseudepigrapha

2 Baruch

13:9–10	226n28

1 Enoch

1–36	71–76
1:6	238
14	37n33
14:9–22	70n3
14:19	73, 137
15:19	3n4
17:5	74, 98, 141
18:13	85
18:15	85
21:3	85
27:2–3	87
48:9	87
52	13
52:6	99
52:6–7	88
67:4–7	88–90, 141
67:13	3n4, 98, 137
89:50–73	80
90:8–9	77
90:26–27	87
91:11–17	76
92:1	79
93:8	76, 81
93:1–10	76
98:3	81, 82
99:2	85
100:5	83
100:7–9	81
102:1	129
102:1–3	82–83
103:5–8	81
103:8	82
104:10–11	85
106:1–18	71
108:2–6	84–85

2 Enoch

10:20	3n4

4 Ezra

7:6–11	127–28, 141
7:31	13
7:77	235
7:83–84	235
13	37, 37n37
13:2	128
13:3	129
13:6–7	129
13:9–11	128–29
13:10–11	3n4, 137

Jubilees

4:18	76
9:15	69n1
16:6–8	43

Liber antiquitatum biblicarum

1:497–500	204
12:8–9	205

Life of Adam and Eve

49	12, 13
49:3	236n67

Letter of Aristeas

87	210

Psalms of Solomon

9:5	235
17:8	240

Sibylline Oracles

2:196–205	3n4
2:196–213	123
2:286–89	18
2:252–54	3n4, 141, 174
2:315–17	18
3:46–62	121–22
3:80–92	122–23
3:54	3n4, 137
3:689–90	236n67
3:77	118
3:80–92	238
3:776	118
3:84–87	3n4
3:287	119
3:504–7	119
3:542–44	119
3:669–74	120
3:702–6	120
3:798–99	120
4:42–43	123
4:80–82	124
4:159–92	123n203, 124–26
4:175	237n70
7:13	13

8:411	18

Testament of Benjamin

10:8	226n28

Testament of Levi

3	13
4:1	238
9:14	147n10, 155

Testament of Naphtali

3:4–5	179n131

Testament of Abraham

1:1–2	43
13:11	150n21, 211

Testament of Isaac

5:21–25	3n4, 142, 174
7	12
7:3	13

Testament of Moses

10:7	80

Dead Sea Scrolls

Thanksgiving Hymns[a] (1QH[a])

III.29–33	137
IV.13–14	99, 102
IV.16	224
VI.3–4	97
X.26	97
XI.27b–37	97–100, 102, 141, 183, 185, 186
XI.29–21	3n4
XI.29–37	97
XI.32	100
XII.33	97
XIII.15b–19	102–3
XIV.16b–19a	100, 141
XVI.11–20	100, 141
XVI.12	137

War Scroll (1QM)

I.1–2a	108
XI.9	108
XI.10–11	108
XVI.15	224
XVI.15–XVII.3	109–11
XVII.8–9	109–11

Pesher Habakkuk (1QpHab)

1–2	94
XII.13	94
IX.9–X.13	94–96

Rule of the Community (1QS)

I.16–18	107
I.17	224
II.7–8	104–5
II.8	107
IV.11–14	105–6
IV.20b–22a	106, 141
VIII.4	107
VIII.5–8	205

Damascus Document (CD)

II.5b–6	92
V.13	93
XVI.2–4	92
XX.13–15	92
XX.27	224

Florilegium (4QFlor)

II.1	224

Philo

De Abrahamo

1	210
145	214

De aeternitate mundi

110	238

De agricultura

1.13	214

De cherubim

130	214

Legum allegoriae

3.189	214

De vita Mosis

2.10–12	179n131
2.219	210
2.48.263	236n67

Quaestiones et solutions in Genesin

4.10	43
4.54	43

De specialibus legibus

1.53	155
4.28	210

Josephus

Life

304	214

Antiquities

1.203	224
1.205	42
3.9.1	154
12.3.3	154

ANCIENT DOCUMENT INDEX

14.362	214
17.264	209
17.275	214

Jewish War

6.288–289	120

New Testament

Matthew

3:9	143
3:10	131
3:11	7, 21–22, 106, 131–44, 170, 175, 244
3:11–12	4
3:18	131
4:24	161
5:13	155
5:18	237
5:19	197
7:19	188
8:19	195
8:12	82
9:2	176
9:3	176
9:29	176
10:34	40, 160–61
10:37–38	195
10:39	195
11:2–6	134, 159, 171
11:23–24	181
11:24	40
13:40	181
13:42	82
13:50	82
21:25	158
22:13	82
24:35	237
24:37–39	179–81
24:42–44	237
28:19	173

Mark

1:4	146
1:7	175
1:8	131–132, 135, 148n32, 173–74
1:9–11	146
1:12	176
2:5	176
2:7	176n124
4:11	176
4:21–25	154n34
8:12	176
8:31	164
9:11–13	133
9:12	164
9:21	145, 167
9:31	164
9:33–50	152–53
9:40–48	61
9:43	195
9:43–48	1, 151
9:44	87
9:47	195
9:48	87
9:49	1, 1n1, 2, 7, 10, 14, 16, 19, 138n28, 139, 145, 146–59, 164–65, 168, 172, 174, 177, 182–83, 188, 193, 195–98, 245–46
9:50	155
10:33	164
10:38	164, 167–69, 171, 187
10:39b	174
11:30	158
10:38	160–61
10:48	172
12:34	197
13	13, 22
13:15–16	180
13:24–27	
13:31	237
14:36	163n83

Luke

2:25	145
3:8	143
3:9	131
3:16	7, 21–22, 106, 131–44, 244
3:16–17	4, 7
3:17	131
4:38	160
5:20	176
5:21	176n124
7:18–23	134, 159, 171
7:47–48	176
8:37	160
8:45	160
9:51	186–87
9:54	16
10:12	40
12:12–15	162
12:49	6, 16, 145, 187, 193, 195–96
12:49–50	2, 14, 16, 19, 21, 23n73, 138n28, 139, 158–77, 182–83, 187, 196, 196n210, 197, 198, 245
12:50	160
12:51	160
12:51–53	167
12:55	186n165
13:7–9	188
13:32	176
14:34	155
17:26–30	145, 164
17:26–32	177–83, 198, 245
17:28–32	40
19:43	160
20:4	158
21:33	237
22:25	167
22:42	163n83
22:63	160
23:31	145, 183–88, 198, 226n27, 245

John

1:21	133
1:29–34	146
1:33	131–32, 173
3:17	212

Acts

1:5	131–32, 136, 173, 175n117
2:3–4	135–36
7:57	160
11:16	131, 175n117
11:18	132, 173
15:11	212
18:5	160
19:1–7	138
28:8	160

Romans

1	5
3–4	216
5:9	212
12:12	200
12:20	11, 199
14:10	216

1 Corinthians

3:9–15	96n103
3:10	205
3:10–15	2, 6–7, 12–14, 19–22, 96, 117n180, 144, 144n50, 200–217, 221, 223, 225, 237, 244–46
3:12	144
3:12–15	6, 207
3:13	11, 59, 150n21, 199–200, 206n30, n31, 207
3:15	5, 8, 8n21, 116n173, 157, 157n58, 199–217
4:4–5	216
5:5	216

ANCIENT DOCUMENT INDEX

15:1–2	212, 216

2 Corinthians

1:14	202
5:10	206
5:14	161

Galatians

2–3	216
3:11	94
3:13	18, 18n58, 215
4:3	237
4:11	202

Ephesians

2:8	2:12
2:20–22	202

Philippians

1:23	161
2:16	202

Colossians

1:24	172
2:8	237
2:20	237
4:6	155

1 Thessalonians

3:5	202
5:3	237

2 Thessalonians

1:8	11, 14, 22, 199, 200
2:8	129

1 Timothy

2:15	212

Titus

3:5	212

Hebrews

11:33	116
12:9	11

1 Peter

1:5–7	19
1:6	220n6
1:7	2, 210, 218–27, 240, 242
2:4–8	202
2:5–6	225
3	213
3:15	220
3:20	213
4:4	220
4:17–18	225–26
4:12	2, 22–23, 107, 115, 218–27
4:12–13	19, 219–27
4:17–18	184

2 Peter

2:5–7	180n131
2:17	82
3	6, 13, 85, 227
3:3–4	227
3:4	232
3:5–7	233–236
3:7	2n2, 11, 11n37
3:10	2, 240–42, 246
3:10–12	154
3:10–13	22, 236–43, 246
3:10–15	218
3:14	240, 242
3:16	200
3:19–14	2n2

Jude

7	102
14	82
23	157, 201, 209n36

Revelation

1:15	241
3:3	237
3:18	241
6:13	237
8	169
8:5	169
8:5–8	169
8:7	169
8:8	169
8:9–10	119n189
9:13	119n189
9:15	129
16:15	237
18:9	223–24
18:18	223–24
20:9	11

Rabbinic Literature

Baba Batra

84a	73

b. Berakot

17a	214n46

b. Megillah

9a–9b	111n154

b. Sotah

21a	213

Genesis Rabah

51:8	42n37
78:1	38

Hagigah

14a	38n37

Targumic Texts

Fragmentary Targum

Deut 32:34	235

Targum Neofiti

Gen 3:16	213

Apostolic Fathers

Epistle of Barnabas

16:5	80n52

1 Clement

9.4	213
12.1	213
58.2	213
21.6	240–42

2 Clement

3.3	213
16.3	241–247, 247

Didache

16.5	18–20, 150n21, 216, 224–25, 247

Shepherd of Hermas, Similitudes

9.12.3	213

Shepeherd of Hermas, Visions

3.3.5	213
3.8.3	213
3.13.3	238
4.2.4	213
4.3.4	216

Ignatius, *To the Ephesians*

18.2	146

Ignatius, *To the Magnesians*

10	155n43

Ignatius, *To the Smyrnaeans*

4.2	191

Martyrdom of Polycarp

15.2	247

Polycarp, *To the Philippians*

1.3	213

New Testament Apocrypha

Gospel of the Hebrews

2	146

Gospel of Thomas

7	192n187
10	145, 160–61, 195
16	145, 192n187, 195
82	2, 11, 16, 19, 172n109, 189–98, 245

Gospel of Philip

35	147n10
16	155

Apocalypse of Paul

31–36	3n4

Apocalypse of Peter

6	3n4, 174

www.ingramcontent.com/pod-product-compliance
Lightning Source LLC
Chambersburg PA
CBHW061433300426
44114CB00014B/1666